Exploring World Geography Part 2

To Our Grandchildren:
May you live in a better world, and may you help it be so,
until we all live together in the better place God has in store for us.

Exploring World Geography Part 2
Ray Notgrass

ISBN 978-1-60999-155-5

First Edition (2022 Printing).

Previous Page: Lake Louise, Canada, by Braxton Stuntz on Unsplash
Front Cover Image: Shanghai, China, by chuyuss / Shutterstock.com

Cover design by Mary Evelyn McCurdy
Interior design by John Notgrass
Maps by Sean Killen and John Notgrass
Literary introductions by Bethany Poore

Printed in the United States of America

975 Roaring River Road
Gainesboro, TN 38562
1-800-211-8793
notgrass.com

Mo Chu River, Punakha, Bhutan

Part 2 Table of Contents

Church of St. Alexander Nevsky (Russian Orthodox) Tashkent, Uzbekistan

Kosrae, Federated States of Micronesia

20 Southeast Asia 535

21 Australia and New Zealand 567

Lucky Bay, Western Australia

Salisbury Plain, South Georgia Island

Mustard Field, Bangladesh

16 South Asia

Because of its geographic location, Afghanistan has a complex history that is marked by recurring invasions of foreign powers. Mountainous Nepal, positioned between rivals India and China, is a surprising field for Christian evangelism and is seeing significant changes for women. The Siddis of India are a fascinating ethnic study and introduce a discussion of that country's caste system. We look at the first accomplishment of the greatest geographic challenge in the world, the climbing of Mt. Everest. The worldview lesson considers God's worldview as expressed in the first three chapters of Genesis.

Memory Verse

Memorize Jeremiah 29:11-13 by the end of the unit.

Books Used

The Bible
Exploring World Geography Gazetteer
Boys Without Names

Project (Choose One)

1) Write a 250-300 word essay on one of the following topics:
   - Write a script for a five-minute television newscast reporting the successful climbing of Mt. Everest by Sir Edmund Hillary and Tenzing Norgay. Present this newscast to your family. (See Lesson 79.)
   - Describe a big goal you would like to achieve, why you would like to achieve it, and outline the steps you would need to take to accomplish it. (See Lesson 79).
2) Draw or paint a scene in the Himalayas with Mt. Everest in the center. (See Lesson 79.)
3) Research Afghan rugs and shawls. Draw and color the pattern for one or, if you have the interest, create a small Afghan rug or shawl. (See Lesson 76.)

Literature

Boys Without Names is a novel by Kashmira Sheth set in India. Gopal is the eldest son in a warm, secure family living in their beloved, quiet, out-of-the-way village. Disappointing yield from their farm and mounting debt push them to try a completely different way of life in the city of Mumbai. Their world is turned upside down as they try to cope with a strange, fast-paced new culture. Gopal is a strong protagonist who brings his creativity, caring, and bright intelligence to meet each challenge.

Kashmira Sheth was born in Gujarat, India. She moved to the United States at age seventeen when her uncle, a professor at Iowa State University, sponsored her to attend college there. Though she trained and worked as a microbiologist, she discovered writing as a second career. She was inspired by beloved Indian literature of her childhood, family memories, and discovering English literature along with her daughters. She researched extensively to make *Boys Without Names* portray the real dynamics and strong grip of industries that exploit child laborers.

Plan to finish *Boys Without Names* by the end of the next unit.

Schoolchildren in Afghanistan (2019)

76 Trouble and Hope in Afghanistan

Mursal, a twelve-year-old girl, can get an education.

This is remarkable in Afghanistan where children, especially girls, are not always able to go to school. Warfare, whether between the private armies of warlords or involving the armed forces of many nations, often interrupts normal life. Landmines left from previous conflicts make the simple act of walking to school dangerous. Having to flee one's community as refugees when fighting gets close leaves education behind. Economic need leads many parents to keep their children at home so they can generate income or help with crops in a country that has limited economic opportunities. Sometimes a father will withdraw his daughter from school for her to get married. A formal education is not valued in a country where half the men and three-fourths of the women cannot read or write.

In conflicts between groups of adults, children are often innocent victims, "collateral damage" as the euphemistic phrase expresses it. But the children of Afghanistan deserve a chance at a better life than what many of them have known, and education can provide that chance. Education gives children hope and gives the country hope. As one Afghan teacher put it, "If we, as the adults, work for these children, they could become the doctor or the lawyer or the engineer who will fix this country."

Geography and History

Afghanistan is a landlocked country in southern Asia, slightly smaller than Texas. It lies north of Pakistan, east of Iran, and south of three countries in Central Asia: Tajikistan, Turkmenistan, and Uzbekistan. Afghanistan is three-quarters mountainous but has plains in the far north and deserts in the south. In the mountains winter temperatures can dip to -13°, while on the plains summer highs can reach 115°. Differences between daytime and nighttime temperatures in the same place can also be extreme.

The Khyber Pass is the main access route through the mountains between Afghanistan and Pakistan. It has played an important role in many of the invasions of Afghanistan and in military movements between Afghanistan and points south.

The Pashtuns are the largest single ethnic group. This overall group has some sixty tribal groups that sometimes don't get along with each other. Pashtun values include family honor, generous hospitality, personal bravery, and vengeance for insults. The country is 99% Muslim.

View of Kabul (1842)

The area we know as Afghanistan has been traversed by traders and conquered by empire-builders for millennia. One historian called it the hub of the ancient world. It was part of the Persian Empire until Alexander the Great conquered it in the 300s BC. At his death his successors the Seleucids maintained the presence of Greek culture. The Mauryan Empire of India established control over southern Afghanistan and brought Buddhist teaching into the area. In the 200s BC Kushan nomads created an empire that built commerce and brought a culture that lasted until the 200s AD.

Following a period of fragmented rule, in the 600s Arab Muslims invaded and brought Islam. Various Muslim dynasties ruled until the Mongols led by Genghis Khan established dominance about 1220. The later conqueror Timur and his successors ruled from the 1300s until the early 1500s, when the Mughals took over. In 1747 indigenous Pashtuns called the Durrani and led by Ahmad Shah established unified rule that continued until 1978. The Durrani ruled an area that was much the same as modern Afghanistan.

In the 1800s Russia and Britain competed for control of Central and Southern Asia in what came to be known as the Great Game (think about the worldview that would call competition to control land and peoples, involving the cost of many lives, a game). However, Afghan rulers were generally able to resist the efforts of both countries most of the time except for brief periods when Britain ruled. Britain and Afghanistan fought a series of wars for control in the 1800s and early 1900s. In 1893 Britain drew the Durand Line in the Hindu Kush mountains. This established Britain's sphere of authority but it divided Pashtun tribes. Today the line serves as the border between Afghanistan and Pakistan in practical terms, although no Afghan government has accepted the Durand Line. In 1919 the Treaty of Rawalpindi ended the last British-Afghan conflict and established the independence of Afghanistan, but then Afghanistan had to maintain a balance between British interests to the south and Communist Soviet interests to the north.

After World War II and the establishment of Pakistan, conflicts grew between Afghanistan and Pakistan. Those governing Afghanistan drew toward the Soviet orbit for protection and assistance while the U.S. aided Pakistan. A Communist revolt overthrew the sitting leader of Afghanistan in 1978. The next year, anti-Communist fighters in several Afghan tribes, called the mujahideen, threatened another revolt; so the Soviet Union sent in 80,000 troops to defend the government. This resulted in a long guerilla war that lasted nine years between tribal forces on one side and Afghan government and Soviet troops on the other. The United States gave assistance to the mujahideen because they were anti-Communists, even though they were Muslims. In 1988 the Soviet Union backed a new government that was Communist but weak. The last Soviet troops left the next year. This ended the conflict that

Scene in Kabul After the 1978 Revolution

At Left: Afghan Mujahideen (1985)
Above: Mujahideen Praying (1987)

cost thousands of lives and caused 5 to 6 million people to become refugees.

Mujahideen fighters then entered Kabul, the capital, and declared an Islamic state. Continued conflict between the government and tribal factions led to a civil war in 1992. One fundamentalist Islamic group, the Taliban, gained control of much of the country in 1996 and established a repressive regime. Many mujahideen joined the Taliban, which allowed the terrorist group al-Qaeda to use Afghanistan as a base for their activities. A coalition of tribes called the Northern Alliance opposed the Taliban.

The United States government accused Osama bin Laden of overseeing the terrorist attacks on the U.S. on September 11, 2001. After the Afghan government refused to surrender Osama bin Laden, who was then living in Afghanistan, the U.S. and its allies invaded the country in late 2001 and fought the Taliban with the help of the Northern Alliance. The Taliban government collapsed, but Taliban and al-Qaeda leaders escaped capture. U.S. and allied occupation forces supported the founding of a new government that supported the United States. The allies have enjoyed military successes against the radical Muslims, but the situation in Afghanistan has not improved to the point that all U.S. and allied troops can leave.

Afghanistan is sometimes called the graveyard of empires. History has indeed shown that Afghanistan is difficult to rule, both for geographic reasons and because of the fierce independence of the native tribes and warlords. However, empires tend to wind up in graveyards (think Alexander, the Mongols, the British, and the Soviets, among many others). The difficulties of ruling Afghanistan have contributed to this pattern, although Afghanistan was not the only factor in the demise of these empires.

Afghanistan's 46-Mile Border with China

In addition to Afghanistan's long borders with Iran, Pakistan, and three Central Asian countries, Afghanistan has a 46-mile border with China that is at the end of a tiny finger of land 185 miles long that points to the northeast. The finger of land is the Wakhan Corridor within the Hindu Kush mountains. The border lies among the Hindu Kush, Karakorum, and Pamir mountain ranges.

Most of the border is mountainous, but the small Wakhijir Pass cuts through the mountains about 16,400 feet above sea level. The pass was once a part of the Silk Road. Marco Polo traveled through this pass in the late 1200s. It now serves as almost the only way to get through the mountains between China and Afghanistan.

In the Great Game that saw Russia and Great Britain competing for control of Central Asia,

Panj River and Pamir Mountains in the Wakhan Corridor

Russia lay to the north of Afghanistan and British-controlled India lay to the south (before the creation of Pakistan). In 1895 Britain and Russia agreed between themselves to adjust Afghanistan's borders by creating the finger of land to keep the two empires separate. The significance of the corridor and border with China today is that it keeps the fighting in Afghanistan separate from the unrest among the Muslim Uyghur people in the Xinjiang province of China (see Lesson 93).

Another significant fact about the border is that it creates the largest time difference in the world between two neighboring countries. All of China, which is about as wide as the United States, uses only one time zone (imagine when it is 6 a.m. in New York City it is also 6 a.m. in San Francisco). In 1949 Communist leader Mao Zedong, in an attempt to enhance national unity, declared that all of China would use the time that Beijing is on. So when a traveler steps from China into Afghanistan, he steps into a time zone three and a half hours earlier. Afghanistan, along with neighboring Iran and nearby India, maintains a time that is on the half hour compared with most of the rest of the world. Just be sure your watch or phone adjusts easily to such a change if you ever travel through the Wakhijir Pass.

The Country That Needs Fixing

Education is not the only unstable aspect of Afghan life. Centuries of war have cost countless lives and done significant damage to the land and the economy. Afghanistan used to grow its own food; now it must import much of the food that the people need. Leadership that has focused on political power, military control, and the imposition of Islamic fundamentalism instead of improving the welfare of the people has contributed to low life expectancy and a high infant mortality rate. Trafficking in illegal drugs, primarily because of the poppies grown in the country which are used to produce heroin, is a serious problem. Two major earthquakes in northeast Afghanistan in 1998 killed six thousand people and left many more thousands homeless.

The United States and its allies have helped in building the country's infrastructure (including roads, hospitals, and schools), and the new constitution has given women more rights and freedoms. Some gemstone mining takes place; and the production of oil, natural gas, coal, and other minerals holds promise. Improvement has definitely taken place. Perhaps the coming generation will indeed learn and adopt a new and better way that will help Afghanistan enjoy a promising future.

The prophet Jeremiah told the Jews who would be captives in Babylon that God had in mind for them a future and a hope if they would seek Him:

"For I know the plans that I have for you," declares the Lord,
"plans for welfare and not for calamity to give you a future and a hope.
Then you will call upon Me and come and pray to Me,
and I will listen to you. You will seek Me and find Me
when you search for Me with all your heart."
Jeremiah 29:11-13

Assignments for Lesson 76

Gazetteer	Study the map of South Asia and read the entry for Afghanistan (pages 140-141).
Worldview	Copy this question in your notebook and write your answer: What is one lesson you get from the accounts of Creation and the sin of Adam and Eve in Genesis 1-3?
Project	Choose your project for this unit and start working on it. Plan to finish it by the end of this unit.
Literature	Begin reading *Boys Without Names*. Plan to finish it by the end of the next unit.
Student Review	Answer the questions for Lesson 76.

Annapurna Nature Reserve, Nepal

77 A Fertile Field in the Mountains of Nepal

Tej Rokka is a native Christian evangelist in Nepal. He has seen a huge increase in the number of people coming to Christ in the twenty-first century in that predominantly Hindu country. Nepal has one of the fastest-growing Christian populations in the world.

Why Is This Happening?

Rokka can identify several reasons why people are turning to Christ in Nepal. After years of political instability, Nepal has become a constitutional democracy with multiple political parties. The bad news is that Maoist and Communist parties have come into power and lead the country. The good news is that the country has moved away from official Hinduism and toward being an officially secular state. This means that evangelism is more acceptable and can take place more openly, even though it is still technically illegal. The secular national government does not exert strong control over the countryside, where 85% of the people live in villages and small market centers.

Another factor is the poverty and social caste system in Nepal. About one-fourth of the people live below the generally accepted poverty line. Although Nepal officially outlawed the caste system in 2001, people in rural areas still practice it. Many feel hopeless until they hear about the One who came to free the oppressed.

A third, more immediate reason for the growth in Christianity is the devastating 2015 earthquake that hit Nepal. The 7.8 temblor, centered about fifty miles northwest of the capital Kathmandu, along with the aftershocks, killed almost 9,000 people, injured many thousands more, and destroyed over 600,000 homes and businesses in and near Kathmandu. While political squabbling and government red tape

Members of Kathmandu's Gyaneshwor Church celebrate a baptism in the 1960s.

have slowed the response through official channels, Christian relief organizations were able to get help to hurting people, who noticed where it came from and started asking questions.

Census numbers show the dramatic growth of Christianity. The 1951 census listed no Christians in the country. Ten years later, the number was 458. By 2001 the census accounted for almost 102,000 Christians. By 2011 the number had tripled to over 375,000, and some observers believe the actual count is even higher. Nepal has a population of about 29 million, 80% of whom are Hindu, 10% Buddhist, 4.5% Muslim, and 1.5% Christian. Although Christians are still a small fraction of the population, the growth in their actual numbers is nothing less than phenomenal. Tej Rokka and other evangelists and Christian teachers are amazed to see what God is doing. Churches are starting and growing in many places throughout Nepal.

Political Changes for Women

Religion is not the only aspect of life that is changing in Nepal. Women are becoming more involved in politics. In 2017 the country conducted elections for local offices for the first time in twenty years. About 20,000 women ran as candidates, and many of them won.

Sita Chhaudry was one. Twenty years earlier her parents, members of a low social caste, sold her into virtual slavery as a ten-year-old to a wealthy family for $50 a year as a *kamlari*. A kamlari was a servant who cooked, cleaned, and looked after children. Kamlaris were sometimes physically and mentally abused by their so-called employers (actually owners). They were not allowed to sleep inside the house where they worked. The bargaining between parents and potential masters for the services of kamlaris took place every year. The practice was a significant source of income for many poor Nepali families.

Sita endured this treatment for years, as did many other children. Because she did, Sita never learned to read and write. Nepal's supreme court eventually outlawed the practice, and the government has settled many former kamlaris into communities to help them start over. Unfortunately, the government put many of these communities on poor land that often floods. Sita is now a local official who wants to use her new position to help bring about justice and a decent life for all Nepali, especially those in her own community.

The Geography of Nepal

Nepal is a long, narrow, landlocked country. It stretches about five hundred miles from east to west and between 90 and 150 miles north to south. It is slightly larger than New York state. Nepal has three distinct geographic regions. The Tarai is a 15-20 mile wide band of fertile Ganges River lowland plains in the south. Almost half the population lives in this region. Just to the north is a hilly region where most of the rest of the people live. The remaining portion, by far the largest, consists of the Himalayas, the tallest mountain range in the world. The Himalaya range includes Mt. Everest, the tallest mountain in the world, which sits astride the border between Nepal and Tibet. Eight of the world's ten highest mountains are in Nepal. About three-fourths of the land area of Nepal is hilly or mountainous.

Nepal is one of the poorest countries in the world. The country has relatively few miles of road, few vehicles, and one airline. Most transportation

Village of Saldang, Nepal

Yaks Carrying Loads in Nepal

occurs on footpaths that follow rivers. Farming takes place on a little more than one-fourth of Nepal's land area, but about three-fourths of the population depends on subsistence agriculture for their income. Farming accounts for one-third of the country's gross domestic product. Much of the industry in Nepal involves food processing. Another one-fourth of the land is forested, including the sides of many mountains. The country has few natural resources, and it imports much more than it exports. About 30% of Nepal's GDP comes from remittances that Nepali workers in other countries (mostly India) send home. Nepal has the highest rate of remittances in the world. The fact that so many men work out of the country is one reason why the door is open for women to take part in politics as mentioned earlier. Tourism is a relatively small but growing part of the economy. It began to grow after Sir Edmund Hillary and his Nepali guide Tenzing Norgay climbed Mt. Everest in 1953; and others wanted to do likewise.

The most important factor in Nepal's geography is its location, sandwiched between India and Tibet. But what is Tibet? That is a contentious issue on the world stage. Tibetans and many others in the world believe Tibet is (or should be) an independent country. China, on the other hand, believes that Tibet is part of China and refers to it as an autonomous region within China.

Thus, Nepal lies between India and China, two bitter enemies. Nepal likes to keep both sides happy. Most Nepali exports go to India, and India and Nepal have a mutual defense agreement. Nepal also trades with China as well as the United States and a few other countries. China is pursuing the expansion of its influence in Nepal by funding projects that it says will help Nepal, even as the projects also help expand China's international influence and get them closer to having an unimpeded route to the Indian Ocean.

Nepal is a land of rich ethnic diversity. Many groups live within its borders, and most of these groups are related to people in either India or Tibet. The Newar people are the largest group; Newar and Nepal are different forms of the same word in Nepali. One smaller group is the Sherpas (probably from Tibet), many of whom work as guides and burden bearers for mountain climbers. Tenzing Norgay was a Sherpa. Another group is the Gurkhas (or Gorkhas) who are related to the people of northern India. Many Gurkhas have served in the British or Indian armies.

Nepal is also a land of diverse languages. Only a little less than half of the population speaks Nepali, the official language, as their primary language, although many speak it as a second language. The 2011 census listed 123 languages that at least some people reported as their mother tongue. The literacy rate is about 75% of men, 57% of women. The average life expectancy is about 71 years.

Besides the large Hindu and Buddhist presence mentioned earlier (Buddha was born there, and the blending of the two belief systems is common), many Nepali also worship local so-called gods and spirits and consult shamans to solve problems. Some polygamy (also called polygyny) and polyandry (a woman having more than one husband, who are often brothers) takes place.

A centuries-old practice in Nepal is identifying and worshiping a Hindu child goddess, called the Kumari Devi. Hindu religious leaders interview and examine candidates from the Newar people, the earliest known residents of the Kathmandu Valley, and declare a girl sometimes as young as five years old to be a goddess. After an elaborate initiation ritual, the girl lives an almost completely isolated life until she relinquishes the title in her teen years. Sometimes more than one girl at a time can be goddesses. In the Kathmandu Valley, most people believe in the influence of planets, karma, and a host of gods as part of their worldview.

Birthplace of Buddha, Lumbini, Nepal

The Backstory

Small independent kingdoms flourished in what is now Nepal until the 1700s. Then the Gurkha king launched a military campaign to unify the kingdoms under his rule. He took the title of the king of Nepal.

In the early 1800s, Nepali rulers attempted to expand into British-held India. This led to a war between Nepal and Great Britain in 1814. The two countries concluded a peace treaty in 1816, and the countries became political allies even though Nepal was never a British colony. Even today, the British army includes a brigade of Gurkha soldiers.

In 1846 Jung Bahadur seized power from the Nepalese king and took the title or family name of Rana. He declared that a Rana would be prime minister from then on. However, when Great Britain withdrew from India in 1947, the Rana lost an important support for its rule. Other Nepali who wanted to gain power began a revolution that culminated in 1951 and returned actual power to the king. The king established a cabinet government and allowed political parties to form. This system continued until 1960, when a long pattern of political instability began. The king banned political parties; but then he reinstated them in 1990 and declared Nepal to be a constitutional monarchy and a multiparty democracy.

Supporters of the late Chinese Communist dictator Mao Zedong started an uprising in Nepal in 1996, which began a ten-year civil war. In 2002 the king reclaimed power and dissolved the government, but fighting continued. The warring factions signed a peace treaty in 2006 and devised a new constitution. In 2008 the Constituent Assembly abolished the monarchy, declared a federal democracy, and elected the country's first president. However, the process of writing a new constitution dragged on until late 2015 (only finishing after the earthquake, which many believe was the impetus for finishing it). A coalition of Communist and Maoist parties came together in 2014, and they now hold a substantial majority of seats in what is now a two-house legislature.

Since Nepal separates India and China, Nepal is literally caught in the middle. The country is poor, and the Communist-led government is not likely to make things much better, although Chinese investment might raise the standard of living for some. The country is heavily Hindu, but as the country gains more Christians, it also gains more hope for a brighter future.

The believers who began spreading the gospel in Nepal echoed Paul's desire to preach the gospel where others had not gone before.

And thus I aspired to preach the gospel, not where Christ was already named, so that I would not build on another man's foundation; but as it is written, "They who had no news of Him shall see, and they who have not heard shall understand."
Romans 15:20-21

Assignments for Lesson 77

Gazetteer Read the entries for Bangladesh, Bhutan, and Nepal (pages 142, 143, and 147).

Worldview Copy this question in your notebook and write your answer: What attempts to describe the origin of the universe, other than the description in Genesis, seem to be widely popular in our culture today?

Project Continue working on your project.

Literature Continue reading *Boys Without Names.*

Student Review Answer the questions for Lesson 77.

Sidi Saiyyed Mosque, Ahmedabad, India

78 The Siddis of India

Malik Raihan Habshi, known as Ikhlas Khan, became chief minister in the Sultanate of Bijapur in Western India. This portrait is from about 1650.

You know about African Americans. In this lesson you will learn about African Indians.

Along the western coast of India, in the dense forests that are the home of black panthers and hornbill birds, live about 20,000 descendants of Bantu East Africans. Most of their ancestors came as slaves, brought by Muslims as early as the 600s, and then later by the Portuguese and British when they began developing a presence in India. Some ancestors of modern Siddis came to the Indian subcontinent as free merchants, sailors, and mercenaries.

These migrant people became known as Habshis, the Persian word for Abyssinians (Ethiopia was once known as Abyssinia). The later term that Indians applied to them was Siddis, possibly derived from Arabic. The term Siddi is now used for all those of African descent who live in India. Most of the Siddis have been Muslims, which is still true today, although some converted to Hinduism. A Siddi leader oversaw construction of the mosque pictured above in the 1570s.

A few Siddi achieved high status as generals, admirals, and local rulers, while some became architects and city planners. Some of the structures they built and lived in still stand. Africans who moved from Africa to other continents, whether by choice or by force, are called the African Diaspora.

The accomplishments of those who became successful and prominent in India speak well of their talents and determination to overcome their humble origins there.

However, most of the Siddis were slaves. When the British Empire abolished slavery in the early 1800s, the Siddis fled to the jungles out of fear that they would be recaptured. There they were able to hide, and other Indians were satisfied to leave them there in the villages that they formed.

The Siddis are a fascinating mix of African and Indian characteristics. Their physical features are African. Their music and dances are African. Their dress and language are Indian. Many of their first names reflect African, British, or Portuguese background; but their last names are Indian.

As slaves and the descendants of slaves, the Siddis have occupied the lowest ranks in India's caste system. As you will learn below, those whose work is particularly distasteful to Indians are considered untouchable. Their work and dark skin cause many Indians to treat them with prejudice. Most Indians are happy with them living in the forest apart from Indian society—unless Indians decide that they want the land that the Siddis live on and sometimes own. Then the Indians want the Siddis to move further into the jungle.

Siddis in Gujarat, India (2015)

Siddis in Mumbai, India (1912)

The Caste System

The caste system is a ranking of groups in society. It is the product of the Hindu religion and is part of the Hindu worldview. The Hindus believe that the castes sprang from various parts of the body of Brahma, the Hindu god of creation.

Castes are based largely on the occupations of those in the castes. The highest caste is the Brahmans (also spelled Brahmins). These are the priests and teachers of sacred knowledge. Hindus believe that Brahmans sprang from Brahma's head. It is not surprising that Brahmans hold the highest caste, since they are the ones who devised and established the system.

The second highest caste is the Kshatriyas, the ruling class and warriors, who are said to have come from Brahma's arms. Third are the Vaishyas, the farmers, merchants, and so forth, who sprang from Brahma's thighs. Finally are the Shudras, the servants who perform menial labor and who are said to have come from Brahma's feet. A fifth group, the Dalits, is outside of the four categories and contains those who perform work considered especially disgusting, such as removing dead bodies or cleaning latrines. These are considered the "untouchables" and are the origin of the term outcasts (outcastes).

These are the main caste groups, but they are broken down into about 3,000 castes and 25,000 subcastes, according to specific occupations. People almost never move from one caste to another but stay in the caste into which they were born. Theirs was the caste of their parents, and it will be the caste of their children. Other factors such as race, where a person lives, and the individual's language and culture can also contribute to a person's caste. The caste worldview also influences followers of other religions in India.

Government leaders paid tribute to B. R. Ambedkar in 2009 at India's Parliament House and took this photo beneath his portrait.

The caste system keeps members of the different castes separate in society. Traditionally, the members of different castes did not marry one another, eat together, or even share the same water wells. The Dalits were subjected to severe and sometimes cruel discrimination. The people in a caste often share their last name. British colonial rulers in India maintained the caste system to make the ordering of society easier and generally appointed members of the higher castes to hold government positions.

When India became independent of Great Britain in 1947, the new constitution outlawed discrimination based on caste, although discrimination still persisted in practice. India passed additional laws that gave more rights and protections to people of the lower castes and set quotas for jobs and college admissions. Urban areas do not have the same pattern of discrimination as rural areas do.

A few Dalits have taken advantage of these laws and have been successful in crossing social lines. For instance, B. R. Ambedkar was a Dalit who wrote the new Indian constitution. K. R. Narayanan, also a Dalit, was elected president of India in 1997. Jitan Ram Manjhi was elected chief minister of Bihar province in 2014. He is a member of the lowest Dalit caste, called ratcatchers. He grew up on a farm where his parents were laborers and where he literally caught and ate rats. But he took advantage of the opportunity to get an education and was able to open doors for himself. Old habits die hard, however, and the structuring of society on the basis of castes still takes place to a considerable degree.

What Is the Answer?

God created all people with equal value in His eyes. Mankind, however, has not generally looked on people that way. We have tended to discriminate between groups on the basis of race or other characteristics. Modern Siddis have generally not received the opportunities that many other people in India have had simply because of their race and background. Is it possible for Siddis to have brighter prospects?

A few Indians have tried to open a door for Siddi youth by encouraging their skills in sports, especially soccer. Other Indians have encouraged the Siddi to learn, develop, and take pride in their distinct culture. Meanwhile, most Siddis still work as farmers and manual laborers. Since most Siddis are poor, they cannot obtain an education that would give them greater opportunities. Playing soccer and appreciating one's heritage do not usually go far in paying the bills.

Perhaps by studying this situation in another country and culture, we can see more clearly how harmful prejudice is, and how official policies and social norms can treat people unfairly. But change comes slowly. Would you want to change the system? How would you go about it? Are there ways that Americans can do a better job of treating all people fairly and equally in our own country?

The fact is that geography affects people's lives. Where people live, their ethnic background, and certain physical characteristics can make a huge difference in their lives and in the prospects for a better life for them and their children.

Paul says that in Christ there is a renewal in which such traits do not matter.

. . . a renewal in which there is no distinction between Greek and Jew, circumcised and uncircumcised, barbarian, Scythian, slave and freeman, but Christ is all, and in all.
Colossians 3:11

Assignments for Lesson 78

Gazetteer Read the entries for India, Pakistan, and Sri Lanka (pages 144, 148, and 149).

Worldview Copy this question in your notebook and write your answer: What might have been the value of the account of creation in Genesis for the nation of Israel? What is its value for the church today?

Project Continue working on your project.

Literature Continue reading *Boys Without Names.*

Student Review Answer the questions for Lesson 78.

Mount Everest

79 "Because It's There"

It is without a doubt the greatest challenge to humankind that the earth's geography presents: scaling the 29,035 feet to the summit of the highest mountain in the world, Mt. Everest in the Himalayas. The mountain sits on the border between Tibet and Nepal. Over four thousand people have accomplished this feat. In 1953, however, no one had; but it was not because they hadn't tried.

The Roof of the World

The Himalayan Mountain range includes the highest mountains in the world. About 110 peaks rise above 24,000 feet. The range extends for some 1,550 miles and varies between 125 and 250 miles wide. Geographers consider the Himalayas to be part of an extended mountain system that stretches from North Africa to the Pacific coast of Southeast Asia. The Himalayas divide the Tibetan plateau to the north from the plains of India to the south. Along their summits is the border between Tibet and Nepal, but several countries claim parts of the range.

The name comes from the Sanskrit words *hima* (snow) and *alaya* (abode). At the highest reaches, snow and glaciers are year-round fixtures. However, some parts of the region descend to more typical levels. The lands of the range include valleys, rivers, and farming areas. Some parts of southern Nepal are subtropical and are only about one thousand feet above sea level.

The British named the highest mountain in the world in 1865 for George Everest, a Welsh surveyor who worked in southern Asia in the 1800s. Even this massive mountain is subject to the forces of geography. A powerful 2005 earthquake in the region moved Everest just over an inch to the southwest, although it did not affect its height.

The peak of Mt. Everest sits at the cruising altitude of jets. It pokes up two-thirds of the way through the earth's atmosphere. At that altitude the temperature is cold, the weather is unpredictable and dangerous, and the oxygen level is low. Oxygen masks are a necessity (or almost, as we will see below).

The Assault on Everest

Although local people have long looked upon Mt. Everest with awe and reverence, and even though Westerners had known about it for generations, no one was known to have tried to climb it until an attempt by British climbers in 1921, who had to turn back because of bad weather. After World War

II, interest grew in scaling the mountain's heights. Several countries organized expeditions to do so. In 1952 a Swiss team led by Richard Lambert and Sherpa *sirdar* (head porter) Tenzing Norgay fell about 825 feet short. The British team was next in line. Other countries had arranged efforts that were scheduled to follow, so the British had a strong desire to get the job done.

We must not think that scaling Mt. Everest is something an individual climber does on his own. The efforts are carefully planned expeditions that often involve hundreds of people. The Royal Geographical Society and the Alpine Club of London sponsored the 1953 British expedition. British military officer Sir John Hunt led the expedition with precise military planning. The personnel included several hundred porters who carried equipment and supplies, twenty Sherpa climbers, and ten British climbers, all of whom had extensive experience at mountain climbing. The team planned their route carefully based on earlier experience and set up camps along the way as places to rest and to store supplies. Each British or British Commonwealth climber was paired with a skilled Sherpa climber. The first two pairs failed to reach the summit, the second getting to within three hundred feet of the peak.

The third climber in the group was Edmund Hillary of New Zealand. Hillary, 33, had begun climbing mountains in his homeland when he was in high school. By profession he was a beekeeper like his father, but he was an active climber. He trained for this climb on the glacier peaks of New Zealand. It was his fourth climbing expedition in two years.

Accompanying Hillary was Tenzing Norgay, 38. The Sherpa are an ethnic group of about 150,000 who live mostly in Nepal but also in Tibet and India. Tenzing was born in Tibet and had lived in Nepal, but at the time he was living in India. Many Sherpa became skilled at mountain climbing. Tenzing had been involved in six previous attempts at Everest, including the 1952 Swiss expedition.

Edmund Hillary and Tenzing Norgay After Their Historic Climb (1953)

The climbing season at Everest lasts for only two months, and the best time is the last half of May. This is the time that the British chose. On May 29, 1953, Hillary and Tenzing took on the last few feet of the climb, including a forty-foot sheer cliff that became known as the Hillary Step. About 11:30 that morning, Edmund Hillary became the first person known to step onto the peak of Mt. Everest. Tenzing followed a few moments later. In proper British fashion, Hillary shook hands with Tenzing. In reply, Tenzing hugged Hillary, who hugged him back. They spent about fifteen minutes on the summit, then they started back down the mountain.

As Hillary and Tenzing descended, others took the news ahead of them. A runner carried the news from the base camp to a radio post. Queen Elizabeth II learned of the accomplishment on June 1, the day before her coronation as queen. The news became generally known around the world on the day of her coronation as ruler of the United Kingdom and the British Commonwealth of Nations, which includes New Zealand. Later that year, Queen Elizabeth II knighted Hillary. Because Tenzing was not a citizen of a commonwealth nation, he received the George Medal, which is given to those demonstrating great bravery.

Afterwards

Hillary was not prepared for the avalanche of publicity that followed the historic feat. He tried to use his fame and the financial rewards that came to him through books and other endeavors to help others. Hillary organized the Himalayan Trust, which improved the lives of the Sherpa people by building schools, hospitals, and airfields. He also continued his adventures, revisiting the Everest region several times. He reached the South Pole in 1958. In 1967 he was part of the first team to climb to the peak of Mount Herschel in Antarctica. Hillary died in 2008.

Tenzing Norgay continued to live in India and became the chief instructor at the Mountain School in Darjeeling. He frequently traveled to other countries as an ambassador of goodwill. Tenzing died in 1986. Hillary and Tenzing remained good friends for the rest of their lives, and their sons became close friends in later years.

The accomplishment of Tenzing and Hillary opened the door for a significant increase in the number of people attempting the climb. The first American to reach the top, James Whittaker, did so in 1963. In 1975 the Japanese climber Tabei Junko became the first woman to accomplish the feat. Two climbers in 1978 reached the top without using additional oxygen. Around 800 people attempt to climb the mountain every year. Unfortunately, over 300 people have lost their lives making the attempt.

Sherpa guides are much in demand. Namche, a town in Nepal at an elevation of over 11,000 feet, has become quite wealthy because of the influx of climbers who go there and are able to spend significant amounts of money attempting to climb Everest. The town even has Internet access.

Namche, Nepal

Why?

View from Mount Everest (2013)

In a 1974 interview, Sir Edmund Hillary said, "I think I mainly climb mountains because I get a great deal of enjoyment out of it. . . I think that all mountaineers do get a great deal of satisfaction out of overcoming some challenge which they think is very difficult for them. . . . Not until we were about 50 feet from the top was I ever completely convinced that we were actually going to reach the summit. . . . I think my first thought on reaching the summit—of course, I was very, very pleased to be there, naturally—but my first thought was one—a little bit of surprise. I was a little bit surprised that here I was, Ed Hillary on top of Mt. Everest. After all, this is the ambition of most mountaineers."

After the failed 1921 British attempt to climb Mt. Everest, a journalist asked George Leigh Mallory, a member of the expedition, why he wanted to climb it. He answered:

"Because it's there."

As the psalmist considered the mountains around Jerusalem, he knew from where his help came.

I will lift up my eyes to the mountains;
From where shall my help come?
My help comes from the Lord, Who made heaven and earth.
He will not allow your foot to slip. . . .
Psalm 121:1-3a

Assignments for Lesson 79

Gazetteer	Read the entries for Iran, Maldives, and the British Indian Ocean Territory (pages 145, 146, and 150). Read "The Faces of South Asia" (pages 295-297). There are no questions on this photo essay in the *Student Review Book.*
Geography	Complete the map skills assignment for Unit 16 in the *Student Review Book.*
Project	Continue working on your project.
Literature	Continue reading *Boys Without Names.*
Student Review	Answer the questions for Lesson 79.

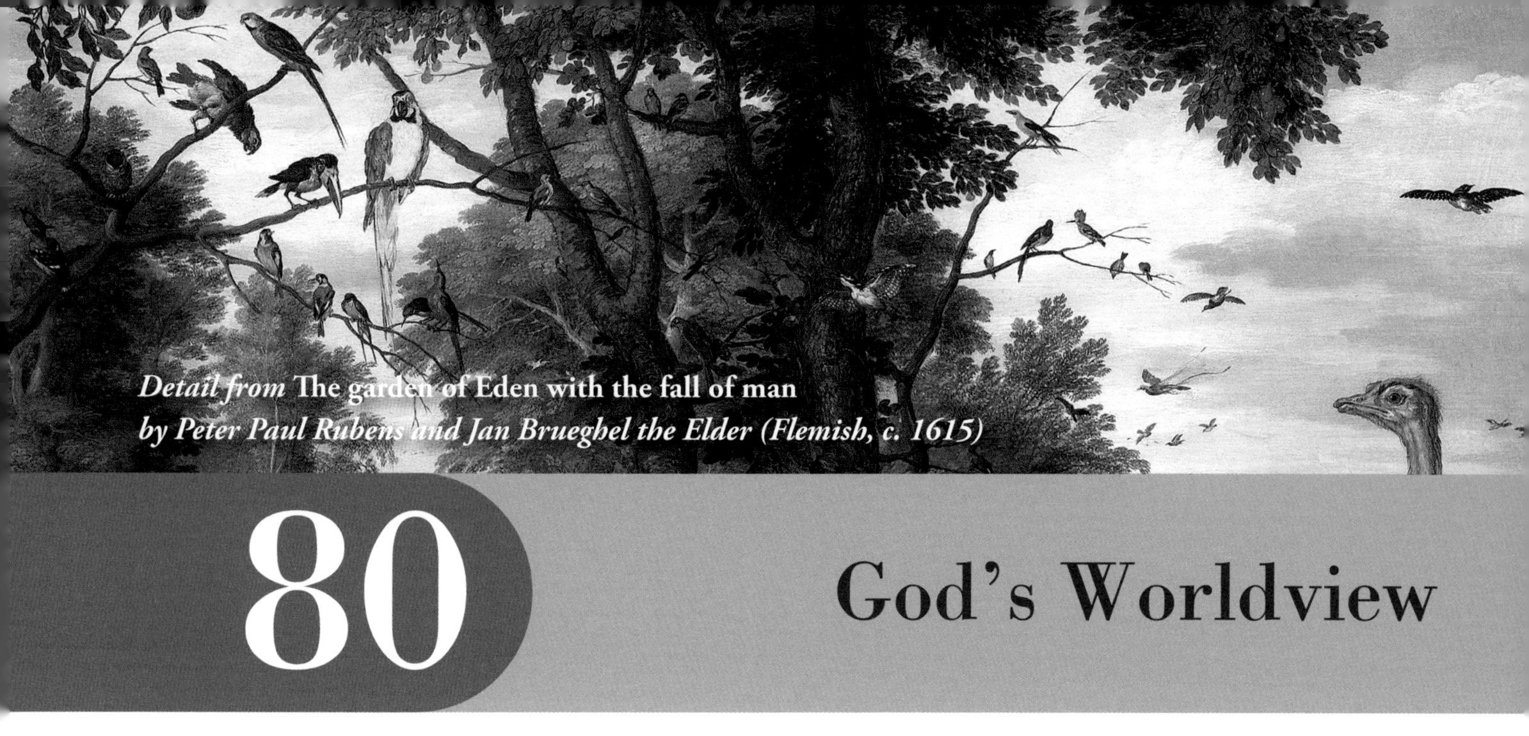
Detail from The garden of Eden with the fall of man
by Peter Paul Rubens and Jan Brueghel the Elder (Flemish, c. 1615)

80 God's Worldview

Knowing your past helps you know who you are. No one stands on his own. We all stand on the shoulders of the many generations who have gone before us. The lives and decisions of our parents, grandparents, and earlier ancestors help to shape our own thoughts and lives.

The most important factor in our past is what God has done in the world and especially in the lives of those who have gone before us. Understanding the story of our past is an important element of our worldview. Our history gives us a way of understanding the world in which we live—in other words, our worldview.

God inspired Moses to write the narrative of Genesis to tell the people of Israel about their past. The Lord wanted them to understand how they came to be His people. In telling this story, He started with Creation. By doing this the Lord gave Israel the big picture of what He had done. God provided Israel with the basis for the worldview they needed as God's people. Remembering what God had done in their history was going to be important for them as they lived in the land God had promised to give them. Even this was a significant aspect of the worldview they were to have: they were to understand that God had given them the land in which they were living.

God's View of the World—His Worldview

The opening passages of Genesis were God's first lessons in giving Israel the understanding they were to have about God, their world, and themselves. Here are some key points from those passages.

God created everything out of nothing. He gives life to all living things, including people. God wanted Israel to understand this fact so that they would know whom to thank and praise and on whom they depended. Everything that Israel possessed and everything they experienced was the result of what He brought into existence.

He also wanted them to understand that this account of Creation was different from the many creation stories of the nations around them, stories that the Israelites might well have known. Every culture has a story about how the world began. These stories are attempts that people have made to explain their origins. Some of these stories involve cosmic battles between spiritual forces or conflicts among the gods or the actions of primordial animals, which result in the creation of our world. Sometimes in these pagan stories the physical creation is a byproduct of those conflicts and actions. In addition, some of the religions of the nations around Israel taught that people should worship the sun, moon, and stars. Genesis teaches that God created the sun, moon, and stars; and thus He is superior to them. Genesis told Israel that God created the world with purpose, in an orderly fashion, and by the power of His mere word. That's the kind of God they had.

Geography is from God. He created the expanse of the heavens, the waters and the dry land of the earth, and everything that lives on the earth. Genesis 1 uses the phrase "after their kind" repeatedly. The different living aspects of Creation are all part of Creation; but God created them separately, and they are different from each other. God's will was for the living things to multiply and fill the earth, both on land and in the sea.

Light is an important part of Creation. When God said, "Let there be light," he was referring to the entire electromagnetic spectrum; but of course, the ancient Israelites would not have understood this. The emphasis on light in Creation reveals how important light is to the ongoing functioning of God's world.

The pinnacle of God's Creation is mankind. Of all that God created, only man bears the divine image. God created humankind male and female; in Scripture, gender is this simple and clear-cut. These realities—our divine image, our maleness or femaleness—are part of our identity as humans, and so they are part of who we should understand ourselves to be. God gave mankind the role of ruling over the other living things. This gives humankind rights, but it also gives them responsibilities. The Israelites were to understand themselves to be stewards of all that God gave them.

On every day of Creation, God saw that what He had created that day was good. In summary, on the sixth day, God's worldview was that everything He created was very good. It was beautiful, orderly, and complete. It worked together. It had purpose and gave honor to Him.

Another element of God's worldview is that rest is good. Rest is part of the order and rhythm of life. God ceased from work on the seventh day, which set that day apart and made it holy. God expanded on this concept when He gave Israel the

This view of the sun rising over Sri Lanka is taken from a mountain known in English as Adam's Peak. At the top of the mountain is a depression shaped like a larger-than-life footprint. Some Buddhists associate the "footprint" with the Buddha. Some Hindus associate it with the god Shiva. Some Muslims associate it with Adam, suggesting he made it when he left the Garden of Eden.

Ten Commandments. At that time He told Israel to remember the sabbath day to keep it holy (Exodus 20:8-11). They were not to do work that profited or advanced themselves. This taught the Israelites the worldview that God was in charge and that He provided for them. They were to work as they lived out their lives, but their success and well-being ultimately did not depend on their work; it depended on God, and they could depend on Him.

Genesis 2 returns to the theme of the significance of man in God's creation. Genesis 1 describes God creating the different aspects of Creation, culminating in the creation of man. Genesis 2 focuses on man by first describing God creating him and then describing the rest of God's Creation. God provided man a place to live and food to eat.

Then the account emphasizes the special, unique relationship that man and woman have. The Lord brought to Adam every living creature, but none was a suitable helper for him. At that point God created woman from one of Adam's ribs. When God brought the woman to Adam, Adam exclaimed, "Ah, now this—this is the one who is part of me. This is woman." As God took the woman out of man and made them separate persons, in marriage God brings the man and the woman back together and makes them one. This taught the Israelites that the husband-wife relationship is God-made, beautiful, and unique. This account taught them that marriage is to be completely open with nothing hidden and nothing to be ashamed of. It is a relationship that makes the two into one. The husband and wife are to cherish and protect each other and to rejoice in their relationship.

The View of Sin in Genesis

The Garden was wonderful for Adam and Eve, but evil was a reality even then. The Garden contained the tree of the knowledge of good and evil. Adam and Eve were not to eat from it. Even though people were made in God's image, and even though God gave people a special status and important responsibilities, God put limits on their behavior. There was something that people were not to do. It was best for them not to do it, and they were to trust God in this and to be content enjoying what God gave them and doing what He told them to do.

The serpent was an influence for evil. Genesis says nothing about the origin of Satan. The Bible gives hints elsewhere, but perhaps that is something we should not spend a lot of time trying to figure out. Satan is real and his influence is present even in the best of circumstances; that is probably all we need to know. Satan showed himself to be deceptive and the father of lies.

Eve and Adam gave in to temptation and sinned. Genesis teaches us that sin is primarily a heart problem. For instance, after the serpent tempted Eve with its lies, Eve "saw" the tree in a different way and desired what it offered her. Cain murdered Abel because Cain became angry that God did not regard his sacrifice. Later, men's pridefulness led them to build the tower at Babel. We see sin's devastating consequences in the lives of Abraham's descendants. Sin is not a matter of people being social misfits; sin is rebellion against God. Sin comes from the decision to exalt and follow self instead of God.

Sin is real; it matters; it has many consequences. Most importantly, sin separates people from fellowship with God. It says that a person does not really trust God. Sin takes away beautiful innocence. It leads to defensiveness and blaming others; notice how Adam and Eve both blamed others for their actions: Adam blamed Eve, and Eve blamed the serpent. Sin had negative consequences for Adam and Eve for the rest of their lives and for the lives of others who followed them. It made life harder for them. Sin appears to be a shortcut to a better way, but in reality it is the road to bitterness and difficulty.

Despite man's sin, God provided for them. He made clothing for Adam and Eve instead of leaving them to fend for themselves. God is relentlessly good, even in the face of sin.

How Genesis Provides a Worldview

The first few chapters of Genesis gave Israel (and give us) an understanding of the world. The passage tells how Creation began. It tells of God's power, goodness, and discipline. We learn how sin entered the world and how its consequences extend into the lives of the people who commit sin as well as others who are affected by it. Later, we learn the origin of the rainbow after the flood and the origin of languages after the Tower of Babel. God created a world that was very good, but the influence of Satan and the sin of human beings besmirched that world and led to separation from God, murder, widespread unrighteousness, and the separation of people from each other through the introduction of many languages and the scattering of people over the face of the earth.

But Genesis also tells us how God began the process that He had planned even before the creation of the world to redeem mankind from sin and death (Genesis 3:15). Through His choosing of Abraham and later the people of Israel, and through His sending Jesus to be the Redeemer of Israel and the world, God provided a way to restore people to a relationship with Him and to give them everlasting life after this life ends.

Even though this world brings sin, hardship, separation from God, and divisions among people, God provides hope that does not disappoint (Romans 5:5). As we consider the world in which we live, we must acknowledge the realities of sin and death but also the realities of God's love and hope. This is how Genesis gives us an accurate worldview.

And I will put enmity
Between you and the woman,
And between your seed and her seed;
He shall bruise you on the head,
And you shall bruise him on the heel.
Genesis 3:15

Assignments for Lesson 80

Worldview	Recite or write the memory verse for this unit.
Project	Finish your project for this unit.
Literature	Continue reading *Boys Without Names.*
Student Review	Answer the questions for Lesson 80. Take the quiz for Unit 16.

Carpet Workshop, Bukhara, Uzbekistan

17 Central Asia

This unit discusses Central Asia, the region that lies between the Caspian Sea and China. We examine the story of the Aral Sea, which almost dried up due to poor Soviet environmental policy. Kazakhstan is the largest country in Central Asia, and its geographic location is strategically important. Tajikistan is the poorest former Soviet country. The story of Turkmenistan reveals the wide impact of the Turkish people. The worldview lesson presents one Christian viewpoint on the environment.

Memory Verse

Memorize Psalm 146:2-3 by the end of the unit.

Books Used

The Bible
Exploring World Geography Gazetteer
Boys Without Names

Project (Choose One)

1) Write a 250-300 word essay on one of the following topics:
   - Which state and national policy do you think is best: government regulation of the environment, or allowing businesses, individuals, and private groups to maintain the environment with minimal government oversight? Give reasons for your answer.
   - Research and write a report about what life is like in the desert climate of Central Asia.
2) Prepare or buy your favorite bread. When you serve it to your family, give a brief talk on what you think Jesus meant when He said, "I am the bread of life" (John 6:35).
3) Research and write a report on the distinctive aspects of Turkish culture.

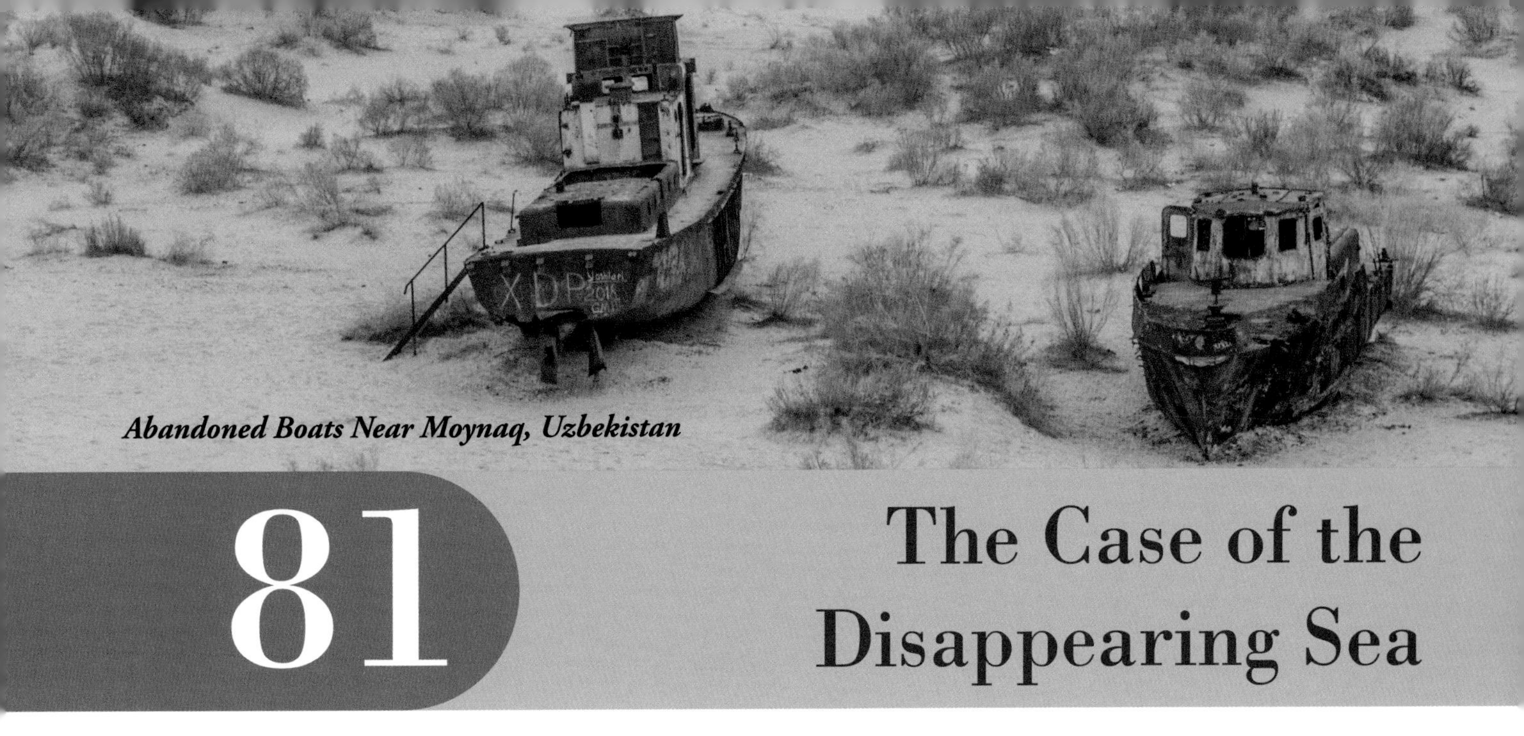
Abandoned Boats Near Moynaq, Uzbekistan

81 The Case of the Disappearing Sea

Khojabay used to be a fisherman who lived by the sea. Today, he lives in a desert.

Khojabay didn't move. Instead, the sea all but disappeared.

The once-thriving fishing village of Tastubek had ninety households around 1980. By the mid-1990s, only nine families lived there.

Similar scenes fill the surrounding landscape for hundreds of miles. Countless fishing boats lie empty, abandoned, and useless on dry ground, miles from the current shoreline. Fish canneries that once processed abundant catches are empty and unused. Cities that were popular with tourists are nearly ghost towns, as residents moved away to find work and a healthier environment.

These scenes are the human evidence of what scientists have called one of the worst environmental disasters of the twentieth century. But rather than blame capitalists, we have to place the responsibility for this tragedy on Soviet Communists.

The Sea That Was

The Aral Sea lies in Central Asia. It straddles the border between Kazakhstan to the north and Uzbekistan to the south. These now-independent countries were once part of the Soviet Union.

At one time, the Aral Sea was the fourth largest inland body of water in the world, behind the Caspian Sea between Europe and Asia, Lake Superior in North America, and Lake Victoria in Africa. It covered about 26,300 square miles, a little more than the size of West Virginia. The sea stretched 270 miles north to south and 180 miles east to west. Several varieties of fish lived in abundance in its waters. This bounty enabled an active fishing industry which included bustling fish canneries on shore. Several seaside towns were popular tourist destinations in a mostly desert region of the world.

The area only receives about four inches of rain annually. Dry, windy conditions cause significant surface evaporation. The sea was fed mainly by two rivers, the Amu Darya and the Syr Darya. Water that the Aral received from these rivers maintained the sea's level despite the dry climate.

The Great Communist Idea

In the 1950s, the central planners of the Soviet Union decided that they had a better idea than the way God created the region. The government decided to develop the mostly unused land around the Aral for agricultural production, primarily for

Cotton remains a major crop in Uzbekistan.

cotton. They diverted water from the two rivers to irrigate the crops. With this the problems began.

Without the inflow from the rivers, the Aral Sea began to dry up. With less water, the concentration of salt increased. The pesticides and chemical fertilizers that farmers used in growing their crops washed into the sea and poisoned the waters. Because of these factors, the fish population decreased. Khojabay's last catch, in 1976, was a net full of dead fish. With fewer fish, the fishing industry died.

The sea produced zero commercial fishing in 1987. As the sea receded from what had been seaside towns, shipping and tourism dried up (pardon the expression). Shops closed. Abandoned port facilities and hundreds of ships languished on dry ground, what used to be coastal areas. People cut up some ships to sell as scrap metal.

By 1990, the rivers had virtually dried up and some 80-90% of the surface area of the Aral Sea had disappeared. The Aral Sea had become the Aral Desert. The one-time seaside city of Aralsk in Kazakhstan, with a population of about 40,000, now stood twelve miles from the shore of the Aral. The saline level of the water that remained was three times higher than it had once been, killing off more fish and making the water largely unusable. Even the tips of the grass that grew in the region became salty. This in turn negatively affected the camels that people relied on for transportation.

This was not all. The pesticides and fertilizers that farmers had used on the agricultural fields also poisoned the land. Dust storms blew toxic dust into people's faces and homes. The region saw a marked

The town of Kantubek on Vozrozhdeniya Island was home to the Soviet biological weapons facility.

Images taken from space in 1985 and 2020 dramatically illustrate the shrinking of the Aral Sea.

increase in cancer, anemia, kidney disease, and infant mortality.

This was not all. The weather changed dramatically without the moderating effect of the sea. The climate was hotter in the summer and colder in the winter. This was another reason why people moved away from the area.

This was not all. The Aral Sea contained many islands. One, Vozrozhdeniya Island, became a peninsula connected to the mainland. This allowed easier access to the one-time island, but therein was a problem. The island had been a secret testing site during the Cold War, where the Soviets worked on biological weapons. This included tests on bubonic plague and anthrax bacteria. A team from the United States cleaned up the remnants of the biological weapons—supposedly—in 2002.

The Aral Sea became two much smaller bodies of water, the North Aral Sea and the South Aral Sea. In sum, the Soviet policy destroyed the Aral Sea ecosystem and with it the way of life of the people who lived there. With the fall of Communism, Kazakhstan, Uzbekistan, and the other Central Asian republics were left with the responsibility of determining what to do about the situation.

Turning the Tide

In 2005 the World Bank funded the construction of the eight-mile-long Kokaral Dam that created a reservoir which in turn fed the North Aral Sea and increased its size. The Kazakh government helped with replenishing fish. As a result, the fishing industry and fish processing operations have begun to recover. Later generations of families that once made their living from fishing are returning to the industry and finding it profitable. Tourists are also returning and shops are reopening, although the people involved in these activities are perhaps not fully addressing the continuing environmental dangers.

The renewal of life in the Aral Sea region is slow and uneven. Although the North Aral Sea is recovering, the South Aral Sea is still largely dead.

The efforts to increase the flow of water have led to conflicts over where the water goes and what parts of the affected region should benefit from the projects.

The interaction between people and geography can be helpful, but it can also definitely be harmful. People must be wise and respectful of God's Creation because the way that humans treat the earth can have effects that last for generations.

By God's mercy, recovery is sometimes possible. The prophet Isaiah described a renewal of the land of Israel in this way:

The scorched land will become a pool
And the thirsty ground springs of water;
In the haunt of jackals, its resting place,
Grass becomes reeds and rushes.
Isaiah 35:7

Assignments for Lesson 81

Gazetteer Study the map of Central Asia and read the entries for Kazakhstan and Uzbekistan (pages 151, 152, and 156).

Worldview Copy this question in your notebook and write your answer: What do you think God meant when he said, "Be fruitful and multiply, and fill the earth, and subdue it; and rule over the fish of the sea and over the birds of the sky and over every living thing that moves on the earth" (Genesis 1:28)?

Project Choose your project for this unit and start working on it. Plan to finish it by the end of this unit.

Literature Continue reading *Boys Without Names*. Plan to finish it by the end of the this unit.

Student Review Answer the questions for Lesson 81.

Charyn Canyon, Kazakhstan

82 Kazakhstan: Who Are They?

In June 2019 Kassym-Jomart Tokayev was re-elected president of Kazakhstan. He had been in office since the previous March, when the incumbent, Nursultan Nazarbayev, had resigned after exercising power for thirty years. Tokayev, whom Nazarbayev had chosen to be his successor, received just over 70% of the vote; his nearest challenger received about 16%. The election was marked by anti-Nazarbayev demonstrations and charges of corruption and vote fraud.

Even after he left office, Nazarbayev continued to exercise considerable influence as chairman of the Security Council; and he retained the constitutional title of Leader of the Nation. His daughter held the position of speaker of the upper house of the country's parliament. Many observers believed that she was preparing to assume the presidency at some point. However, in 2020 Tokayev replaced Ms. Nazarbayeva, who had become connected with scandals, with someone more loyal to him.

Nazarbayev had become head of the Communist Party of Kazakhstan in 1989, when it was still part of the Union of Soviet Socialist Republics. The following year, the Kazakh Supreme Soviet named him president. The Soviet Union disintegrated in 1991, and Kazakhstan became an independent, officially non-Communist country which Nazarbayev continued to lead. In the first election after independence, Nazarbayev received 95% of the vote, surprisingly low since he was the only candidate on the ballot.

At first, Nazarbayev appeared to be open and accommodating to change and reform; but as his presidency continued he became more authoritarian. Over the years, the government changed Kazakhstan's constitution and various laws so that Nazarbayev could stay in office well past the two-term constitutional limit. In his last election victory, he received 98% of the vote. Nazarbayev is genuinely loved and respected by many of the people, a large percentage of whom have never known any other leader. On the other hand, a vocal segment of the Kazakh people condemn his practices and demand greater rights and a genuine democracy.

As just one indication of the former leader's power and influence in the country, the day that Tokayev assumed the presidency from Nursultan Nazarbayev, the new president announced that the name of the capital city was being changed. The new name? Nursultan.

This swirl of activities in the highest levels of government reflects questions about the country as a whole. For instance, what kind of country is Kazakhstan? It is officially considered to be a republic. To be sure, the country holds elections; but the government functions in ways that seem more authoritarian than democratic. This conflict illustrates how it is difficult to pin down what Kazakhstan is and who the Kazakh people are.

The Land

The land area of Kazakhstan is about four times the size of Texas. It is the largest country in Central Asia and the largest former Soviet republic besides Russia. It is the largest landlocked country in the world. Kazakhstan and Azerbaijan are considered to be in both Europe and Asia. They are the only two landlocked countries that are on two continents.

Kazakhstan is strategically located. It borders China on the southeast and extends to the Caspian Sea in the west. The Caspian Sea is the largest inland body of water in the world. Kazakhstan shares a long border with Russia on its north and east. Historically Kazakhstan has had close ties with Russia, which we will discuss in more detail below; but the Kazakh government does not want just to be a satellite of Russia again. Kazakhstan also wants to develop stronger economic ties with China. This desire involves an intricate diplomatic dance, since China and Russia have long had uneasy relations. Kazakhstan's past and present raise another question about Kazakhstan's identity: Is it an ally of Russia or China? Kazakhstan would like to be both.

Kazakhstan is mostly vast, flat steppes and deserts, with smaller areas of hills and low mountains. Parts of the land around the Caspian Sea are below sea level. Only about 3% of the country's land area is woodlands. The country's main geographic advantage is its abundant store of natural resources, which include petroleum, natural gas, coal, iron, and other metals. Kazakhstan exports oil to China and to Europe through Russia. It has the largest economy in Central Asia. Kazakhstan is a member of the Russian-led Eurasian Economic Union, which is Russia's attempt to have a European Union-like alliance of nations. The Soviet Union built its space program launch facilities (their equivalent of Cape Canaveral) in Kazakhstan, and it now rents those facilities from the Kazakh government.

The bitter contrast to this wealth of the land is that Kazakhstan suffers from serious environmental issues. The Soviet Union had nuclear missile test ranges there, which led to atomic radiation impact. Soviet and later Kazakh controls on pollution were lax, which means that there is significant pollution from industry and from the use of agricultural chemicals. The Caspian Sea has significant pollution issues also.

So, is Kazakhstan a wealthy nation or a nation with significant problems? It is both.

Aktau, Kazakhstan, on the Caspian Sea

These women are celebrating Kazakh culture at an ethnic festival in Almaty in 2019.

The People

Arabs brought Islam to the region from the west in the 700s and 800s. Mongols from the east led by Genghis Khan swept over the area in the 1200s. The Uzbeks who came in the 1400s were known as wandering Uzbeks. The word *kazakh* means to wander, so Kazakhstan means land of the wanderers. The Kazakhs are a mix of Turkic, Mongol, and Persian backgrounds.

Imperial Russia took an interest in Kazakhstan in the 1700s and came to dominate it in the 1800s. Some 400,000 Slavic Russians moved into the land during the nineteenth century. About one million Slavs, Germans, and Jews came in the first part of the 1900s. These movements drove Kazakhs off the best farmlands and frequently into poverty.

Kazakhstan became a Soviet Socialist Republic in 1936, but the policies of Joseph Stalin led to widespread famine as most food produced there was shipped to Russia. An estimated 1.5 million people died from famine and disease; many Kazakhs fled to China, Uzbekistan, and Turkmenistan. In the 1950s the Soviet government began the Virgin and Idle Lands project to increase agricultural production in Kazakhstan. One impact of this was the diversion of water from the Aral Sea to irrigate farmland, as discussed in Lesson 81. Another impact was a large influx of Russian and Ukrainian Slavs to work the land. More Kazakhs emigrated to other countries during this time. The result of these people movements was that when the country became independent of the Soviet Union, Kazakhs had become a minority in their own country.

Following independence, many ethnic Russians left and moved to Russia, many members of other ethnic minorities emigrated also, and about one million Kazakhs who had been living elsewhere moved to Kazakhstan. The result has been that the population of the country is now about two-thirds Kazakh and about one-fifth Russian. Kazakhs are still a majority of urban dwellers, but most of the growth in urban population has come from the immigration of foreigners and not Kazakhs moving from rural areas.

About 70% of Kazakhs are Muslim, but for centuries the Islam practiced there was not as strict as that typical in Arab countries. This changed as Muslim schools became more common in cities before the Russian takeover in the 1900s. Under Communism the U.S.S.R. discouraged any religious activity. After the fall of Communism, Kazakhs have enjoyed freedom of religion. About one-fourth of the population is Russian Orthodox. Over half of the population is under thirty, and one-fourth is under fifteen.

So, who are the Kazakhs?

The Languages

The languages of Kazakhstan open another window on the influence of Russia and the complexity of life for the Kazakh people. Kazakh is a Turkic language that used to be written in Arabic characters. During the period of Russian domination, the Russian language became the common language of government and education. The Soviets applied the Latin alphabet to Russian in Kazakhstan in the 1920s but then shifted to Cyrillic (the alphabet of the Russian language in Russia) in 1940. This emphasis on Russian meant that many Kazakhs grew up not knowing the Kazakh language.

But Cyrillic uses 42 characters, which moderns find cumbersome on keyboards and digital devices. In 1989, Nazarbayev announced that Kazakh (using the Cyrillic alphabet) would be the official language of the country, although the constitution adopted in 1995 officially acknowledged Russian as well. In 2017 Nazarbayev declared that Kazakh would henceforth use the Latin alphabet, which English and other Western languages employ, as a way to minimize Russian influence. These changes had significant impact in government communication, education (how teachers taught and the textbooks they used), and the media. Fluency in Kazakh is increasing, but in the 2009 census, 62% of the people said they were fluent in Kazakh while 85% said they were fluent in Russian.

So what language is the way people really communicate in Kazakhstan?

The culture of any people that has had significant interaction with others will be a mixture of features drawn from several influences. Changes imposed by government decree but that don't really change the hearts of the people add an additional level of complexity. The history and geography of Kazakhstan have led the people of Kazakhstan to live with this complexity every day.

While the New Testament teaches believers to respect their governing authorities, the Bible also reminds us that they are not the source of our salvation.

I will praise the Lord while I live;
I will sing praises to my God while I have my being.
Do not trust in princes,
In mortal man, in whom there is no salvation.
Psalm 146:2-3

Assignments for Lesson 82

Gazetteer	Read the entry for Kyrgyzstan (page 153). Read the photo essay on "Architecture in Kyrgyzstan" (pages 298-301).
Worldview	Copy the following question in your notebook and write your answer: In what ways have people cared for the environment well, and in what ways have people harmed the environment?
Project	Continue working on your project.
Literature	Continue reading *Boys Without Names.*
Student Review	Answer the questions for Lesson 82.

Market in Dushanbe, Tajikistan (2010)

83 Of Bread and Poverty: Tajikistan

The center of every meal in Tajikistan is the flatbread called *non*, also spelled *naan*. The cook molds the dough into a thick, flat circle about a foot in diameter. She imprints one side with a tool that leaves a beautiful design and then lightly covers the dough with a mixture of salt and water. To bake it she slaps the finished loaf onto the vertical side of a clay firepit oven called a *tanur*. The high temperature from the fire holds the loaf in place. When it has baked, the cook removes the non and places it on a dish in the center of the table.

Tajikistanis treat non with great respect. A meal is incomplete without non. While the diner is eating, he must keep the imprinted side up; turning it upside down is considered disrespectful. One does not cover a piece of non with anything except another piece. If someone finds a piece of non on the street, he or she will pick it up, kiss it, touch it to the forehead three times, and then place it on a high ledge for birds to eat. Tajikistanis believe that to leave it on the street would be disrespectful and invite bad luck.

Non speaks of many important things for Tajikistanis. It speaks of tradition, beauty, cultural pride, and a sense of accomplishment. It also speaks of necessity: the country is so poor that this simple bread constitutes a major portion of the nation's diet.

Preparing Non in Tajikistan

The Setting

Tajikistan is a landlocked, mountainous country in Central Asia, slightly smaller than Wisconsin. Its highest point, Ismoil Somoni Peak, is 24,590 feet in elevation. The mountain derives its name from the ruler of that land around 900 AD. When Tajikistan was part of the Soviet Union, the mountain was the highest point in the entire Union. At that time it was called Communism Peak.

Unlike most of its Turkic neighbors, the Tajikistani people are a Persian ethnic group, and the Tajik language is derived from Persian. Eighty-four percent of the population are Tajik and speak Tajik as their native language. Ninety-eight percent of the people are Muslim.

The Influence of Russia

Leaders of the Russian Empire had an interest in expanding into Central Asia from at least the time of Peter the Great in the early 1700s. A more concerted effort took place in the 1860s for several reasons. Russia wanted to have greater security for its frontiers. They feared an invasion of Central Asia by the British, who already controlled India. With the loss of cotton imports because of the American Civil War, Russia wanted to grow cotton in Central Asia to provide a more stable source. Like most empires, Russia wanted to control more land and expand its trade. As Russia gained control of much of Central Asia, it called the region Turkistan because of the predominance of Turkic people.

After the Communist revolution in Russia in 1917, the Communist government defeated scattered resistance movements in Central Asia. In 1924 the Communist government made Tajikistan an autonomous part of Uzbekistan, but in 1929 Tajikistan became a full-fledged Soviet Socialist Republic. Moscow's main interest in Tajikistan was the cotton production there.

As the Soviet Union was dissolving in 1991, Tajikistan declared its independence. The last communist leader, Rakhmon Nabiyev, became the first leader of the independent country. However, reform groups opposed his rule; and a bitter civil war erupted and lasted from 1992 until 1997. Between 50,000 and 100,000 people died in the conflict. Over one million Tajikistanis became refugees in neighboring countries or were internally displaced persons within Tajikistan itself. Nabiyev's government fell, and the country endured several years of unstable government. Rebels gained control of large areas of the country while the government continued to depend on troops sent from Russia. Finally in 1997, the government of Tajikistan signed a peace agreement in Moscow with a coalition of Islamic and secular opposition leaders. Some opposition groups rejected the agreement and continued to work against the government, but in 1999 the government put coalition leaders into positions in the government and integrated opposition forces into the country's military.

Poverty

Tajikistan is the poorest of the former Soviet republics. Cotton growing is the dominant agricultural activity, although farmers are diversifying into wheat and other crops. Tajikistan imports 70% of its food. Some mining takes place, and researchers are hopeful that oil reserves might be found there.

The country has two major sources of income, neither of which is an indication of economic health. About one million Tajikistanis work in other countries (90% in Russia, most of the rest in Kazakhstan), and many of them send part of their pay home in what are called remittances. Many people who live and work in the United States do the same thing for their families who live in other countries. Many immigrants to America in the late 1800s and early 1900s either sent money to their families in the home country or saved money to pay for family members' passage to America.

It is good that so many people are so self-sacrificing, but the situation is not healthy in several ways. For one, it tears families apart. For

Fann Mountains, Tajikistan

another, Tajikistanis have to depend on political and economic conditions in other countries and are not able to be productive in their own country. It is also not economically healthy for the country to be so dependent on this source of revenue. These remittances account for about 35% of Tajikistan's gross domestic product (GDP).

A second unhealthy economic activity is the illegal drug trade. Tajikistan is on one of the world's busiest drug trafficking routes. Opiates produced in Afghanistan to the south find their way to buyers in Russia to the north and in Eastern Europe. Some illegal poppy growing and drug consumption take place within the country. At least part of the drug trafficking activity finances Islamic militants based in Afghanistan. Estimates of the economic significance of drug trafficking in Tajikistan range between 30% and 50% of GDP.

The government has been in the hands of Emomali Rahmon, the former director of a state-run farm, and his family since 1994. Rahmon has arranged to be president for life and for his son to assume the presidency when they decide the time is right. Few doubt that Rahmon and his associates have profited from the cotton and drug trades. Rahmon and his associates dangle hard memories of the civil war in front of Tajikistani voters to discourage thoughts of potentially risky change and thus keep themselves in power.

Some positive trends are taking place. An Italian company is building a hydroelectric facility on the Vakhsh River that has already begun producing electricity. When completed in 2026, the facility will include the world's tallest dam and will enable Tajikistan to become energy independent.

And Tajikistan has not completely missed the modern age. Its 8.6 million people use a total of 9.4 million cell phones.

Rudaki Park in Dushanbe is named after a Persian poet named Rudaki (c. 859-941), who was born in what became Tajikistan.

Tajikistan and Geography

The geographic position of Tajikistan in the world led to its ethnic and linguistic makeup. Its geographic proximity to Russia led to a long and often troubled relationship with that country. Its land produced cotton, which has been both a boon and a bane to Tajikistanis. The people of the country have seen strife and bloodshed, but they also have created objects of beauty and pride, such as non bread. All of these realities indicate again how much geography influences everyone's lives.

The prophet Amos recognized the One who is the Author of our world, the world's geographic features, and its inhabitants.

For behold, He who forms mountains and creates the wind
And declares to man what are His thoughts,
He who makes dawn into darkness
And treads on the high places of the earth,
The Lord God of hosts is His name.
Amos 4:13

Assignments for Lesson 83

Gazetteer Read the entry for Tajikistan (page 154).

Worldview Copy this question in your notebook and write your answer: Why do you think some people, from ancient times to the present, have attributed a spiritual nature to the created world?

Project Continue working on your project.

Literature Continue reading *Boys Without Names*.

Student Review Answer the questions for Lesson 83.

Ashgabat, Turkmenistan

84 Turkic, Turkmenistan, and the Turks

Turkic.

Turkmenistan.

Turkey.

Turkish.

Is there a connection? *Yes.*

Isn't it just that Turks are from Turkey? *No, it's not that simple.*

Islam Spreads East, Turkic Spreads West

The Altai Mountain range lies in Central and East Asia where China, Mongolia, Kazakhstan, and Russia come together. The Altai region gave rise to three related but distinct people groups and languages: Mongolian, Manchu-Tungus, and Turkic.

In the decades following the life of Muhammad, Arab Muslims spread their religion east to the Altai region. A few centuries later, Turkic-speaking people spread west from their homeland, across the Central Asian steppes and along the Silk Road that ran between China and the West. Turkic people were the origin of the country we now call Turkmenistan; but they continued to spread further west into Iran, the Caucasus region, and ultimately into Anatolia—what we call Turkey today. The spread of Turkic peoples across Central and Western Asia is why the two countries of Turkey and Turkmenistan, even though they are over a thousand miles apart, have names that are somewhat similar and why they share ethnic and language backgrounds.

The Seljuk Turks brought Turkic culture to Anatolia. A Seljuk army defeated a Byzantine force at the Battle of Manzikert in eastern Anatolia in 1071. This opened Anatolia to Seljuk conquest and began the downfall of Byzantium, the remnant of the Eastern Roman Empire. The Ottoman Turks were a later group that exercised great political and cultural power for centuries in what we call Turkey.

Turkic people developed into the groups we now call Azerbaijanis, Kazakhs, Kyrgyz, Tatars, Uzbeks, the Turks of Turkey, and the Uyghurs of western China. We will have more to say about this last group in Lesson 92. As immigration of Turkic people has continued, groups of Turkish people from Turkey have come to live in Europe and the United States.

Turkic people have a common history and language heritage, and almost all of them share the religious heritage of Islam. People who speak the various Turkic languages might be the largest language group you've never heard of. An estimated 140 million people in the world speak a Turkic language. This makes them one of the world's ten

Traditional Clothing and Instruments in Turkmenistan

largest language groups. In terms of language, culture, religion, and ethnicity, the influence of Turkic people in the world, especially in Turkey and Central Asia, has been immense.

All Show, No Substance

However, the people of Turkmenistan have not always been their own masters. The area we now call Turkmenistan has been ruled at different times through the centuries by the Persians, Alexander the Great, Muslim invaders, Mongol invaders, and Russians (both in the mid-1800s and after the Russian Revolution of 1917). It once occupied an important place on the Silk Road.

Turkmenistan is the second largest country in Central Asia. It is slightly larger than California. However, its land area is nine-tenths desert with a few mountain ranges. The population is about 5.4 million, whereas the population of California is about 40 million. Some 80-85% of the population is Turkmen; most of the rest are Russians, Uzbeks, Kazakhs, and Tatars. Almost 90% of the people are Muslims. Because of Russian influence there, about 9% of the people are Russian Orthodox. A few are Buddhists and followers of folk religions.

The chief natural resource of the country consists of its oil and natural gas reserves. Exports of oil and gas have brought considerable wealth into the country, but most of that wealth has wound up in the hands of the rulers while the everyday people have benefited hardly at all.

Turkmenistan was part of the Union of Soviet Socialist Republics from 1925 until 1991, when it declared independence. Unfortunately, the people of Turkmenistan are still not their own masters. Saparmurad Niyazov, chairman of the Turkmenistan Supreme Soviet during Communist days, became the leader of independent Turkmenistan; and the country traded one repressive, tyrannical form of government for another. In 1993 Niyazov took the name Turkmenbashi, which means "Leader of the Turkmen." (In the 1920s the leader of the first Republic of Turkey, Mustafa Kemal, had taken a similar step in assuming the name Kemal Ataturk, "Father of the Turks.")

Niyazov developed a personality cult that approached a religion. He had himself declared

Darvaza Gas Crater

Certainly one of the most unusual geographic features anywhere in the world is the Darvaza gas crater. Located in Turkmenistan's Karakum Desert about 160 miles from the nation's capital of Ashgabat, the pit is about 226 feet across and about 65 feet deep.

The story is that in 1971, when Turkmenistan was part of the Soviet Union, Soviet geologists were looking for oil deposits in this area that is rich with oil and natural gas. They began drilling in the desert at this site, which turned out to be a cavernous pocket of natural gas. The weight of their equipment caused the surface to collapse, which triggered cave-ins here and in several nearby places. No one was injured in the collapse.

However, natural gas began escaping from the crater. The methane in natural gas displaces oxygen in the air, which makes breathing difficult. Animals in the area began to die. The risk of methane explosions increased dramatically. The scientists' solution was to set the crater on fire. They expected the natural gas to burn off in a few weeks. This is a common procedure at a drilling site where all the natural gas at the site cannot be recovered. But the scientists did not know how much natural gas was there.

The crater is still burning. The government has declared that something should be done, but nothing has been done. Hundreds of tourists, such as these taking selfies, visit the site each year. Tour companies take groups there.

President Niyazov ordered the construction of this Monument of Neutrality in 1998 in the center of the capital. The statue of Niyazov on top rotated to follow the sun each day. President Berdimuhamedow had the monument moved to the suburbs of town. The statue no longer rotates.

president for life. He spent lavishly on impressive-looking public projects and numerous gold statues of himself. Meanwhile, the people spent hours in lines waiting for the chance to buy scarce groceries. Niyazov died in 2006. His successor, Gurbanguly Berdimuhamedow, rolled back some of Niyazov's excesses, such as the renaming of days of the week and months of the year for himself and family members; but the government continues to be secretive, abusive, and corrupt. Elections are a farce.

A Lesson in People and Geography

The Turkic peoples of the world and Turkmenistan in particular give us yet another example of the connection between geography and people. Pre-modern emigrants from one place in the world spread their culture and language across a broad sweep of the globe that includes many places and affects millions of people.

Modern Turkmenistan reminds us of the value of natural resources in a given geographic location but also reminds us that a few people can mishandle and selfishly use a resource that could help many people. Proverbs speaks of this danger:

Like a roaring lion and a rushing bear
Is a wicked ruler over a poor people.
Proverbs 28:15

Assignments for Lesson 84

Gazetteer Read the entry for Turkmenistan (page 155).

Geography Complete the map skills assignment for Unit 17 in the *Student Review Book.*

Project Continue working on your project.

Literature Continue reading *Boys Without Names.*

Student Review Answer the questions for Lesson 84.

Summer Flowers in Kyrgyzstan

85 The Environment

The story of the Aral Sea in Lesson 81 leads us to think about a worldview regarding the environment. This lesson is not the setting for a discussion of options regarding government policy about the environment. Here we want to concentrate on the worldview issue of how we should see the environment from a spiritual perspective.

The Earth Is the Lord's

God created the world, and it belongs to Him. He created it by His decree; His creation of it was purposeful and not an accident.

> *It is He who made the earth by His power,*
> *Who established the world by His wisdom;*
> *And by His understanding He has stretched out the heavens.*
>
> *Jeremiah 10:12*

> *The earth is the Lord's, and all it contains,*
> *The world, and those who dwell in it.*
> *For He has founded it upon the seas*
> *And established it upon the rivers.*
>
> *Psalm 24:1-2*

He's Got the Whole World In His Hands

God not only created the world; He also maintains loving, wise, active control over it. Just after God brought the devastating flood upon the earth, He promised that He would never again destroy every living thing the way He had:

> *While the earth remains,*
> *Seedtime and harvest,*
> *And cold and heat,*
> *And summer and winter,*
> *And day and night*
> *Shall not cease.*
>
> *Genesis 8:22*

God knows and cares for every aspect of creation. He says:

> *For every beast of the forest is Mine,*
> *The cattle on a thousand hills.*
> *I know every bird of the mountains,*
> *And everything that moves in the field is Mine.*
>
> *Psalm 50:10-11*

The Bible tells us that God guides what we usually call the natural forces of the earth. Psalm 147 says that it is God:

Who covers the heavens with clouds,
Who provides rain for the earth,
Who makes grass to grow on the mountains.
He gives to the beast its food,
And to the young ravens which cry. (verses 8-9)

He gives snow like wool;
He scatters the frost like ashes.
He casts forth His ice as fragments;
Who can stand before His cold?
He sends forth His word and melts them;
He causes His wind to blow and the waters to flow. (verses 16-18)

Jesus taught that God causes the sun to rise and sends the rain (Matthew 5:45). God feeds the birds of the air and clothes the grass of the field with its beauty (Matthew 6:26, 30).

To be sure, the earth experiences earthquakes, volcanoes, tsunamis, local floods, and other events that we consider to be disasters. But even when our environment seems out of control, God still has everything under His control. We can trust that everything that happens is within God's sovereign and loving power and rule.

Responsible Human Interaction with the Environment

God created and sustains this amazing and marvelous world to support life, including human life. The atmosphere and its composition, the plants and animals that live on the earth, the earth's temperature and weather, and the sources of fuel we find on and in the earth support human life.

Golden Eagle, Kyrgyzstan

God commanded humans to "be fruitful and multiply, and fill the earth, and subdue it; and rule over the fish of the sea and over the birds of the sky and over every living thing that moves on the earth" (Genesis 1:28). It logically follows that humans should exercise this rulership wisely and responsibly; in other words, in a godly manner.

Since God created the earth to provide for human needs, we must utilize and manage it to this end. We should employ the natural resources of our environment for sustaining and bettering mankind, while at the same time doing all we can to see that resources are available for future generations as the Lord wills. Any natural resources that people use up or render useless, such as land, water, and minerals, will not be available to later generations. On the other hand, later generations can easily continue to use other resources, such as wind and solar power.

The Bible is not a textbook on environmental science, but it does provide principles for being responsible stewards of the environment. Exodus 23:11 required giving farmland a sabbath year. This let the land lie fallow and regain its nutrients for growing crops in future years. Leviticus 9:12 tells what to do with the waste of animals that the priest sacrificed, and Deuteronomy 23:12-13 gives instructions about disposing of human waste. Such laws enabled the Israelites to be good stewards of the land God gave them to possess.

Unfortunately, people have not always lived responsibly regarding the environment. For instance, in medieval Europe people practiced unsanitary habits, which led to the outbreak of disease from time to time. Too often people have utilized the earth's resources with little or no regard for the impact of such activity on the environment, on people alive at the time, or on future generations. Strip mining, clear-cut logging without replanting, and irresponsible disposal of waste and chemicals are examples of such activities. Look at pictures of smog-covered Los Angeles in the 1960s and you will see what we mean.

Humans have an impact on the environment, just as humans have an impact on other aspects of life on our planet, including society, government, and religion. Our world is different in material terms because of farming, mining, construction, energy production, and other activities. Not all human activity with regard to the environment is bad, but human interaction with the environment should be wise. When it is not, environmental consequences result which impact other people.

Harmful human activity can be a result of ignorance, or carelessness, or defiant disregard for information we know to be true. Carelessness and disregard are sins to be repented of. People should be willing to correct ignorance as soon as possible.

Climate Change?

A topic of considerable discussion and debate is the role of mankind in what is called climate change, formerly called global warming. Some observers believe that humans are causing the earth's temperature to rise. They predict catastrophic consequences if we do not reverse this trend. Other scientists acknowledge that the earth's average temperature has increased slightly over a period of many years but believe that the cause of this rise is not clear. It might simply be a long-term temperature swing that the earth experiences from time to time. Scientists base their predictions of future effects on computer-generated models and scenarios, which may or may not be accurate.

Mankind has definitely impacted the earth's environment, as noted above. Whether mankind's activity has caused changes to the environment that will bring about catastrophic results is a question about which people have very different and very strongly held beliefs. We know enough, however, to act wisely regarding the earth and its environment. If more conclusive evidence emerges, we need to respond appropriately and not reject it merely because it supports the political views of people with whom we disagree.

A Godly Perspective About the Earth

We can go to either of two erroneous extremes in thinking about the earth and its environment. One is to attribute a spiritual nature to it. In Greek mythology, the earth was the manifestation of the goddess Gaia, who was involved in producing all living things. She was the rough equivalent of the Roman goddess Terra. This belief was the origin of the idea of Mother Earth and Mother Nature. Some pagan thought holds that the earth has its own spirit. Some folk religions include the idea of the earth as the mother of living things. Some people use the terms Mother Earth and Mother Nature today to attribute a spiritual aspect to our world without admitting the existence of God. Such beliefs beg the question of where the earth came from.

The other extreme is to see the world as merely a disposable resource that man can exploit for his short-term material profit and comfort without any regard for environmental consequences or long-term impact. Human-caused environmental catastrophes demonstrate the short-sightedness and dangers of poor environmental management. How people handle the environment has consequences on others, including those who are yet unborn.

A Christian worldview of our physical world and environment involves seeing the earth not as a resource to exploit without regard for the consequences, nor as a deity to be worshiped, but as a gift from God that provides what we have to have to live physically and deserves and needs our wise and godly management and care. Doing this will honor God, provide for current needs, and insure resources for those who will inhabit the earth after us as the Lord wills. People will face difficult choices regarding environmental policies, and this worldview will not eliminate all of those hard decisions. However, this outlook will provide people with some principles they can use when facing specific issues.

In his sermon at Lystra, Paul says that in the ordering of the environment, God:

. . . did not leave Himself without witness, in that He did good and gave you rains from heaven and fruitful seasons, satisfying your hearts with food and gladness.
Acts 14:17

Assignments for Lesson 85

Worldview Recite or write the memory verse for this unit.

Project Finish your project for this unit.

Literature Finish reading *Boys Without Names*. Read the literary analysis and answer the questions in the *Student Review Book*.

Student Review Answer the questions for Lesson 85.
Take the quiz for Unit 17.

Mongolia

18 East Asia

The island nation of Taiwan has an unusual background and an unusual geographic setting. The peninsula of Korea has been divided for decades and has developed two different forms of government and ways of life. Rice-growing is a major agricultural and economic activity that began in Asia and has spread to many places throughout the world. The Japanese tea ceremony is a cultural activity that reflects a particular worldview. The worldview lesson surveys meditative religions that reflect worldviews quite different from the worldview of most Christians.

Memory Verse Memorize John 12:24 by the end of the unit.

Books Used

The Bible
Exploring World Geography Gazetteer
Revolution Is Not a Dinner Party

Project (Choose One)

1) Write a 250-300 word essay on one of the following topics:
   - What do you think is the solution to the dilemma of Taiwan's relationship with China, and what role should the United States play in this solution? (See Lesson 86.)
   - If you were living in North Korea, what would you do and why? (See Lesson 87.)
2) Conduct a Japanese tea ceremony—less elaborate than the real thing, and definitely not taking four hours or having everyone drink from the same bowl, but make it a formal occasion with definite elements. Feel free to prompt participants on what they should do and when.
3) Write a 5-to-8 minute play about someone trying to escape from North Korea.

Literature

Author Ying Chang Compestine based her historical novel, *Revolution Is Not a Dinner Party,* on her own childhood during China's Cultural Revolution. Her heroine, Ling, is the daughter of respected doctors working in Wuhan, China. As the revolutionary regime, led by Chairman Mao Zedong, aggressively removes everything considered to be old-order, security and freedom crumble on every side. Ling and her family must hold onto strength and hope as the revolution turns their world upside down.

Ying Chang Compestine was born in Wuhan, China, in 1963. She completed her college education in China, then went on to graduate studies in the United States, where she now lives. She writes young adult fiction and picture books celebrating Chinese culture. *Revolution is Not a Dinner Party* has won numerous awards. Compestine is also a respected food writer, with many articles and cookbooks to her credit.

Before you read this book, read the background material in the *Student Review Book* starting on page 66.

Plan to finish *Revolution Is Not a Dinner Party* by the end of Unit 19.

Islands in the South China Sea, as Seen from Hong Kong

86 One China, Two Chinas

China was in chaos. Around 1900 the weakening Qing (Manchu) dynasty had little control over the country. Foreign powers had carved China into spheres of influence. The United States, which did not have a sphere of influence, urged China to adopt an Open Door policy that would give all nations, including the U.S., access to trade with the tottering giant. One group of Chinese attempted to use force to reassert Chinese power and culture against foreign domination and to drive out the "foreign devils." Meanwhile, Chinese warlords fought each other.

Nationalists and Communists

In 1911 a group of revolutionary nationalists overthrew the dynasty and declared the Republic of China (ROC). Dr. Sun Yat-sen served briefly as president and organized the Kuomintang (KMT) or Nationalist Party to govern the country. Chiang Kai-shek was leader of the party's military force. Sun died in 1925, and by 1928 Chiang was in firm control of the party and functioned as the leader of the nation.

Mao Zedong helped to organize the Communist Party in China in the 1920s and eventually became its leader. At first the Communists and the KMT tried to work together, but both sides wanted to control China so they could not be allies. Civil war ensued until 1934, when the KMT emerged victorious. The Communists withdrew to the far northwest of China to regroup. The two sides stopped attacking each other during World War II in order to fight the invading Japanese, but at the end of that conflict in 1945 civil war between the Nationalists and the Communists began again. This time the Communists emerged victorious. Chiang and his army and other supporters, a total of about 1.5 million people, withdrew to the island of Taiwan off the coast of the mainland. On the mainland,

Mao Zedong (left) and Chiang Kai-shek (right) celebrate victory over Japan in 1945.

Mao declared the People's Republic of China (PRC) on October 1, 1949.

Chiang vowed to invade the Chinese mainland one day and retake control of all of China. Nationalist forces have never attempted this, although a few shots were fired between the two sides in the 1950s. The United States and most other nations recognized the government on Taiwan, known as the Republic of China (ROC) as the legitimate government of all of China and Chiang as the legitimate ruler of China. The United Nations (with the influence of the United States playing a major role) declared that the Nationalist government was entitled to membership while the Communist PRC government was not. Meanwhile the Soviet Union recognized Mao as the real ruler of China.

The Diplomatic Dance

For decades most of the world, again with significant influence from the U.S., operated under the diplomatic fiction that the de facto rulers of the most populous country in the world did not really exist. To recognize the Communist PRC, the thinking went, would be to legitimize the Communists and to turn our backs on an anti-Communist ally. The U.S. sold weapons to the Chiang government and engaged in trade that helped build Taiwan into a major world economic power. However, the Chiang-KMT government was corrupt and oppressive and allowed no other political party to exist out of fear of Communist influence. The American position was that, yes, Chiang is a dictator, but he's our dictator in the fight against Communism.

Meanwhile, the PRC on the mainland descended into Communist darkness. Millions of people died as a result of Communist oppression and the failures of a centrally planned economy. Program after program, such as the Great Leap Forward and the Cultural Revolution, cost many lives and only proved the failures of the Communist Marxist system. With help from the Soviet Union, however (until 1960, when the two Communist giants went their separate ways), the PRC military grew stronger and posed a threat to other nations. The PRC insisted that Taiwan was really part of China but never tried to invade the island to end the ROC. A Communist Chinese invasion of Taiwan would have risked a direct confrontation with Taiwan's major ally, the United States. An invasion of the mainland by the ROC would have almost certainly met defeat at the hands of the much stronger PRC.

Chiang Kai-shek at a parade in Taiwan in 1966.

A New Era

This diplomatic stalemate continued until the 1970s. In 1971 the United Nations (with U.S. support) rescinded ROC membership and gave the seat for China to the PRC. The next year, U.S. President Richard Nixon began a thaw in U.S.-PRC relations by visiting Communist China and meeting with Mao and other leaders. Chiang died in 1975 and Mao died the next year. Mao's successors initiated policies that opened trade relations with other countries and moved the PRC economy toward capitalism. However, the Communist government continued to act with authoritarian repression toward its own people.

In 1979 the United States and the People's Republic of China established full diplomatic relations. The U.S. ended diplomatic relations with the ROC government on Taiwan. The U.S. government declared, "The United States of America acknowledges the Chinese position that there is but one China and Taiwan is part of China." However, in the language of diplomacy "acknowledging" something falls short of "recognizing" it. A short time later, Congress passed and President Jimmy Carter signed the Taiwan Relations Act. This law confirmed America's unofficial relations with the government on Taiwan and left open the possibility that the U.S. would defend Taiwan if the PRC invaded. This position is appropriately called strategic ambiguity. The thinking behind it assumed that the possibility of American military involvement would keep the PRC from invading Taiwan.

In 1992 representatives of the PRC and the ROC announced an "understanding," which is called the 1992 Consensus. The two governments agreed that there is only one China, but they allow each other different interpretations of what that means. Both agreed that Taiwan is part of China, but they disagreed about which is the legitimate government of all of China. An informal part of the Consensus was Taiwan's commitment not to seek independence. The PRC has not promised that it will never move militarily against Taiwan in order to reclaim it. Meanwhile, Taiwan has deepened relationships with Japan and several European countries as well as the United States.

President Jimmy Carter (left), former President Richard Nixon (center), and Chinese Vice Premier Deng Xiaoping (right) at a White House dinner in 1979.

A More Intricate Dance

The diplomatic dance has only gotten more intricate. Almost all world governments except those of about fifteen small countries now recognize the PRC and do not have formal relations with the government on Taiwan. However, Taiwan has a robust capitalist economy and carries on trade with many countries. For many years the PRC did not even receive direct flights from Taiwan. People who wanted to travel between the PRC and the ROC had to fly first to Japan or some other country that received and dispatched flights to both places before completing their itinerary. Today, Taiwan's biggest trading partner is—are you ready?—the People's Republic of China.

And yet, the PRC is very sensitive about how other nations refer to Taiwan. As the PRC has developed trade relations with other countries and extended its Belt and Road Initiative (described in Lesson 93), the Communists have insisted that countries wanting to trade with it support its position on Taiwan. On Taiwan meanwhile, the government finally allowed other political parties to form in the 1980s. One party, the Democratic Progressive Party (DPP), has won some national elections. In 2019 the DPP leader of Taiwan, Tsai Ing-wen, criticized the 1992 Consensus, concluded when the KMT Party was in power. Tsai said that the 1992 Consensus did not express the consensus of the Taiwanese people today. Thus the diplomatic status of that agreement is unclear. Tsai won re-election as president in January 2020, strengthening her position and the DPP's position with regard to keeping separate from the PRC.

Ironically, since the KMT accepted the one China position in 1992, that party's position is actually closer to the PRC's stance than is the DPP's position, which leaves open the possibility of independence for Taiwan. After all, in the thinking of the KMT, why declare independence from the country of which you believe you are the legitimate government? On the other hand, if you are merely an ethnically different province, as the DPP sees it, independence makes more sense. Beijing has declared that it will fight if Taiwan declares independence. The KMT is a much weaker party now than it was when it ruled Taiwan.

The DPP rejects the "one country, two systems" approach that Beijing claims to support for Hong Kong and says it would apply to Taiwan. We say "claims" because this was the premise for the PRC assuming sovereignty over Hong Kong in 1999, but the facts since then indicate that the PRC government wants very much to bring Hong Kong under Beijing's domination the way it controls the rest of China. (You will learn more about Hong Kong in Lesson 92.) The DPP is skeptical that the Communist government would leave Taiwan alone and let it continue to operate as it now does.

Background

Portuguese explorers saw the island of Taiwan in 1544 and named it Formosa (Portuguese for beautiful). Many in the West called it Formosa for centuries. The island had no central government until China assumed control in the late 1600s. When Japan won the Sino-Japanese War in 1895, China was forced to give Taiwan to Japan. The Japanese ruled Taiwan until Japan lost World War II, at which time China (under the Nationalist government) again took control of it.

The central two-thirds of Taiwan is mountainous, with slopes extending east and west. The west has the most land that is suitable for farming and the greater part of the population, while the east has little room for either farming or people. The island is 60% urban, with several cities having a population of over one million each.

The Taiwan Strait separating mainland China from Taiwan is about one hundred miles wide. The government on Taiwan claims sovereignty over about ninety other small islands around it, including one about a mile off the coast of mainland China. The ROC, the PRC, Japan, and other countries

Tsai Ing-wen won election to a second term as President of the Republic of China in 2020.

dispute the ownership of other islands in the South China Sea.

The ethnic groups on Taiwan include an aboriginal people and some long-standing groups that immigrated there from the Asian mainland centuries ago. The majority of the population is ethnically Han Chinese, but many of them identify themselves as Taiwanese. In a 2018 survey of the residents of Taiwan, 55% said they are exclusively Taiwanese, 38% saw themselves as both Taiwanese and Chinese, and 4% said they are Chinese. A large majority of people on the islands see themselves as having a separate identity from the population of the PRC, even if they do share the same ethnic background. Thus, the people living on Taiwan in 1949 didn't necessarily welcome the fleeing Chinese with open arms as long lost brothers. This skeptical attitude increased as the native Taiwanese saw the immigrants running the island as they saw fit and even practicing some discrimination against the people who had been living there.

Over the last seventy years, however, the people of Taiwan have worked out their identity and relationships with each other to a great degree. This distinction between Chinese and Taiwanese is understandable given the island's history. Both the PRC and the ROC have threatened each other. Taiwan has seen significant economic and political growth on its own without reference to China. And the people of Taiwan have learned to cherish the freedom that they have.

A Tiny Piece of Geography

The People's Republic of China covers about 3.7 million square miles and has a population of about 1.4 billion people. Taiwan covers about 14,000 square miles—about the size of Massachusetts, Connecticut, and Rhode Island combined—and is home to about 24 million people. It would seem that Communist China can get along just fine without Taiwan. The Taiwan issue is a question that affects a large part of the world directly or indirectly. It plays a role in U.S.-China relations. What's all the furor about a relatively tiny piece of geography? Why the big push by Beijing to take it over?

As we will explore further in Lesson 144, land ownership means power. China is interested (1) for economic reasons. Taiwan would add considerably to China's wealth. But (2) it is also a matter of geographic pride. To let Taiwan go its own way would be admitting that Taiwan is no longer part of China, and countries do not give up territory lightly. It would also be an admission that the Communist government of China lost that little bit of the war back in the 1940s. (3) The question of Taiwan is part of the web of issues related to the South China Sea. The PRC wants control over as much of the geography of that area as it believes it is entitled, which is most of it. We will explore the South China Sea in Lesson 96

Questions

What will the government on Taiwan do? The island has a functioning democracy and a strong capitalist economy. The Communist government of China believes that it is entitled to govern Taiwan and is increasing economic, political, and to some degree military pressure to bring that about. Will Taiwan declare independence at the risk of a major confrontation with Beijing? Or will the current stalemate continue?

What will the PRC do politically, economically, and militarily toward Taiwan? Ancient Chinese military strategist Sun Tzu wrote in *The Art of War* that the best way to defeat your enemy is if you do not have to fight at all. This appears to be the strategy of the PRC government: bring such economic and political pressure to bear that the enemy (Taiwan and its allies) will recognize Beijing's sovereignty over Taiwan without a shot being fired. At least, that is what the PRC hopes will happen.

Tea Fields in Taiwan

The relationship between Taiwan and China is complicated, and their relationships with other countries in the world are complicated also. The status quo involves a certain amount of diplomatic posturing, but it has proven to be workable. The danger in an unstable situation is that one side or the other, in trying to bring stability, might create more instability or even bring about war.

The question that the psalmist asked about the nations concerned their rebellion against God, but the phrases have application in this situation:

Why are the nations in an uproar
And the peoples devising a vain thing?
Psalm 2:1

Assignments for Lesson 86

Gazetteer Study the map of East Asia and read the entry for Taiwan (pages 157 and 164).

Worldview Copy this question in your notebook and write your answer: How do you know what reality is?

Project Choose your project for this unit and start working on it. Plan to finish it by the end of this unit.

Literature Begin reading *Revolution Is Not a Dinner Party.* Plan to finish by the end of Unit 19.

Student Review Answer the questions for Lesson 86.

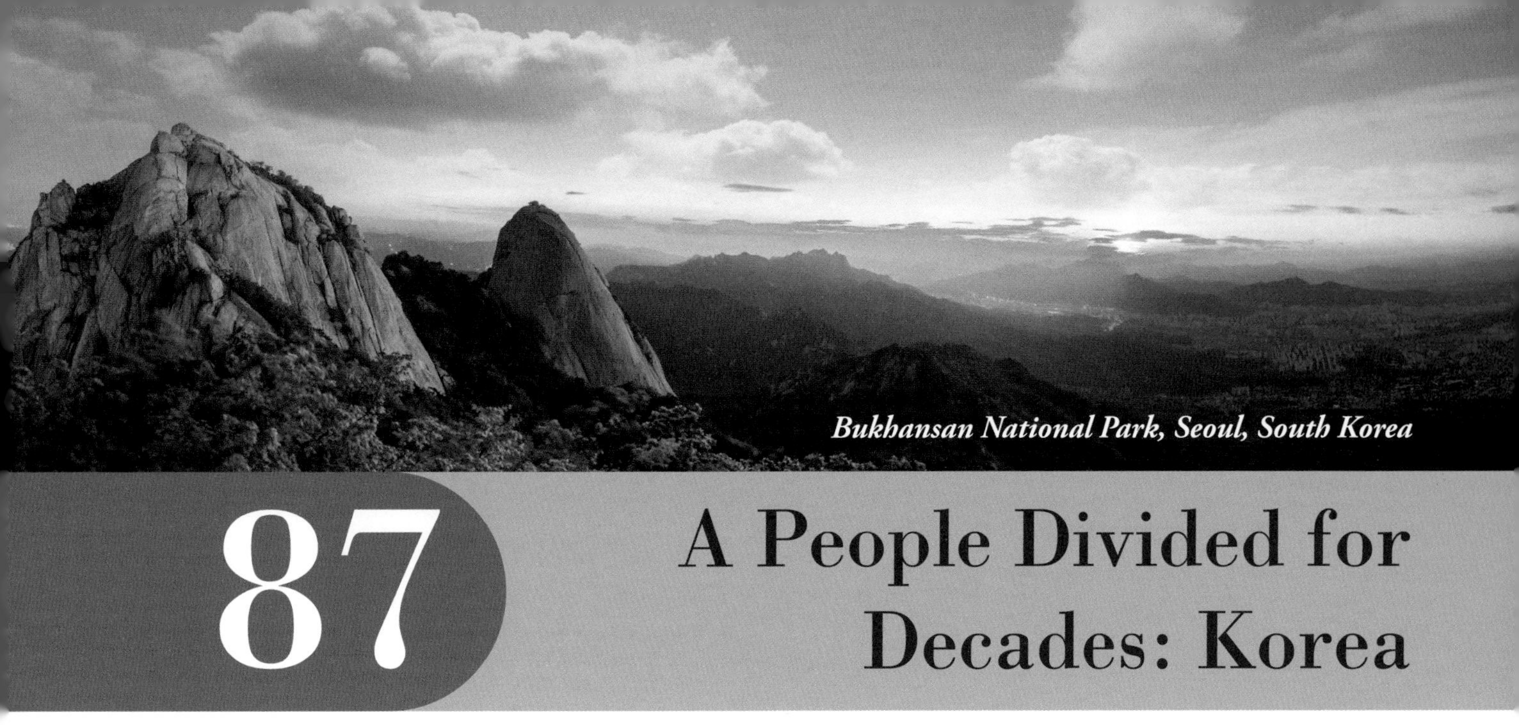
Bukhansan National Park, Seoul, South Korea

87 A People Divided for Decades: Korea

Eunsun (not her real name) had been alone in her bare, cold apartment for a week. The electricity did not work and she had no heat. Her mother and her older sister had gone to a nearby city to find food. They had already sold all of their furniture in order to get money for food, but that food had run out. Her father had died a few weeks earlier.

The only things on the walls of the apartment were two portraits: one of the late Kim Il Sung, the "Great Leader" and "Eternal President," and the other of his son, Kim Jong Il, the "Dear Leader" and current ruler. Every home had these two pictures in it. Citizens had been trained to consider the pictures sacred. Not having them prominently displayed risked severe punishment, imprisonment, or death.

Eunsun expected to die, so she found some paper and wrote out her will. Then she lay down to sleep. Her mother and sister returned a few hours later with no food. Eunsun was eleven years old.

The Most Important Geographic Feature

Korea is a mountainous peninsula about the size of Utah that juts south from the landmass of China. A small portion of its northeastern border touches Russia. Korea lies west of Japan. The Korea Strait that separates Korea from Japan is about 120 miles wide. Korea has mild summers and cold winters. It has limited mineral resources. Most of the people live in the western part of the peninsula where the land is flatter. However, the most important geographic feature of Korea is one that is invisible.

The 38th parallel of latitude is the approximate location of the slightly waving line that divides North and South Korea. This line of division existed before the Korean War, but it hardened into place as the official division at the time of the ceasefire that ended the fighting in that war in 1953. A Demilitarized Zone (DMZ) extends about 1.2 miles on either side of the line. No fighting is to take place in this zone, but each country heavily guards itself just outside of the DMZ. No peace treaty bringing an official end to the war has ever been signed.

Two Ways of Life

The DMZ does not just separate two countries. It separates two philosophies, two worldviews, two ways of life. South Korea (officially the Republic of Korea) is a western-style democracy that practices free market capitalism. From being flat on its back economically in the early 1960s, South Korea has developed a remarkably productive economy, the twelfth largest in the world. The government that ruled for decades following World War II tended to repress any dissent out of fear of Communist revolution. In more recent years, however, multiple political parties have been able to function openly. About a fourth of the people claim to be Christians, and Christianity continues to grow rapidly. However, significant numbers of people follow Confucianism, Buddhism, and folk religions as well.

South Korea has about 51.4 million people. North Korea, which is slightly more than half of the peninsula, has about 25.3 million people. Almost all those who live in North Korea are ethnically Korean. South Korea is predominantly Korean, but also has some residents from other ethnic groups. Koreans have about three hundred family names (which are usually stated before the given name), but the vast majority of people share a mere handful of family names. A few families attained prominence centuries ago, so many Koreans use the names of those families, whether or not they are actually related. Families that go by the names of Kim, Lee, or Park account for about 45% of the population; about 20% have the family name of Kim.

By contrast, the government of North Korea (officially the Democratic People's Republic of Korea or DPRK, even though it is neither democratic nor a republic) is a brutal, oppressive, hereditary Communist tyranny. Its people live in poverty and misery. Any dissent and any expression of religious faith are ruthlessly repressed and punished. Public executions of political enemies, carried out even in front of school children, are common. The government brainwashes the people to believe that theirs is the greatest country in the world and that they have nothing to envy. Accurate economic figures are not easy to obtain, but the British magazine *The Economist* ranked the North Korean economy as the 101st largest in the world—quite a difference between the two systems that function on what is largely the same land. The leaders of North Korea live in luxury while the people live in

Monument to North Korean Leaders Kim Il Sung and Kim Jong Il in Pyongyang, North Korea

poverty. Meanwhile, the government spends billions on developing missiles and nuclear weapons that the government uses to threaten South Korea, Japan, and the United States.

Travel in North Korea is restricted and regulated even within the country. Authorities can check a person's cell phone for foreign songs or to see if the person has called a number in a foreign country. People can be sent to prison for watching a foreign television program—if they own a television and if the electricity works.

The North Korean government operates on fear: fear of punishment and fear of an attack by what it calls blond, big-nosed, bloodthirsty Americans. School children and factory workers regularly take part in self-criticism sessions, and people are encouraged to denounce others as disloyal to the regime. Such denunciations are a good way to get back at someone the accuser doesn't like, and it makes the accused person guilty until proven innocent. The entire charade results in the people living in fear, intimidation, and quiet submission.

The government of North Korea feeds its people lies. It claims that the leader—whether first Communist leader Kim Il Sung, his son and successor Kim Jong Il, or his son and successor Kim Jong Un (called the "Supreme Leader")—as providing bountifully for the people. Before each meal, a family bows their heads and one family member says, "Thank you, Respected Father Leader Kim Il Sung [or whatever the name and title are at the time], for our food." Every school day children sing songs and chant phrases that give praise to the leader. Children receive treats on the leader's birthday. Government propaganda portrays Kim Il Sung as personally winning World War II (he was in the war but did not win it). Supposedly legitimate news reports describe the leader as accomplishing such amazing things as being able to shoot a gun at three years old, to ride a horse at five, and to score a hole in one the first time he played golf. Any criticism of the government can result in beatings, imprisonment, or execution.

The horrible economic conditions that result from the North Korean centrally planned economy give rise to a system of black markets and to those with power playing favorites and demanding bribes to distribute the rare goods and services that people want. Students will offer gifts to their teachers in order to obtain good marks.

Background

The Korean peninsula lies in a geographic position that has resulted in strong ethnic and cultural influences from China and Japan, but its position has also enabled an identity that is separate from those influences as well. Throughout history Korean dynasties ruled the peninsula for periods interspersed among periods of rule by invading Mongolians, Chinese, and Japanese. The desire of the Korean people to live separately from foreign influence led it to acquire the nickname of "The Hermit Kingdom" in the 1700s.

In the early 1900s, Japan and Russia both wanted to expand their influence in Asia. Their conflict over control of Korea and Manchuria came to a head in the Russo-Japanese War of 1904-1905. Japan won the war, and in 1910 began ruling all of Korea as a colony. Japan sought to wipe out Korean language

People in South Korea attach items such as flags, photographs, and Buddhist prayer ribbons to fences along the DMZ expressing their desires for peace and unification.

and culture. During that time, rival Communist and Nationalist groups formed in Korea as they did in China, each wanting to take control of the peninsula.

Japan continued to rule Korea during World War II. In the closing days of that conflict, when Japan's defeat was imminent, American forces moved into the south of the peninsula and supported the Nationalists, while Soviet Russian forces took over the north supporting the Communists. An uneasy agreement between the U.S. and the Soviets divided the peninsula at the 38th parallel. Korea thus became a theater in the developing Cold War. In 1948 two governments formed on the peninsula, the ROK supported by the United States in the south and the DPRK supported by the Soviet Union and Communist China in the north. Both claimed to be the legitimate government for the entire peninsula and threatened to take steps to make that happen.

On June 25, 1950, North Korean troops, supported by Communist China and the Soviet Union, swept south across the 38th parallel. The United Nations authorized a multinational military response led by the United States and General Douglas MacArthur. The DPRK troops pushed South Korean and U.S. and Allied troops to the extreme southern tip of the peninsula. Then MacArthur executed a daring invasion along the western coast that cut the invaders' supply lines, forcing them to retreat. Allied troops then pushed North Korean troops north almost to the Chinese border, at which point Chinese troops poured into Korea and pushed the Allies back to the 38th parallel. At this juncture the ceasefire took effect on July 27, 1953; and the standoff has remained to this day.

Bridge Over the Tumen River Between China and North Korea

The Rest of Eunsun's Story

In the 1990s the fall of the Soviet Union meant that North Korea lost a major trading partner. Then a severe famine hit. Despite the loss of untold numbers of lives, the North Korean government did not change its policies or try to help its people. It was during this crisis that Eunsun's mother decided to attempt an escape to China with her two daughters, hoping eventually to make it to South Korea. Imagine wanting to escape to China because life there was so much better than in North Korea!

Their attempt began a nine-year saga. Their first attempt failed. The second, a year later, involved walking across the frozen Tumen River into China in February in the early morning hours when fewer North Korean guards patrolled the border. The guards were authorized to shoot on sight. Many people on the Chinese side of the border were Korean or spoke Korean, but friends to help were rare. Eunsun and her family lived lives of misery, drudgery, and abuse. They were captured and returned to North Korea, but managed to escape again. This time they made it to a larger city where they were able to obtain fake identification papers with the right payment.

Eunsun's sister was eventually able to establish her life in Shanghai, China. After several years of difficulty, Eunsun and her mother decided to seek refuge in South Korea. But this meant secretly crossing the border into Mongolia, walking into the Gobi Desert and surrendering themselves to the Mongolian army, and trusting that the soldiers would take them to the South Korean embassy in keeping with an agreement between the two countries. They and other escapees received help from Christian missionaries and from South Koreans. They had to pay a large fee, but working with disreputable smugglers carried much greater risk. Eunsun and her mother did make it to the South Korean embassy along with a small group of other escapees, but they still had hurdles to jump. South Korea is constantly on the lookout for spies from North Korea, whom they fear might come disguised as refugees. As a result, Eunsun and her mother were detained for over a month and had to endure harsh questioning. Then they had to wait in a camp until their turn to leave came. Finally they flew to Seoul, the capital of South Korea.

Things were better for them in South Korea, but their problems were not over. They had to endure more questioning before they were free to build their lives. South Korean government workers taught them about capitalism and how to live in South Korea. They had come only a few hundred miles from where they had lived, but they had entered an entirely new world. Eunsun enrolled in school. She used a computer. She began to learn a little English. The South Korean government gives financial assistance to refugees, but Eusun and her mother still had to find a way to support themselves in their new country. Unfortunately, many South Koreans look down on people from the North, so they had to deal with prejudice and discrimination.

Eunsun was able to call her sister in Shanghai and tell her that she and their mother were safe in Seoul. Eventually they became South Korean citizens. After nine years, Eunsun no longer had to fear being discovered and sent back to North Korea. They were able to rent an apartment. In the apartment were goods that a church group had donated to help newcomers. Eunsun decided to go back to school full-time and her mother became a nanny. Eunsun's sister came to live with them for a while, but she had married a Chinese man and they had a child, so the sister wanted to return to Shanghai.

Eunsun was able to visit Paris, France, and the Scandinavian countries. She received a scholarship to study in the United States. When she left North Korea, she did not know that Europe existed. Eunsun has since married and has devoted her life to raising awareness about the brutal regime that controls North Korea with the hope that one day it will end and Korea will be reunited.

An estimated one thousand North Koreans successfully escape that country each year. The number of those who are not successful and who are returned to face the consequences is unknown.

The Geopolitics of Korea

The geopolitical issues involved in the Korean standoff are complicated, to say the least. China supports North Korea and is its largest trading partner, but China does not endorse all of North Korea's policies and actions. The alternative that China sees to having to put up with North Korea would be a united Korean peninsula with a strong capitalist economy and democratic government that would be an ally of the United States with thousands of American troops stationed on its border. Meanwhile, Japan is officially an ally of South Korea; but because of Japan's harsh treatment of Korea in the past, relations between the two countries have been chilly.

North Korea and South Korea each still say that they want to unite the peninsula under one government. They eye each other warily, although they each know that war between them would be devastating for both and could possibly draw China and the United States into direct military conflict, the impact of which is beyond estimation. North Korea occasionally rattles its missile saber but does not indicate an imminent attack on the South. South Korea knows that any economic recovery in the North would come largely at its expense.

The real price of the continued standoff and the continued existence of the North Korean regime is paid by the oppressed, impoverished people of North Korea. They are prisoners in their own land. May they one day be truly free.

Revelation speaks of the self-delusion that characterized the city of Laodicea:

. . . . you say, "I am rich, and have become wealthy, and have need of nothing," and you do not know that you are wretched and miserable and poor and blind and naked. . . .
Revelation 3:17

Assignments for Lesson 87

Gazetteer Read the entries for North Korea and South Korea (pages 160-161). Read the document "Perspectives on Korea" (pages 302-308) and answer the questions in the *Student Review Book*.

Worldview Copy this question in your notebook and write your answer: What do you know about Hinduism, Buddhism, and New Age thought?

Project Continue working on your project.

Literature Continue reading *Revolution Is Not a Dinner Party.*

Student Review Answer the questions for Lesson 87.

Transplanting Rice in Thailand

88 Rice-Growing: It Has to Be Just Right

The geography has to be just right.

Ninety percent of all the rice that is grown in the world is grown in East Asia and Southeast Asia. About the same percentage of all the rice that is consumed in the world is consumed in the same region. People around the world eat rice; in fact, between one-half and two-thirds of the world's population depends on rice as a staple in their diet. Farmers in Africa, the Middle East, Europe, the Caribbean, Latin America, and the United States grow rice; but for the lives of the people in East Asia, for their diet, and for the economy of East Asia, rice is absolutely essential.

Except for a few varieties that grow in the uplands, rice requires a certain kind of geography.

Growing in Water

Rice is an annual grass. Farmers plant rice seeds in special beds. When the seedlings are three to six weeks old, workers transplant them to the rice field or paddy (the word comes from the Malay word for rice). The rice plants sit in 2 to 4 inches of water. Farmers have found that the water helps the rice grow and creates higher yields, and it also suppresses the growth of weeds. So planting involves slogging through the water and mud to plant the rows of rice. Labor is intense to get the seedlings in the paddies and start them growing.

The need for water is why the geography of rice fields is so important. Coastal plains, tidal deltas, and river basins—all large, flat areas with fresh water immediately available—are ideal places to grow rice. Workers irrigate the fields by directing the water flow into them or by using pumps or (in advanced countries) water lines. Farmers build levees around the fields to hold the water in.

But even in mountainous areas, people can grow rice. When the geography does not provide large, flat areas, the people build them. In such places, workers carve out terraces up the mountainside, build retaining walls to hold in the water and to keep the soil from washing away, and pump water up the mountain and into the paddies thus formed. Growers have found that depending on rainfall to provide the needed water results in a much lower yield.

Harvesting and Processing

In 100 to 150 days, the rice plants grow to about four feet high. Workers harvest the rice, again slogging through the mud to do so, from mid-September into October. The window for harvesting to avoid spoilage is relatively brief, so the work at harvest time is again intense. The rice is dried, then comes threshing. Workers pour the rice into large bags that can weigh about one hundred pounds each, which they lift onto trucks to be taken for further processing.

When workers process the rice to remove only the outer husks from the kernel, the product is called brown rice. When workers also remove the bran layer, the result is white rice. Farmers use the by-products of the plants as livestock feed.

Advanced rice farming involves the use of machinery, but machines are expensive and cut down the number of workers needed. Millions of rice growers in East Asia do all of the work by hand. Manually harvesting a hectare (about 2.47 acres) of rice requires about 40 to 80 hours of labor. The world's rice farmers grow about 500 million tons of rice annually.

A Changing Field

Like just about every aspect of traditional life, rice growing faces modern challenges. Most people in East Asia have eaten rice at every meal for centuries. However, as East Asia develops a more industrialized economy, the size of the middle- and upper-income population grows. These people can afford a more varied diet, which tends to reduce the demand for rice to some degree.

On the other hand, about fifty million new rice consumers are born every year, so overall the market for rice is growing. Meanwhile, the same trend of industrialization draws people from rural areas to work in the cities, which reduces the workforce that is available for rice production. For some strange reason, some people believe that working in a factory is preferable to the hot, hard, back-breaking work of growing rice. Thus farmers need to produce more rice using a smaller pool of workers.

Rice Terraces in Vietnam

Rice Terraces in China

People Movements, Rice Movements

As best we can tell, domesticated rice production began in China thousands of years ago and spread to India and Southeast Asia. From these locations, rice growing has spread around the world. Africans have grown their own variety of rice for millennia; but people introduced Asian varieties into East Africa at some point after the time of Christ, and it spread across the continent.

Rice production extended to southern Iraq and Egypt, and Muslims carried it to Spain in the 900s and later to Italy. Spanish explorers introduced it to Mexico in the 1520s and the Portuguese brought it to Brazil, along with African slaves who knew how to grow it. Rice and bean dishes from West Africa became staples in the New World.

Slaves from Africa grew rice in coastal Georgia and South Carolina. For winnowing they used sweetgrass baskets, the making of which was another skill they brought from Africa. Rice growing died out in the coastal southern United States in the early 20th century, but it had developed by the mid-1800s in Arkansas, Louisiana, East Texas, and later Mississippi. Rice farming began in California during the Gold Rush of the 1850s, when some 40,000 Chinese laborers immigrated there and grew rice for their own consumption. Commercial production began in the state in 1912. By the way, what native nations in America, especially the Ojibwe, grow in marshes and call wild rice is not related to the rice grown in East Asia.

The British introduced rice growing in Australia, based on their knowledge of it in India and in their American colonies. Rice production in Australia benefits from cheap land, and in the mid-1900s Australia began exporting rice to Japan; but rice farming in Australia is highly dependent on the availability of often-scarce water. Thus the production and consumption of rice have become essential elements of the lives of many people in many different parts of the world.

Several times during His ministry, Jesus used farming imagery in His teaching. He told the parable of the sower, which compares what grows in different kinds of soil to the different ways that people respond to His message. He also taught the parable of the tares, illustrating how God will separate the righteous from the unrighteous on the day of judgment. Jesus compared the kingdom of God to the man who sowed seed in faith without really understanding how God made it grow. The Lord also used a farming analogy to describe His own sacrificial death for others:

Truly, truly, I say to you, unless a grain of wheat falls into the earth and dies, it remains alone; but if it dies, it bears much fruit.
John 12:24

Assignments for Lesson 88

Gazetteer Read the entry for Mongolia (page 162).

Worldview Copy this question in your notebook and write your answer: How does faith in Jesus help you through hardship?

Project Continue working on your project.

Literature Continue reading *Revolution Is Not a Dinner Party.*

Student Review Answer the questions for Lesson 88.

Japanese Tea Ceremony

89 More Than a Cup of Tea

It is a ceremony that requires years of training to perform well, days of preparation to attend to all of the details, and hours to accomplish. The Japanese tea ceremony is a social event, an artistic event, an interpersonal event, and a worldview event.

Not long after people discovered how to steep tea leaves in boiling water, drinking the concoction became a deeply meaningful, philosophical, and even (to some people) spiritual event. Around the 1300s or 1400s, long after travelers had brought tea from China to Japan, the Japanese developed a standard, elaborate tea ceremony. The Japanese still observe a tea ceremony today. The Japanese word for the ceremony is *chanoyu*, which means "hot water for tea"; but the ceremony involves much more than boiling water. It is actually a complex, intricate event that accomplishes several purposes.

In keeping with the Zen Buddhist worldview, the ceremony enables people to withdraw from everyday life and focus on a single event that involves just a few people as well as the numerous details of that event.

Since the host invites certain guests and chooses utensils, bowls, and decorations appropriate for those guests, each ceremony is a unique event never to be repeated. A proper tea ceremony can be very expensive.

In earlier years the ceremony was a way for members of Japan's ruling class, warrior class (or samurai), and merchant class to create and strengthen interpersonal relationships. We would say today that participants are able to network through taking part in the ceremony.

How the Tea Ceremony Takes Place

The tea ceremony continues to follow essentially the same pattern that hosts have been using for hundreds of years. The host invites his guests several weeks in advance (in the late 1800s, women began hosting ceremonies). He arranges every detail—the decorations, the utensils, the food items served, and so forth—with a view toward connecting with and honoring those who will attend.

In general, the host selects valuable, often older utensils made in China. In the early years the Japanese considered Chinese items to be the highest form of artistic expression. Later, another approach developed that involved the use of simple, even rustic, utensils made in Korea or Japan as a way to express spontaneity and a lack of careful planning. Japanese-made utensils are more common today. Life is like these utensils: some things we successfully prepare for in minute detail, while other things

Gathering for Tea *by Toyohara Chikanobu (Japanese, c. 1893)*

happen without planning and even despite our best preparations.

At the appointed hour, the host welcomes and invites each guest into the tea room, which is about twelve or sixteen feet square and has a small doorway (which encourages an attitude of humility). Each guest removes his or her shoes and enters the room. They wash their hands to represent purity, and in some settings they take their seats according to their social rank. (Sometimes the tea room is located in a garden, and the handwashing takes place outside the room.) Each guest contemplates a Zen saying on a scroll that the host has posted and which is subject to multiple interpretations. As the ceremony begins, the host often serves sweets to balance the sometimes bitter taste of tea.

The host brings in the utensils and cleanses them. The bowls are ceramic, while the teapot to heat the water (usually on a charcoal fire in a pit in the floor) is commonly made of iron. Families often pass these teapots down through many generations. The host then prepares a thick matcha powdered green tea and whips it with a whisk to produce a frothy drink, about the consistency of drinkable yogurt, which he or she pours into a bowl (not a cup). The guest of honor takes a sip, cleans the rim with his or her silk napkin, and passes the bowl to the next guest who does the same, and so forth around the entire group. As each person takes a sip, he or she comments on the beauty of the utensils, the quality of the bowl, the excellent taste of the tea, or other aspects of the setting.

The last guest to drink returns the bowl to the host, who cleans it again. Even the timing and rhythm of handing and receiving the bowl is important in the ceremony. This completes the formal part of the ceremony.

The host then prepares a thin matcha green tea, whips it with the whisk, and pours a separate bowl for each guest. The host might also serve additional confections at this time. Then the host cleans the tea set a third time. The guest of honor usually asks to examine the pots and bowls and praises their craftsmanship. Then the guest of honor passes around the items, which are sometimes priceless antiques, for all to admire. The complete ceremony usually includes twelve courses of food and drink. As each person leaves, the host bows to him or her.

The elements of the ceremony can vary depending on the time of year (which impacts whether a charcoal fire is lit or not), the kind of meal that the host serves, and local traditions. It is not unusual for a ceremony to last as long as four hours.

Is tea the focus of the Japanese tea ceremony? Yes and no. Certainly tea is the central element, but in another sense the focus is on the event itself and tea is just what brings the event about. Details done well, host and guests relating to each other, and the nature of the items used all help make the ceremony memorable. The event helps a person appreciate people and beauty, and the attention to detail communicates the belief that people are worth the effort. People around the world know the Japanese tea ceremony as an expression of what is essentially Japanese, which includes contemplation of the moment, sharing a unique experience—and tea.

We certainly do not ascribe in any way to the Zen Buddhist religion, but the ceremony does raise meaningful questions. Do you ever slow down enough to appreciate the details of a setting or an event? Do you compliment the details of a painting, a craft, a meal, or a home in a way that conveys your awareness of what another person has done? Do you ever focus on a small group of people of which you are a part? Do you ever participate in something while conscious of the fact that you will never do that exact same thing again?

The ABCs of Tea

The Japanese tea ceremony is an illustration of the importance of tea to people around the world. Black tea, white tea, oolong tea, and green tea all come from a single species of plant, the *Camellia sinensis* bush. People first discovered this evergreen bush or tree in the Yunnan province in southwest China, near the border with Burma (Myanmar) and close to Laos, Vietnam, and Thailand. The three main varieties of the tea bush are connected to specific places: the China bush, the Assam bush in northern India, and the Java bush in Indonesia.

Growers have developed over one thousand subvarieties; and as the growing of tea has spread, most of these varieties have become associated with specific geographic places. With the different methods that growers use to prepare tea for market, including the addition of flavors and aromatics, buyers can select from literally thousands of flavors of tea. People also make what they call tea from

Harvesting Tea in Thailand

other plants, such as peppermint, chamomile, lemon balm, and astragalus; but true tea leaves come from some variety of the *C. sinensis* bush.

Variations in tea color and flavor result from differences in the times when workers pick the leaves, how they pick, and how other workers process the tea. How long workers dry the leaves and how finely they crush the leaves also affect the taste. Probably the most important element in the manufacturing process that produces tea for consumers is creating the four basic categories of tea: fermented (oxidized) black tea, unfermented green tea (which does not go through the same chemical reactions as black), partially fermented oolong tea, and white tea (for which less mature tea leaves are steamed or fried).

This Chinese tea bowl from about 1200 AD is inscribed with a wish for "longevity, prosperity, and good health."

A Short History of Tea

The earliest known use of tea was for medicinal purposes. When people began drinking it for pleasure, it was an activity exclusively for the elite. Only as production grew, availability increased, and prices fell did it become something that everyday people enjoyed. Over the centuries, tea houses became popular in China. They served tea and light snacks, and people could gather to socialize, transact business, play games, and listen to poetry and stories. A Japanese Buddhist priest who visited China took some seeds home, and production spread to Japan. Traders also took seeds and plants to other parts of Asia.

Portuguese traders first brought tea to Europe in the late 1500s when they returned to Portugal. Dutch traders brought tea to Amsterdam in 1610; it reached England in 1669. Europeans loved it, and the demand created a huge overseas trade. English traders were not able to break the Dutch monopoly on tea from China, so they pursued tea production in India. At first, the English did not recognize the variety that grew in India and tried to transplant the China bush. In 1848 British businessmen hired Scottish botanist Robert Fortune to disguise himself as a Chinese, go undercover, collect plants, and learn the Chinese method for manufacturing green and black tea in order to develop the industry in India. Fortune was able to smuggle out twenty thousand plants and eighty tea specialists, but the plants failed because India was not hospitable to the Chinese variety. The British finally realized that the Assam variety was simply different, and they developed tea production in India to such an extent that India is now the world's largest producer of tea. As in Asia, in Britain only the upper classes could afford tea at first. Consumption spread throughout society in the 1700s.

The tea craze spawned a related industry. Love for tea encouraged the production of ceramic cups and saucers, with one difference. The Chinese used cups and bowls that did not have handles. The British did not want to hold the hot vessels, so ceramic makers added handles. The British also used milk and sugar (primarily from their West Indies sugar plantations), additions that the Chinese did not use, so British ceramics makers created serving dishes for milk and sugar.

Tea played a role in the American Revolution. Britain controlled the tea market in the American colonies, but Dutch traders smuggled cheaper tea into the colonies. The British East India Company then tried to dump even cheaper tea into the American market to run the Dutch out, but colonists dumped 342 chests of the British tea into Boston harbor as a way to tell the British to quit trying to run the Americans' lives. This conflict indicates how hotly contested the world tea trade had become.

Tea also affected the shipping industry. Majestic clipper ships with their three tall masts of sails came on the scene in the 1840s. They could transport cargo much faster than previous ocean-going vessels, and they made many deliveries of Asian tea to thirsty Europeans and Americans. However, the invention of even faster steamships about the same time cut short the heyday of clipper ships.

In the mid-1880s, a Scottish merchant bought up bankrupt coffee estates in Ceylon (now Sri Lanka) that a blight had ruined. He began growing tea and sold it below the going price with the slogan, "Direct from the tea gardens to the teapot." He was the first to sell tea in individual bags, which offered customers freshness, cleanliness, and consistent weight. The man became a millionaire and won loyal customers in Britain and America. His name was Sir Thomas Lipton. Perhaps you've heard of Lipton Tea.

Tea Ceremony *by Toyohara Kunichika (Japanese, 1883)*

With the restoration of the Meiji dynasty in Japan in 1868, the tea ceremony became less important in the new order of things. However, in just a few years interest in the ceremony rebounded. One reason was that the new industrialist class reinstituted the ceremony to recapture and honor Japanese culture and to network with business and government leaders in ways that earlier elites did. In addition, a new interest in the education of women promoted training them in the fine art of the tea ceremony.

The British devotion to tea is legendary. One story from our family from World War II illustrates this. My father served in Europe during the last half of the war. On one occasion his unit was near a British unit that was tasked with building a pontoon bridge across a river as the Allies pressed toward Germany. The Brits completed the bridge in late afternoon, then they lay down their tools and enjoyed their cups of tea. The Americans were furious and could not understand why they didn't keep moving, to get further along in their ultimate goal—but nope, it was teatime, and the British troops had a previous commitment.

People have continued to expand the production of tea to additional places. About fifty countries around the world now have a tea-producing industry. But tea is more than a product or a drink. For millions of people around the world, tea is part of their culture.

Place and Worldview

Tea involves place: for instance, the places where people grow tea, the differences in tea depending on where the tea grows, and the clays that potters use to make ceramic utensils that have different colors and other characteristics depending on where the clay is mined. The outlook that people have as they drink tea, especially in the Japanese tea ceremony, is an illustration of worldview and culture. While I was working on this lesson, I enjoyed a cup of green tea. It really put me in the mood.

The Lord once appeared to Abraham in the form of three men who came to his home. Abraham welcomed the men with great ceremony and prepared a meal for them.

[Abraham] said, "My Lord, if now I have found favor in Your sight, please do not pass Your servant by. Please let a little water be brought and wash your feet, and rest yourselves under the tree; and I will bring a piece of bread, that you may refresh yourselves; after that you may go on, since you have visited your servant." And they said, "So do, as you have said."
Genesis 18:1-4

Assignments for Lesson 89

Gazetteer	Read the entry for Japan (page 159).
Geography	Complete the map skills assignment for Unit 18 in the *Student Review Book.*
Project	Continue working on your project.
Literature	Continue reading *Revolution Is Not a Dinner Party.*
Student Review	Answer the questions for Lesson 89.

Yin and Yang Symbol at Taoist Temple in Chengdu, China

90 Faith System: Meditative Religions

The believer says, "It seems reasonable to me that God is real."

The atheist says, "It seems reasonable to me that God is not real."

The Hindu says, "What is reason? What is reality?"

In this lesson we examine four perspectives on reality that are truly different worldviews from how Western Christians think and perceive reality.

Yin and Yang

The yin and yang worldview originated in China around the 200s BC. The words represent the north, shady side of a mountain and the south, sunny side. A drawing of a circle with a dark portion and a light portion often represents this philosophy.

This worldview holds that the universe operates on the basis of two complementary forces or principles which together create the whole. These complementary principles are seen in such things as darkness and light, heaven and earth, winter and summer, male and female, passive and active, receiving and giving, even numbers and odd numbers, mountains and valleys, life and death, and so on. In this worldview these opposite principles and their dynamic, ongoing operation sustain and transform the universe. Yin and yang are not simply mutually exclusive opposites; in this worldview they are interdependent forces that explain how the world works.

The yin and yang worldview influenced Chinese thought and is important in taoism (or daoism) that many people in Japan follow. Adherents believe that gaining an understanding of yin and yang helps in medicine, art, the martial arts, knowing lucky and unlucky days, arranging marriages, and many other aspects of life.

This worldview differs from Christianity in significant ways. The belief in yin and yang attempts to explain how the world operates, but it does not provide an answer for how the world started or where it is headed. It has no place for the God of design and purpose. Yin and yang holds to a cyclical view of time whereas Biblical faith holds that our universe had a beginning and will have an end. Christians believe that God controls and operates the universe by His divine will and that trusting Him and praying to Him, not having an understanding of lucky and unlucky days, will best guide our lives. The Bible teaches that God is sovereign, holy, complete, and eternal; whereas evil is a lesser power and is partial and temporary. Good and evil are not balanced in the universe. God is ultimate.

Pura Luhur Uluwatu, Hindu Temple in Bali, Indonesia

Hinduism

Hinduism arose in the region of India probably around 1500-1000 BC. Today about 80% of India is Hindu. A major motivation was probably an attempt to find meaning in a world that had much suffering and much that individuals could not understand. This belief system rejects individuality and holds that all is one. This oneness is called Brahman. Western thought generally emphasizes what is distinct about different parts of our world: stars are not trees, chairs are not humans, and humans differ from one another. To Hindus, separateness is an illusion, what they call maya. The individual soul is the atman, which is part of Brahman. In fact, each atman is Brahman, at once the whole and part of the whole. Every person is the deity and part of the deity.

In this system of thought, the purpose of life is the pursuit of worthy goals, such as devotion to truth, respect for life, and detachment from the world or from things in the world that lessen quality of life. Failing to pursue worthy goals keeps one from being united with Brahman. This pursuit is a major undertaking and is more than one person can do in one lifetime. As a result, the individual soul is reincarnated into many lives. Pursuing this goal results in a person being reincarnated into a higher human caste in life; failing to do so results in being reincarnated into a lower human caste or even as an animal. One's karma—a person's actions or behavior—determines one's caste (have I lost you yet?).

In Hindu thought, reality consists of the oneness of everything. Many paths lead to unity with the oneness; in fact, all paths lead to it to a greater or lesser degree, either more quickly or more slowly. All is good; therefore, there is no truth and there is no lie. Whatever leads an atman to Brahman is worthwhile. A person turns away from separateness and moves closer to oneness by chanting a contentless syllable, such as "om," over and over, or by meditating on nothingness. The goal is to move beyond individual personality and beyond knowledge to experience oneness with Brahman. The individual has no value. Hindu temples and the activities in the temples are intended to help people in this quest for oneness with Brahman.

As you can imagine, Hindu thought is a complex diversity of deities, writings, applications, and teachers. Each one emphasizes different aspects of the philosophy or expounds new meanings in Hindu thought. The search for oneness continues....

Carved out of a cliff in the 700s AD, this statue of Buddha at Leshan, China, is 233 feet tall.

Hindu thought is inconsistent. For instance, Hinduism holds that love is better than hate and that it is better to be enlightened than not to be; however, to define some things as worthwhile and good and other things as not identifies a division or distinction in reality. Thus it appears that even in Hinduism all is not quite one.

More importantly, Hinduism conflicts with the teaching of Scripture. Scripture teaches that God is distinct from the universe He created. The Bible does not dismiss the reality of the material world; instead, Scripture says that it declares the glory of God. Hebrews 9:27 says that "it is appointed for men to die once and after this comes the judgment." This refutes the belief that a person dies many times and goes through reincarnation. The Bible nowhere admits to the possibility of reincarnation. John 3:16 and many other verses teach the value of the individual before God; in other words, the individual is not nothing. Also, the New Testament teaches that Jesus is the way to deal with sin, not meditation and chanting and unity with the oneness.

Buddhism

Buddhism began as an offshoot of Hinduism. Siddharta Gautama was a Hindu born about 563 BC in the region of the Himalayan Mountains. He was the son of a wealthy warrior and lived a privileged and protected life. He married and had one son.

When Gautama was 29 years old, one day he saw, in succession, an old man, a sick man, and a dead man, in that order. This experience prompted him to want to figure out the meaning of life. He left his family and lived as a beggar, trying to determine the cause of suffering. Six years later, while sitting under a tree, Gautama believed that he was "enlightened" with true insight. The Sanskrit word *bodhi* means "enlightenment," and Gautama became known as the Buddha ("the enlightened one"). He organized groups of monks and nuns who were dedicated to following his philosophy.

The Buddha identified what he believed were four truths:

1. Suffering and misery are universal.

2. The cause of suffering is desire.

3. The way to end suffering is to overcome desire. Freedom from desire is the state of nirvana, which involves wanting nothing.

4. The way to escape pain and suffering is to live the Eightfold Path: compassion, knowledge of the good, good intentions, right speech, right conduct, pursuit of a proper livelihood, mindfulness, and meditation.

The ultimate reality in Buddhism is the Void: a state of pure consciousness that overcomes desire. Buddhists believe that the Void is not nothing, but at the same time it is not something. It is something that man cannot adequately name or grasp, but it is reality all the same.

Buddhism accepted some beliefs from Hinduism, such as karma and reincarnation, but Buddha did not believe in the soul or in any deities. Like Hinduism, Buddhism has splintered into many branches and emphases. One branch exalted Buddha to the status of a god. Zen Buddhism, which arose in East Asia (primarily China and Japan), especially encourages a master teacher to guide and instruct new converts.

Buddhism suffers from the errors of Hinduism mentioned above. In addition, the Bible does not say that nothingness is at the heart of reality; God is. Nirvana is an impossible, imagined state; to desire what is good is not wrong. Buddhism's answer to suffering is inadequate also. The Bible says that we can learn from suffering and that it can have redemptive value. Even if one lives a good life (such as the Eightfold Path), one cannot escape all suffering.

Risshakuji, Zen Buddhist Temple in Yamagata, Japan

New Age Thought

A fourth alternative way of looking at reality, one that has arisen in modern times, is called New Age. The premise of New Age thinking accepts evolution and holds that man is evolving to a new age or new stage of existence and thought. The higher consciousness that man is entering will, adherents say, result in a new humanity. This new humanity will supposedly be free of the shackles of religion and perceptions that have limited us.

The emphasis in New Age thought is the self. In New Age thinking, there is no god but the self, and the self is god. Prime reality is the self. There are no standards, no authority, outside of self. Each person makes his or her own reality and truth. New Agers claim mystical experiences that transcend human limits of time, space, and morality (such as the simple issue of what is right and wrong). Some of these experiences are drug-induced, but New Age accepts that "reality," too. New Agers practice channeling, which supposedly gets them in touch with the spirits of dead persons. They also claim to recall experiences that they had in previous lives.

New Age thought is an attempt to have spirituality without religion and without God. Statements and claims by New Agers reach the point of absurdity. For instance, in one of her many books on New Age, actress Shirley MacLaine wrote, "I could legitimately say that I created the Statue of Liberty, chocolate chip cookies, the Beatles, terrorism, and the Vietnam War." In another book she says, "Know that you are God; know that you are the universe." The egotism of such statements is obvious, profound, and ridiculous. Their use of language and terms says that language has no meaning except what they give it. Without a common understanding of language, how can discussion take place?

Moreover, how can there be many realities? People can create their own thought worlds, but those are not necessarily reality. People can claim anything, but New Age claims have no evidence—no reality, if you will—to back them up.

New Age thought is similar in some ways to gnosticism in the early church. *Gnosis* is the Greek word for knowledge. Some persons claimed to have deeper knowledge than what ordinary people had, and as a result they claimed to have a deeper grasp of reality beyond the simple doctrines of Christianity. One purpose of this was to attract a following of people who would look to these "enlightened ones" as spiritual leaders. Paul's letter to the Colossians addresses this kind of thinking, especially his warning against being taken captive through "philosophy and empty deception" (Colossians 2:8). Paul might be warning against similar false teachings in 1 Timothy 6:20, where he warns Timothy about "what is falsely called knowledge." Thus it turns out that New Age thinking is just the old gnostic heresy in new clothes.

A recent manifestation of New Age thought that has gained popularity in the United States is the practice of "mindfulness." The emphasis in this mindfulness trend is on being aware of every moment that occurs and everything in you and around you. Mindfulness teachers encourage meditation to accomplish this. Being aware of what is going on around you and inside you, and not being distracted by petty things from focusing on the more important things, are healthy goals. However, typical mindfulness teaching encourages a nonjudgmental attitude and says nothing about God. Once again, it's all about the self and finding peace solely within you. Mindfulness training for businesses has developed into a multimillion-dollar industry, and its teachers often promote it as a way for workers to be more productive and profitable. But mindfulness is really just Buddhist and New Age thinking dressed up in secular garb. The Bible has many passages that teach healthy, God-centered attitudes, such as Philippians 4:8-9 and Colossians 3:15-16.

Reality in our world is sometimes not pleasant, but it is still reality. Redefining it or claiming a different reality does not make reality go away; if anything, adopting such a worldview just makes dealing with reality more difficult. Reality is that car bearing down on you or the doctor's grim report, but it is also beautiful scenery and the joy of a newborn baby. God created the world as it really exists. Real people live in it and real things happen in it. Jesus came as a real human being into our world to provide answers for the real issues that we have and to teach a different way to see and appreciate reality. Our goal should be to develop a worldview that helps us face reality and be more than conquerors in it. This is what Jesus helps us to do.

Who will separate us from the love of Christ? Will tribulation, or distress, or persecution, or famine, or nakedness, or peril, or sword? Just as it is written, "For Your sake we are being put to death all day long; We were considered as sheep to be slaughtered." But in all these things we overwhelmingly conquer through Him who loved us.
Romans 8:35-37

Assignments for Lesson 90

Worldview	Recite or write the memory verse for this unit.
Project	Finish your project for this unit.
Literature	Continue reading *Revolution Is Not a Dinner Party.*
Student Review	Answer the questions for Lesson 90. Take the quiz for Unit 18.

Wulingyuan Scenic Area, Hunan, China

19 China

This unit discusses several aspects of the People's Republic of China. We first consider the world's largest population, the one-child policy that China followed for many years, and the current status of that law. Then we consider two groups that don't fit into China's repressive system: the persecuted Uyghur Muslims of Xinjiang province and the pro-democracy protesters in Hong Kong. We examine the Belt and Road Initiative, which is China's attempt to expand its influence in many places around the world. Since much of the trade with China takes place using sea containers, we learn about the inventor of "the box," Malcom McLean, and what those containers have meant to international trade. The worldview lesson focuses on atheism.

Memory Verse Memorize Luke 12:15 by the end of the unit.

Books Used

The Bible
Exploring World Geography Gazetteer
Revolution Is Not a Dinner Party

Project (Choose One)

1) Write a 250-300 word essay on one of the following topics:
   - What is the value of life? In your essay, include thoughts about the unborn, the elderly, and the physically and mentally handicapped. (See Lesson 91.)
   - How should Christians respond to persecution? (See Lesson 92.)
2) Draw a diagram of a sea container using accurate dimensions and/or an ocean-going cargo vessel that transports sea containers. (See Lesson 94.)
3) Write a story about taking a long journey to see relatives. Tell how the people would prepare, the journey, what happened in the visit, and how difficult it would be to leave. If you have taken such a journey, you can tell about your own experience. (See Lesson 91.)

Chinese New Year Lantern

91 The People of China

Qin Xua works in a factory in Shenzhen, China, that makes parts for Apple computers. She lives in an apartment that she shares with two friends. Qin is saving money to buy a car. She goes on vacation every year, usually to somewhere in China; but once she went to the western United States.

In mid-January, Qin and almost everyone else in China begins making preparations for the highlight of the year: the celebration of Chinese New Year. She buys her train ticket early (people can buy tickets up to sixty days before departure) to be sure she can make the 200-mile trip to see her parents and extended family. Estimates are that in the peak buying season, Chinese railroads sell about one thousand tickets per second.

The lunar calendar determines the timing of Chinese New Year, so it can fall anytime between mid-January and mid-February. It is a time when families get together. For many who work in urban factories, it might be the only time during the year that they see their family. Families exchange gifts, and the traditional dish involves dumplings. New Year's is also a time for festive and elaborate celebrations, including colorful parades and eye-popping shows. Chinese New Year is something like Christmas, Thanksgiving, New Year's Eve, the Fourth of July, and the Super Bowl rolled into one two-week period. The celebration of New Year's Eve and New Year's Day in China brings the world's largest one-day lighting of fireworks.

About 20-25% of China's people travel for Chinese New Year. This means that about 230-260 million people go somewhere, usually by train. It is the world's largest seasonal human migration. That travel is almost completely from the cities to the rural parts of China and back again. Factories and stores close for at least some of the period surrounding the actual New Year's Day.

A small but increasing percentage of Chinese people travel overseas during Chinese New Year. Perhaps six million people will go abroad either instead of going home or for part of the holiday. This overseas travel is an indication of the growing wealth of many Chinese people.

The Chinese New Year is connected to several worldview issues. The observance began as a time to pray to the gods and to ancestors for good planting and harvest seasons in the coming year. A few days before the celebration begins, the people in each home perform a thorough cleaning to sweep out demons and bad luck. The figures of lions and dragons in the parades are supposed to chase away

Symbols of the Chinese Zodiac, Shanghai, China

bad luck. Most people do not take a shower on New Year's Day to avoid washing away any good luck.

The Chinese zodiac consists of twelve animals: rat, ox, tiger, rabbit, dragon, snake, horse, goat (or sheep, or ram), monkey, rooster, dog, and pig. Each animal has its year, and the twelve signs rotate through a twelve-year period. Each animal supposedly gives its positive traits to people who are born that year. However, of the twelve signs, a person's birth year animal is believed to be his or her unluckiest. A person's best defense against bad luck brought on by the zodiac animal is the color red. Many Chinese believe that the zodiac signs influence their lives in many ways.

It is interesting—and sad—that millions of Chinese have been trained not to believe in the one, true, loving Creator God but to believe in the power of superstitions, luck, and ritual.

Big Numbers

China has the largest population of any country in the world, about 1.4 billion. This population means that China has many workers, a large consumer market, and many people who can serve in the military in time of war. It also means that the country has a great demand for food, healthcare, and public services such as utilities, water, and sewer service.

One region of China, the North China Plain, which lies between Beijing and Shanghai, is about half the size of the United States. The total population of the United States is about 326.5 million. The North China Plain is home to over one billion people.

The Han ethnic group makes up about 92% of the Chinese population. The Han people take their name from the Han dynasty, a family that ruled China from 202 BC to 220 AD, with the exception of a few years during that period. The Han era was a time of great accomplishment and is a source of much pride among the Chinese.

The Han make up the world's largest ethnic group and constitute about 18% of the world's population. In other words, almost one in every five people in the world is a Han Chinese. About fifty-five other ethnic groups also live within China's borders.

China's One-Child Policy

Soon after the Communists took over China in 1949, Communist party propaganda encouraged couples to have many children in order to produce a large workforce and military force that they believed the country needed. The party also condemned the use of contraceptives. The population of China doubled in just over two decades.

Then party leaders began to rethink this policy. The growing population strained the food supply. Many scholars believe that a famine between 1959 and 1961 cost 15 to 30 million lives. In 1979, the

Family in Xi'an, China

Celebrating National Day of the People's Republic of China in Beijing, October 2013

Communist government announced a one-child policy. To slow population growth, the government allowed parents to have only one child.

Most Chinese couples want a male child to carry on the family name and property and to care for them in their old age. This desire and the government's policy led to some harsh consequences. Many Chinese women underwent sterilization procedures after having a male child, sometimes voluntarily and sometimes as the result of governmental pressure. In many cases parents abandoned female children and second children or put them up for adoption; in some cases, they were put to death. When medical technology enabled parents to learn the gender of a child in utero, many parents chose to abort unborn female babies. Sometimes the government forced women to have abortions.

The government refused to provide many services for families that had more than one child, such as health care and allowing their children to go to school. Some second children were hidden away. They often had a hard time finding jobs because, in the eyes of the state, they were not supposed to exist. The government enforced the one-child policy more stringently in urban areas and not as strictly in rural China.

The one-child policy had several other negative consequences. China experienced a slowdown in the growth rate of the work force, and many leaders feared that the country could not supply the workers needed for a growing economy. The older generation became concerned that there would not be enough children to care for them in their old age. The government implemented this policy as longevity was increasing, which meant that there were more older people and fewer young people to support them.

The number of male children compared to female children increased, which meant that fewer girls were available to become wives and mothers. The natural ratio is 105 male births for every 100 female births; in China the rate became 114 males to 100 females. This further slowed the population growth. On the other hand, China offered an increased number of female children for international adoption.

In 2013 China began changing the one-child policy. Now the government allows a couple to have two children. However, some experts doubt whether China can return to the normal male-female balance in its population. The current mandatory retirement age is 55, so that the younger generation can have more jobs. Some have suggested raising the mandatory retirement age to 65, to enable the Chinese economy to have more workers.

The two-child policy is a step toward greater freedom, but the government still does not allow a couple complete freedom in how many children they can have. This entire problem-filled situation shows what happens when the government meddles in matters that are best left to parents and God.

Some governments, when they believe the population is growing too fast for their economy or food supply, attack the problem by trying to control population instead of changing their policies to allow greater economic freedom such as encouraging greater food production or trade.

One way that God showed the value of infants was in Jesus' praise that God had hidden His truths from the wise and intelligent and revealed them to infants. Sometimes those who are wise and intelligent from the world's perspective don't understand God, while little children understand what it means to depend on God.

At that time Jesus said, "I praise You, Father, Lord of heaven and earth, that You have hidden these things from the wise and intelligent and have revealed them to infants."
Matthew 11:25

Assignments for Lesson 91

Gazetteer Study the entry for China (page 158).

Worldview Copy this question in your notebook and write your answer: What do people miss out on when they do not believe in God?

Project Choose your project for this unit and start working on it. Plan to finish it by the end of this unit.

Literature Continue reading *Revolution Is Not a Dinner Party*. Plan to finish by the end of this unit.

Student Review Answer the questions for Lesson 91.

Kashgar, Xinjiang Uyghur Autonomous Region, China

92 People Who Don't Fit

In opposite corners of China, in the far northwest and the extreme southeast, two groups of people don't willingly accept being part of the People's Republic of China. The Uyghur people (pronounced WEE-jer and sometimes spelled Uighur) of Xinjiang Uyghur Autonomous Region and many of the residents of Hong Kong Special Administrative Region are protesting the heavy-handed way that the central government in Beijing is ruling them. In both cases, geography is a major factor.

The Uyghurs

How remote is Xinjiang Province in northwest China? It includes the point on the earth that is farthest away from any sea. The dry Tarim Basin dominates the province. The basin contains the Taklamakan Desert. The name of the desert roughly translates to "the place that one does not leave upon entering." Routes on the Silk Road passed along the northern and southern edges of the basin, but the basin itself is largely impassable.

The Tarim Basin is surrounded by mountain ranges, including the Tien Shan mountains. Some plants and wildflowers in the Tien Shan have never been classified. Colonies of Buddhist monks lived in caves there to escape civilization. Xinjiang Province has several rivers and streams, but the waters of only one reach the sea. The Irtysh River flows west into Kazakhstan and then north into Russia, merging there with the Ob River which flows into the Arctic Ocean. The other streams and rivers of Xinjiang disappear into deserts and salt lakes.

The Qing (Manchu) Dynasty annexed the Xinjiang region in the 1700s, and it has officially been considered part of China almost continuously ever since. However, the Uyghur people who live there are Turkic in origin; the Uyghur language is Turkic; and the Uyghurs are almost all Muslims. Racially they are a mixture of Caucasian and East Asian. In other words, the Uyghurs have much more in common with the Turkic people of Central Asia than they have ever had with the Han Chinese who govern them.

In the 1900s Uyghur writers and political leaders reasserted the Uyghur identity after many years of rule by the Han Chinese. The region came to be called East Turkestan or Chinese Turkestan. Leaders there declared themselves to be the country of East Turkestan in 1949, a name the Uyghurs continue to believe appropriate. Unfortunately for them, Mao Zedong declared the People's Republic of China on October 1 of that year; and the Chinese Communists reabsorbed Xinjiang under their rule.

Some Uyghurs fled to the Soviet Union and into Central Asia. Xinjiang is the largest province-level administrative district in China, larger than Alaska.

During the 1980s, China allowed the Uyghurs to build mosques and practice their faith with some freedom. A resurgence in the use of the Uyghur language occurred. The Uyghurs were emboldened in the 1990s to protest rule by China, but the Beijing government suppressed the demonstrations and other expressions of Uyghur ethnic identity. Tension in the region has grown ever since.

Systematic Violation of Uyghur Human Rights

The Beijing government has instituted a systematic program that is intended to eradicate Uyghur culture, language, and religion. An estimated eleven million Uyghurs live in Xinjiang, and China recognizes the Uyghurs as one of 56 minority ethic groups in the country. The capital, Urumqi, has an estimated population of 2.5 million people. However, Beijing has clamped down on the Uyghurs to foil any religious expression and any thoughts of political separation. The Uyghurs see Chinese rule as illegitimate, while China sees the Uyghurs as posing a threat of separatism and religious radicalism. The following actions have been reliably reported, and this author believes them to be true.

The Beijing government has moved millions of Han Chinese into the Xinjiang Province, especially to work in projects that contribute to the province's economic development. This decreases the percentage of Uyghurs in the population and prevents Uyghurs from holding the best jobs. An estimated 40% of the population of the province is now Han. At the current rate Uyghurs could become a minority of the population in Xinjiang in a few years.

The government is using advanced technology to identify and spy on Uyghurs suspected of being religious extremists or terrorists. The technology includes facial recognition cameras that are posted

Market in Kashgar, China

above streets in cities in the province. The government has released "flocks" of surveillance drones disguised as birds. The drones have wings that flap to make them look even more like birds. With these tools, the government can spy on where people go, what they do, and whom they talk to.

Police can demand to see anyone's cell phone at any time and can detect if the owner has any videos that promote ideas that are not in keeping with Beijing's policies or if he or she has placed calls to or received calls from foreigners. The government has offered cash payments for interethnic marriages (hoping to lessen Uyghur identity and culture) and has forced some women to have abortions.

Beijing's treatment of Uyghur Muslims as potentially disloyal and a political threat parallels how the Roman Empire treated the church in the period before Constantine. In the name of preventing terrorism, the government treats Uyghurs harshly. Thousands of names are added to a watch list regularly, and people are routinely brought in for difficult questioning. The government has outlawed most or all religious activity, including the banning of fasting during the Muslim festival of Ramadan. A group of Uyghurs who went on a pilgrimage to Mecca in 2018 had to wear lanyards that transmitted their location to the Chinese government. Beijing has outlawed or restricted the use of the Uyghur language and the conducting of schools in Uyghur.

Without a doubt the most egregious violation of human rights takes place in the hundreds of "re-education centers" across Xinjiang. To these places that have been compared to concentration camps, the Chinese government has taken over one million Uyghurs suspected of having extremist thoughts. At these centers government workers practice what in other contexts has been called brainwashing. They force the inmates to renounce their Muslim faith, sometimes by forcing them to eat pork and drink alcoholic beverages, both of which violate Muslim doctrine.

Inmates are deprived of sleep and food, interrogated with electric shocks, and kept in isolation. If a family member goes outside of the country, that is sometimes reason enough for a Uyghur to be put in a camp. Inmates must take courses in Mandarin Chinese (the language of the majority of Chinese), receive training in vocational skills, and learn about Chinese law (which they did not know they were violating). Inmates must stay in a center for varying lengths of time, but eight to ten months is not uncommon. The government has reportedly moved Han men into Uyghur homes to fill the roles of husband and father while the man of the house is in a re-education center, assuming he is still alive. Some inmates have been transferred to prisons in other provinces and even in Mongolia.

Beijing has denied much of what detainees have claimed about the centers; but we do know the centers have existed, and the reports out of them have been too consistent to believe anything other than that such harsh treatment has taken place. The government has justified the policy by saying, in effect, "Don't you quarantine someone with an infectious disease? These people have dangerous thoughts or are capable of dangerous thoughts. Won't the country be better off without these dangerous, extremist ideas?" In late 2019 the Chinese government claimed that all of those receiving what it called voluntary vocational training in the re-education centers had completed their studies and that the centers have been closed. No independent confirmation of this claim was possible.

Violence has taken place in connection with the Uyghur situation. Protest demonstrations crushed by authorities have resulted in hundreds being killed. Uyghur terrorists unleashed a knife attack at a train station in the city of Kunming in southeast China in 2014. Over thirty people died and many more were injured. Beijing uses such attacks and fear of them to justify its repression of all Uyghurs.

In other words, the government of China is engaged in what is probably the largest and most brutal systematic, government-sponsored repression of human rights taking place in the world today, except perhaps what is happening in North Korea.

Hong Kong

At the opposite end of China, more unrest has taken place in Hong Kong. Great Britain acquired the valuable port city island of Hong Kong in 1842 as a concession after a war with China over the opium trade. Britain acquired more adjacent territory in succeeding years. The city, which is now about six times the size of the District of Columbia with a population of 7.2 million, developed into a vibrant business center. In 1898 Britain signed a 99-year lease for Hong Kong with China. The city became a haven for refugees fleeing Communist-controlled China and an embarrassment for China as its economy boomed while China's socialist system struggled.

Communist China was not about to renew the lease, so Great Britain and China agreed to end the lease as of 1997, when the city returned to Chinese control. Beijing talked nice about the transfer and said that the arrangement would prove that "one country, two systems" could work and that no significant changes in how Hong Kong operated would be made for fifty years. China hoped that the arrangement would be a model for how Taiwan could be reabsorbed into its control.

However, in 2019 the Chinese government considered a new extradition law that would require persons accused of crimes in Hong Kong to be tried in courts in China. This struck many people in Hong Kong as a significant and dangerous change, and thousands took to the streets of Hong Kong in protest. The demonstrations sometimes turned violent and Chinese authorities brutally repressed them. The proposed extradition law was dropped (although the issue did not go away), but demonstrations continued over other issues. The emphasis of the demonstrations became a demand that freedom and democracy continue in Hong Kong. Pro-democracy candidates won a landslide

Hong Kong Skyline

Protestors in Hong Kong (2019)

victory in district council elections in November 2019, giving them control of 17 out of 18 district councils. The election was seen as a stern rebuke to how the government handled the demonstrations earlier that year. The Communist leader of Hong Kong said that she would "seriously reflect" on the outcome of the elections.

It is apparent to the world that China has at least two large groups of people who don't want to be part of China. After President Richard Nixon opened the door for China to have greater contact with the democratic West, many hoped that political freedom would follow an increase in capitalist activity in China. Thus far, this has not happened.

And yet they call it the People's Republic.

Isaiah uttered woes on those who turned genuine standards of right and wrong upside down for their own evil purposes.

Woe to those who call evil good, and good evil;
Who substitute darkness for light and light for darkness;
Who substitute bitter for sweet and sweet for bitter!
Woe to those who are wise in their own eyes
And clever in their own sight!
Isaiah 5:20-21

Assignments for Lesson 92

Gazetteer	Read "Riding the First Wave of CDC's COVID-19 Response at Ports of Entry" (pages 309-310) and answer the questions about the article in the *Student Review Book.*
Worldview	Copy this question in your notebook and write your answer: What is your outlook on the world and your life when you get up in the morning?
Project	Continue working on your project.
Literature	Continue reading *Revolution Is Not a Dinner Party.*
Student Review	Answer the questions for Lesson 92.

Port of Khorgos

93 China's Belt and Road Initiative

Khorgos, Kazakhstan is a bustling port that each year handles thousands of shipping containers transported internationally. But this port has no water. Khorgos is inland, far from any ocean. All those containers travel by railroad. This rail line has transformed the city of Khorgos. Who built this key trade location?

China.

The Confucius Institute of China offers classes in Chinese language and culture. The people that the institute teaches want to learn Chinese because it opens career opportunities for them. The people there celebrate the Chinese New Year. Chinese construction companies build apartments with dimensions appropriate for Chinese people. Chinese young adults teach Chinese songs to local children. Where is this happening? Nairobi, Kenya, East Africa. Who is paying for it?

China.

A deepwater port at Gwadar in southwest Pakistan on the Arabian Sea can handle large merchant marine vessels as well as naval vessels. Who built this facility and many others like it along the coast of the Indian Ocean?

China.

Do you see a pattern here? You should. There is one.

A Coordinated Plan

In 2013 China announced a sweeping plan to build roads, railroads, bridges, gas and oil pipelines, power plants, and seaports in many locations around the world. The plan ultimately included seventy countries (about one-third of the countries of the world). The plan called for an investment, with both grants and loans, of money equal to almost one trillion U.S. dollars—yes, that's trillion. Taken together, the plan is the largest construction project in the history of the world. No other country has this kind of large-scale vision for how to influence the world.

Observers have called this plan One Belt One Road, or the New Silk Road, or (as has become most common) the Belt and Road Initiative (BRI). The land routes run from China through former Soviet republics in Central Asia, through the Middle East, and into Europe. The route roughly follows in many places the path of the original Silk Road of the Middle Ages. The Silk Road was the route by which traders brought silk, spices, and other commodities from China to Europe and over which people and goods went from Europe to China. The original Silk Road began in the easternmost province of China, which is the heart of the current China. The chain of ports on the Indian Ocean that China is building shadows much of the route that the Chinese seafaring explorer Zheng He followed in the early 1400s.

Many Projects, Many Countries

Here are just a few examples of the many projects in which China is investing its money, people, and prestige:

- China is building a high-speed railroad linking it with Laos and Thailand.
- China has invested in oil projects in Sudan and the United Arab Emirates.
- China has built its first naval and military bases on foreign soil in Djibouti in Africa.
- China built an electric railway between Djibouti and Ethiopia. Djibouti has borrowed from China an amount equal to more than 75% of its gross domestic product.
- China has spent billions building resort facilities—including ports, hotels, resorts, and a golf course—in the Bahamas.

Baha Mar Resort, Bahamas

Chinese Warship in Djibouti

- China has built a port in Santiago, Cuba. China is Cuba's second-largest trading partner.
- A Chinese construction company built a large mosque in Algiers, Algeria, with the world's tallest minaret. China is the largest foreign construction investor in Algeria.
- At one point, a single Chinese dam-building company was engaged in over seventy hydropower projects across Africa; and it was not the only Chinese construction company active on the continent.

What Does China Get For Its Money?

What is China buying for its trillion dollars? Several things. The Chinese government wants to expand its influence in the world economy. This Communist country means business—big business. If companies anywhere want to engage in international trade, increasingly they will have to do business with China.

China is also obtaining reliable sources of energy for its ever-growing number of factories, cars, and homes. This is fuel that does not have to pass through the Strait of Malacca in Malaysia, which is busy and could be a choke point in the event of an international crisis.

Great Mosque of Algiers

China is buying food for its billion-plus population, not only the food itself but land (in Africa, for example) on which farm workers grow food dedicated to the Chinese market.

China is opening new markets for the goods that Chinese factories produce.

China is achieving its long-desired access to Indian Ocean ports, such as by the road it is building from Tibet through Pakistan to the port in Gwadar. Pakistan gets a nice new road and a nice big port, and China gets a broader economic reach and Pakistan's silence regarding issues the Chinese government sees as important.

China is employing hundreds of thousands of its citizens in building these projects. One estimate is that about one million Chinese live in Africa, building infrastructure, growing food, running stores, and doing many other kinds of work. A smaller number of Africans are moving to China for education and training, thus becoming more dependent on Chinese money and the Chinese approach to business and life.

China is also creating places where its growing middle and upper class population can go on vacation, confident that they will feel at home and allowing them to return some of their earnings to Chinese companies.

China is hoping that some of these projects might win over minority ethnic groups in China. For instance, the New Silk Road runs through Uyghur territory, and China hopes that an increasing standard of living and economic dependence on China will discourage any ideas about political revolt. In addition, as the Muslim peoples west of China become more dependent on Chinese trade, they will be less likely to support the Uyghurs in any rebellion.

At the same time, China is buying friends in the international community, and in the United Nations in particular, who commit themselves to supporting

China's policies and actions. For instance, whenever China acts on its claim that Taiwan is really part of China and moves to absorb it, the countries that China has helped economically (and have become dependent on China to some extent) will support China's claim. When China takes clear action to support its claim that the South China Sea is part of its legitimate territorial waters, these countries will either support China or stand by without objection. These countries also commit themselves not to invite for a visit the Dalai Lama of Tibet, the spiritual leader-in-exile of the disputed region that China claims over Tibet's objections. In other words, money talks.

Not Everything Is Rosy

With the increasing influence of the Chinese Communist government around the world, several specific problems have arisen in China's relentless outreach. China has sent many of its own people to work on these projects, which is disappointing to local residents who had hoped to obtain jobs. When contractors do hire local people (usually to perform more menial labor), reports are common that Chinese managers are abusive and racist toward them and that their pay is less than what Chinese workers receive.

The government of China is not concerned about whether another government with which it wants to do business is corrupt or allows human trafficking or engages in other illegal activities. Those are "domestic issues" that don't concern China. In other words, integrity in a business partner is not the most important factor to the Chinese.

Moreover, the influence of most of these Chinese workers, Chinese officials, and groups such as the Confucius Institute will for the most part be godless. Some Christians might be among the immigrants from China, and God can work through them to further the kingdom; but the official stance will be that of atheism.

Map of Key Routes and Locations in the Belt and Road Initiative

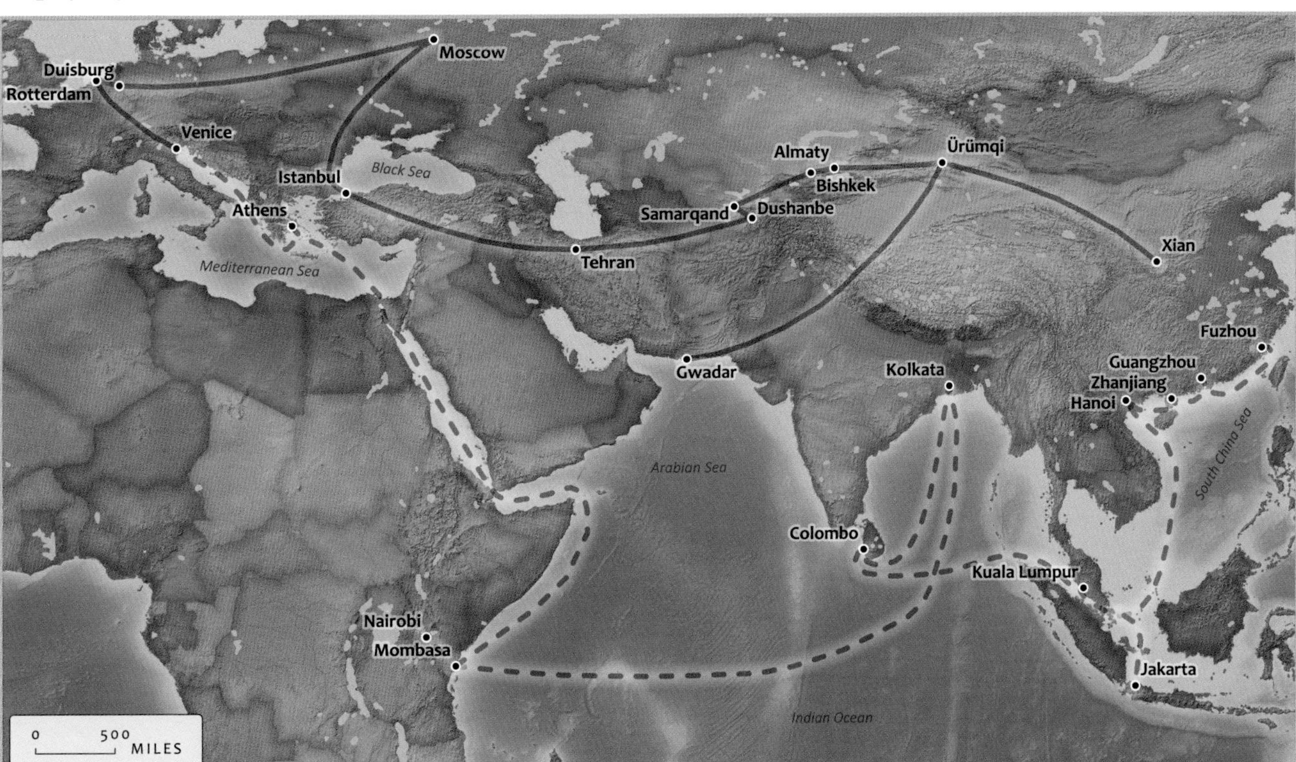

Not Unprecedented?

China attempts to draw a parallel between its policy and that of the United States. The United States and American businesses have invested heavily in many other countries, and the U.S. maintains many military installations around the world. The United States has certainly wanted to influence other countries, but its goal was to encourage democracy and free enterprise. The goals of the American and Chinese policies are very different, but from China's perspective, it is doing nothing different from what the U.S. did throughout the latter half of the twentieth century.

While the United States has been distracted by conflicts in the Middle East for decades, in the Balkans in the 1990s, and by wars on terrorism in Iraq and Afghanistan in the twenty-first century, China has developed its economy and its military to become a major world power. It may well surpass the United States in annual economic output sometime in the next few years. China certainly wants to be seen as the dominant power in Asia and as a country to be reckoned with around the world, but it hopes to achieve this return to greatness by soft, economic power rather than with military power.

A Different Worldview

All of this activity by China comes as a result of a profound shift in the worldview of the Chinese Communist government. During the days of Mao Zedong's leadership in the 1950s and 1960s, China saw almost the entire non-Communist world as the enemy. Chinese propaganda portrayed the United States and other Western nations as "imperialist lackeys and their running dogs." Now China wants those lackeys and dogs to be their trading partners.

Chinese President Xi Jinping Addresses the United Nations General Assembly (2015)

In a sense this is a triumph of capitalism. China's economic success and power have become possible because that nation has largely abandoned traditional Marxist economics and moved toward market economics, although the government's control over the economy is still strong.

We can hope and pray that disciples of Jesus around the world can influence China for the Lord and that its people can come to know religious, intellectual, and political freedom as well as greater economic opportunity. After all, a road runs in both directions.

The prophet Isaiah spoke of a different kind of road, a Highway of Holiness:

A highway will be there, a roadway,
And it will be called the Highway of Holiness.
The unclean will not travel on it,
But it will be for him who walks that way,
And fools will not wander on it.
Isaiah 35:8

Assignments for Lesson 93

Worldview Copy this question in your notebook and write your answer: If I have my truth and you have your truth, how can we have any sort of meaningful discussion?

Project Continue working on your project.

Literature Continue reading *Revolution Is Not a Dinner Party.*

Student Review Answer the questions for Lesson 93.

Container Ship Docked in Rotterdam, The Netherlands

94 Malcom McLean and the Box

On April 26, 1956, Malcom McLean changed the world. At least, he changed the shipping industry. He did it with a box. It was a big, specialized box, to be sure; but basically it was a box.

Getting Things from Point A to Point B

Trade has always depended on transporting goods from one place to another, which is where geography comes in. At various times in history, traders have used ships, wagons, camels, railroads, trucks, barges, airplanes, and tanker ships. Each mode has advantages, and each has potential difficulties. Weather sinks ships, shuts down highways and airports, and stymies railroads. Camels can travel across deserts but can also get sick. Thieves, whether workers within a company or bandits from without, steal goods, which means lower profits. Rough roadways could ruin wagons. Long distances could prevent trade from taking place at all between two otherwise eligible parties.

As a result of these factors, traders, shippers, and inventors are always developing new ways to make trade safer, quicker, and less expensive. The development of refrigerated railcars in 1878 is one example of such innovations. As technology has advanced, larger trucks and airplanes have carried more goods, providing greater cost efficiency.

Transporting manufactured goods or farm products, especially by ships, was for a long time a complicated, labor-intensive process. Workers in a factory made the goods and loaded them by hand onto wagons, railcars, or trucks. A driver would take the goods to a dockyard, which was a flurry of activity. Longshoremen would carry the goods onto a ship on their backs or with small carts and store them among different kinds of goods in the hold of the ship.

When the ship landed at the receiving port, the process was repeated in reverse. Middlemen stored the goods in warehouses and eventually distributed them to retail stores. Transportation, labor, and storage costs added significantly to the retail price that consumers paid. Manufacturing and warehouse facilities were often located as near to the ports as possible to save on costs and so businessmen could transport the merchandise more quickly.

In the first half of the 1900s, the Interstate Commerce Commission oversaw and heavily regulated commercial transportation in the United States. The ICC's primary interest was the railroads. It applied strict rules to the trucking industry as well.

Longshoremen's unions controlled who worked at the docks. Corruption was common, and strikes were crippling. A few companies experimented with shipping containers that they could use on trucks, trains, and ships; but their efforts were inefficient and the containers they developed were not standardized.

Enter Malcom

Malcolm McLean (he later changed the spelling of his first name) was born in 1913 in rural North Carolina. He demonstrated a remarkable entrepreneurial interest early in his life. McLean began acquiring trucks and eventually built one of the largest trucking companies in the U.S.

McLean noticed the inefficiencies of the shipping industry and decided to find a solution. He also noticed the increase in vehicle traffic after World War II and the inadequacy of the country's road system to handle it. One of his early attempts involved putting tractor trailer bodies filled with goods onto unused oil tankers and sending them between ports in the United States instead of along overcrowded highways. This approach ran into regulatory issues and the opposition of existing industries and proved to be an inefficient use of space.

Through a series of business deals, McLean acquired a small shipping business and created the SeaLand Company. He obtained an oil tanker that he named the Ideal-X. McLean acquired the services of engineer Keith Tantlinger to build a steel container that workers could transfer between ships, railroad cars, and trucks. Rather than storing the containers in a hold on the ship, workers stacked the containers on the deck and secured them in place. They developed a crane that could lift the containers on board ship and secure them without any direct human contact.

On April 26, 1956, at the port of Newark, New Jersey, McLean and about one hundred others watched a crane load one container onto the Ideal-X every seven minutes. The ship took less than eight hours to load and departed the same day for Houston, Texas, with its cargo of 58 containers. McLean and his associates flew to Houston and saw the ship arrive six days later. What is more, the shipping cost per unit was considerably lower than with previous methods. McLean's company started receiving orders before the ship even docked in Houston. The revolution had come.

A Combination of Factors

The invention of the sea container was just one of many factors that led to the revolution in shipping. The increased use of computers, the building of thousands of miles of Interstate highways, the deregulation of trucking, and repeated world oil crises, as well as other factors, have contributed to this new industry.

One significant factor in the increased use of sea containers has been the growth of business in Communist China. As a Communist country, China rejected capitalism and pursued a socialist economy that failed miserably. In 1972 President Richard Nixon opened the door to increased contact between China and the West. In 1979 the United States and China established diplomatic relations and began to trade with each other. That year the two countries did about $4 billion worth of business in goods and services.

Drawing of the Ideal-X

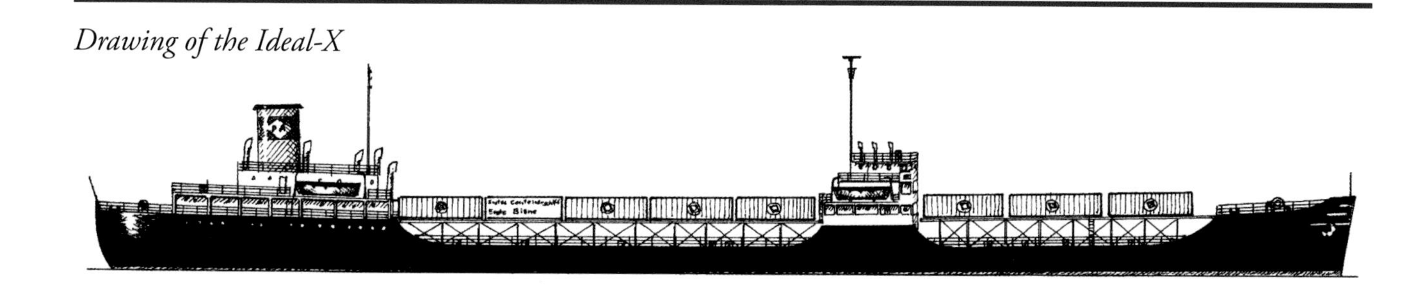

Mao Zedong, China's longtime hardline leader, died in 1976. His eventual successor, Deng Xiaoping (pronounced dung-shaow-ping), began moving the Chinese economy toward what he called "socialism with a Chinese flavor." Deng in effect made a bargain with the Chinese people: "The government will increase your standard of living; you will do what the government tells you to do." The state still owned many businesses, but the government allowed some individual ownership of companies, although they were still under close government oversight. Rather than continuing to reject the hated capitalist West, Chinese companies began doing business with Western companies. China built thousands of factories, recruited millions of its citizens to work in relatively low-paying jobs in those factories, and began turning out consumer goods, computer parts, and heavy industrial machinery.

Now, how would all those goods produced in China reach world markets cheaply and quickly? The sea container! Sea containers enabled China to become supplier to the world. China had the workers who could make the goods relatively inexpensively and a way to deliver the goods to world markets relatively inexpensively. Now containers can deliver raw materials from almost anywhere in the world to a Chinese factory. Workers produce and then package the finished goods into standard containers on standard pallets, which they load into sea containers. The containers are transported to ports, where cranes load them onto container ships. From there they cover the world. It takes about three weeks for a container ship to go from Hong Kong through the Panama Canal to a dock in Germany. Going to the west coast of the United States is a much shorter trip.

In the American port, another crane lifts the sea containers off the ship and onto railcars and trucks. These deliver the containers to distribution centers, where the pallets of goods are touched by human hands perhaps for the first time since the goods were manufactured and loaded into a sea container. From

COSCO and China Shipping were two Chinese companies that merged in 2016. This container ship is crossing the Indian Ocean in 2020. Modern container ships can carry thousands of individual containers.

the distribution center the products are carried by truck to individual stores, where workers place them on the shelves for consumers to buy.

Over time manufacturers agreed to use standard sizes for sea containers. The most typical container is a corrugated steel box measuring eight feet wide, eight feet tall, and twenty feet long with two doors on one end. The floor is usually plywood. Shippers refer to the capacity they fill in terms of TEUs—twenty-foot equivalent units—and FEUs—forty-foot equivalent units (which is simply two twenty-foot units). Containers are numbered, and shippers can keep up with their whereabouts by means of computer tracking systems. Malcom McLean's first transport carried 58 containers. The largest modern container ships can carry over 20,000 each.

A Study in Economic Geography

The growth of maritime shipping because of sea containers is a study in economic geography. This is a subfield of both economics and geography that deals with the impact of geography on economic activity.

The economic impact of sea containers has been enormous. About 90% of global trade is seaborne, and the vast majority of that is accomplished with sea containers. The U.S. and China did $4 billion in trade in 1979. In 2018, the United States imported $557.9 billion from China and exported $179.3 billion to China, and this is only a fraction of world sea trade. Canada, Mexico, and China are our three biggest trading partners.

The largest categories of U.S. imports from China in 2018 were electrical machinery, machinery, furniture and bedding, toys and sports equipment, and plastics. The largest categories of U.S. exports to China were aircraft, machinery, electrical machinery, optical and medical instruments, and vehicles. The vast majority of these imports and exports travels in sea containers. To complete the circle, almost all sea containers are now made in China.

Lower shipping costs can affect where factories are located. Plants do not have to be near ports for manufacturing to be cost-effective when shipping costs are dramatically lower. Companies can even move their facilities to countries where labor costs are cheaper. Another economic consequence of sea containers is that fewer dock workers are needed. This has resulted in numerous layoffs of longshoremen. Modern ports can move seventy containers per hour by crane without their being touched by workers.

Numerous side industries developed because of sea containers. One is the development of new types of cranes to move the containers. Another is the use of retired containers in various ways. They are put to use for businesses, as homes, and, in Odessa, Ukraine, as an outdoor shopping mall (more like a marketplace) with 16,000 vendors.

Sea containers hastened the rise of China as an economic power. In the early 2000s, Rotterdam in the Netherlands was the busiest port in the world. Since then, seven ports in China have surpassed Rotterdam in the amount of freight handled. China has four of the five busiest ports in the world and seven of the top ten. In 2017, 235 million containers passed through Chinese ports. These accounted for one-third of all container trips in the world. The United States was a distant second with 51.4 million container shipments.

Some results of this trend are more ominous. The increased industrialization in China has caused pollution, underemployment (people working at jobs for which they are overqualified), and poverty that often results from economic change, such as when people lose their jobs in economic downturns. In addition, we can see political and military consequences from China's growing economic power. As we have noted in other lessons, China has used its new wealth to buy influence in Asia, Africa, and other parts of the world, sometimes using its power to intimidate other countries. China has made investments in other countries, and those countries sometimes become dependent on Chinese money and agree to support China's political agenda.

Malcom McLean (c. 1960)

The widespread use of sea containers has created new terms, such as containerization and intermodal freight transport, and it has hastened what is called the globalization of the world's economic activity, as the people of many nations have come to depend on what international shipping can bring them.

Come to think of it, with all of these significant changes that have taken place in how the people of the world live, maybe Malcom McLean really did change the world.

Even as average income has increased around the world, and as people have had more helpful products available to them because of world trade, we must also keep in mind Jesus' warning about what our lives are really about:

Then He said to them, "Beware, and be on your guard against every form of greed; for not even when one has an abundance does his life consist of his possessions."
Luke 12:15

Assignments for Lesson 94

Geography Complete the map skills assignment for Unit 19 in the *Student Review Book*.

Project Continue working on your project.

Literature Continue reading *Revolution Is Not a Dinner Party*.

Student Review Answer the questions for Lesson 94.

95 Faith System: Atheism

A person basically has two ways to begin, live, and end each day.

Christians can get up every morning with a purpose that is larger than themselves and that is not merely the product of man's flawed thinking or the faulty, changing trends of society. They can live each day to honor and thank God, who made them and who made the universe. They can live in the salvation from their sins that Jesus made possible by His loving, self-sacrificing death on the cross. They can help, serve, and love others in the name of Jesus to bless others. They can be a light of hope for others and draw others closer to the God who loves them.

Christians can learn a loving, faithful response to what happens to them. They can grow in faith and maturity through times of difficulty. They can appreciate by their own suffering what Jesus suffered for them. They can pray to a loving heavenly Father who hears and answers. They can lie down each night knowing that God has blessed and guided them and has worked in the world for good. They can look toward the end of their lives knowing that they have a goal to live for, a place of rest after their years of service, and a time when the One who judges justly will make all things right and all things new, forever.

Atheists, on the other hand, get up every morning with no purpose or destiny larger than themselves and their perceptions. In their view, their lives, and in fact the entire universe, are the result of mindless chance. Many atheists expend a great deal of energy being angry at God, or the idea of God, and at people who believe in God. They feel superior to believers because they think they have learned the real truth about the world in which they live and have decided not to be captive to the myths and fairy tales to which they say believers' minds are captive.

Atheists do battle to tear down, prevent, and resist the force of faith that has done untold good in the world, though admittedly it has done so imperfectly and with many faults. In every blue sky, in every majestic mountain, in every roaring waterfall, in every beautiful newborn baby, in every loving act and relationship, an atheist can only say, "Hmm. Interesting consequence of the Big Bang." They must live among billions of other people who, they think, are either deluded about the purpose of their lives or who are just as self-motivated as they are.

They might see some value in getting along with others, but only as self-preservation. They recognize no Father to bless them and hear them in prayer.

In their lives they can aspire to achieve only as far as humankind has gone or thinks it can go, and in their view humankind is actually nothing more than evolved apes. When atheists lie down at night, they cannot know why they have lived. When they face the inevitable end of life, they must think that it has all been pointless folly.

A Definition

Atheism is the belief that no deity, no spiritual realm, exists. Atheists believe that the material realm is all that exists and all that has ever existed. This worldview is sometimes called naturalism or secularism. Atheists believe that some form of matter has always existed and that slow, evolutionary change over billions of years has brought about the world we know. At some point, this process of evolutionary change resulted in consciousness, in man's ability to think. To the atheist, the idea of Jesus as the Son of God is just the product of human thoughts and fears and is not true.

One major question that arises among atheists deals with the nature of human thought and will. Some believe that, since the universe is the result of completely material forces, everything that happens and everything that exists is determined by what those material forces do. Those who discuss worldviews often call this a closed system. As a result, such people say, human thought and human will are merely the result of material forces. Thus, the idea of individual choice and the idea that people can think, say, and do things that make any difference in the material course of events are simply illusions. One consequence of this belief is that people cannot be held responsible for their actions. In this view, people are simply machines, acting however their knobs, dials, and brain synapses cause them to act. Consciousness is simply a happy result (happy to whom? for what purpose?) of evolution. It is as though a mixture happens to fall into a bowl (where did the mixture come from? and the bowl?) and at some point that mixture or a later descendant of it decided to blend itself, or build a skyscraper, or believe in God.

To atheists, laws, ethics, and social standards are merely human constructs, concepts that develop over time and in different ways in different places. These maintain order in society as long as people follow them, but trouble arises when they don't or when some want to change those accepted standards and ethics.

One branch of secularism does give special value to mankind as thinking beings. These secular humanists do accept the idea that man's ability to think enables them to make a difference. They believe that man has evolved to the point of having personal

Bezbozhnik *("The Godless") was a publication of the League of Militant Atheists in the Soviet Union from 1922 to 1941.*

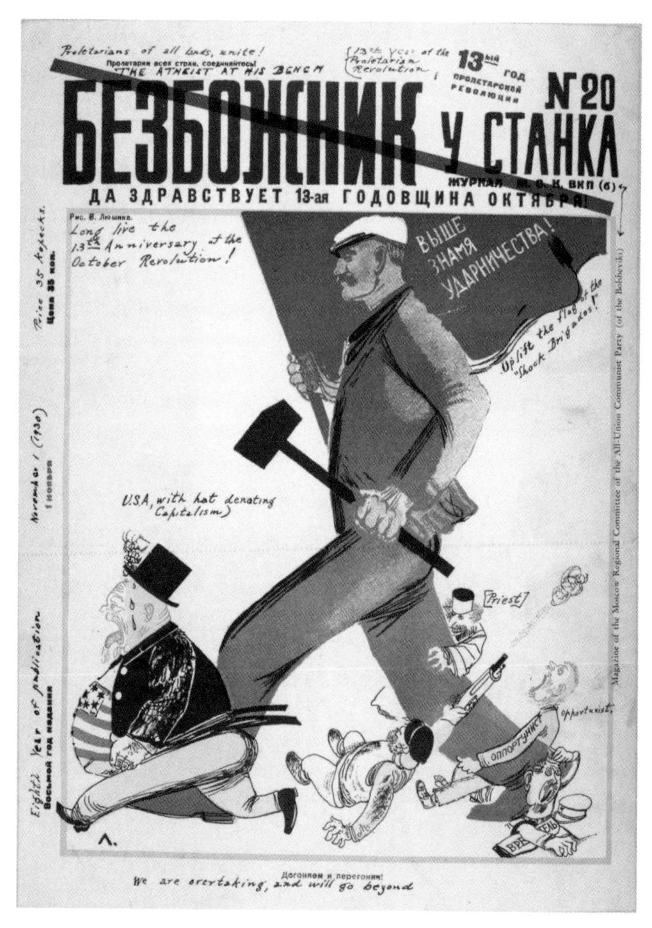

will and so their thoughts and choices actually matter and are not merely the destiny of material forces. But these thoughts and efforts "matter" only to human society and to the physical world. Some atheists admit to the existence of a spiritual realm, that there is more to us than just the physical, but they still deny the existence of God.

The human mind generally recoils against a lack of purpose. This could be the reason why many who deny the existence of God take up a social or political cause to give their lives meaning. Perhaps this is why these activists fight so hard to have others accept their viewpoint as legitimate and (dare we say it?) right.

The two most widely-accepted substitutes for Biblical faith, what we might call secular religion, have been Communism and Naziism. Their proponents were so convinced of the rightness of their ideas that they put to death millions of people who disagreed with them. Those who held these philosophies believed that they could create a heaven on earth (because in their view there was no other heaven for which they could strive). Naziism fell because of its inherent evil. Most expressions of Communism have fallen also, and the adherents of the Communism that remains are changing it to make it more like capitalism. Communism, with its totalitarian government and rejection of God, is evil also. Its presuppositions about economics and human life have been shown to be erroneous.

Secularists for the most part believe that mankind is basically good and can be redeemed by being retrained (indoctrinated) into the truth as they (the secularists) see it. Biblical faith holds that people are sinners and that only faith in Christ can redeem them. Secularists find their "devil" in ideas they disagree with, such as traditional families and religion. People of Biblical faith believe that some human practices, such as family and faith, are good but that sinful humans practice them imperfectly. Secularists believe the source for correcting things is human wisdom. People of faith believe the source for how to correct things is in Scripture.

The Failings of Atheism

Atheism rejects the evidence in the world that God and the spiritual realm exist.

Atheism results in no grand, overarching purpose to the universe, life, and individual lives. Everything just happened; it is all a cosmic accident.

Atheism necessarily concludes that what exists is what ought to be. Atheists cannot determine anything that "should" be, except on the basis of the material actions of the synapses of someone's brain.

However, if the mind is the same as the brain, on what basis can we trust what someone thinks? There is no standard for right and wrong, so who can say whether a thought or action is good or bad, except as society defines it at a particular point in time. If nothing is immoral, then anything is acceptable. In fact, we cannot determine whether materialism is right or wrong, illusion or truth, since any evaluation would be merely the result of the materialistic forces one wants to evaluate.

If we are just deterministic machines, what difference does a worldview make? What difference does anything make? Are education, work, love, and everything else just activities to fill our time?

Some atheists say that law, social behavior, and human expectations are good because they preserve society, but upon what basis does anybody think that society should be preserved? Should the society that existed in the southern United States in the 1950s have been preserved? Should all societies and cultures around the world be preserved just as they are? What do we make of the rebel who wants to make his or her society better (if we can determine a meaning for "better")? We have to have an objective standard that comes from outside of the realm in which we live to know what is good or bad, desirable or objectionable.

The Influence of Atheism

The vast majority of people say that they believe in God. If this is the case, and if atheism is primarily the interest of relatively few writers, philosophers, and celebrities, why bother to try to understand it? There are several reasons:

1) The most populous country in the world, China, along with a few other countries, are officially atheist. Their policies and actions, including their persecution of religious people, flow from this worldview. Understanding this worldview helps us understand why those governments do what they do.

2) Ideas have consequences. Unbelievers and people who act without regard to the existence of God do things that believers find hard to understand. Understanding the godless mindset (not agreeing with it, but understanding it) will help believers see why some people act and speak the way they do.

3) Atheism influences what you see and hear. In previous times school curriculum, even public school curriculum, generally assumed the existence of God. Today much curriculum does not assume the existence of God. This feeds the minds of millions of school children and college students. Atheism is the worldview of many of the writers of popular literature, movies, and television as well as celebrities. This influences their lifestyles. The media rarely mention God except to ridicule the idea or to point to failings of people who say they believe in God. Consider that some 40% of Americans say they attend church services every week. Do 40% of the characters in movies and television programs attend church services every week? Do even 20%? Are there any? The consequences of this pounding away on people's minds with the assumption of atheism, and the acceptance of this "practical atheism" by many people, are the trends we see in society.

Nihilism and Existentialism

Nihilism is the rejection of any and every worldview. Not only does this outlook deny the existence of God, it denies the possibility of knowing anything and even denies (or at least questions) the reality of existence. In the nihilist view, the world doesn't really exist; every person only thinks it exists. At some point the state of mind that questions the existence of the mind shades off into mental instability.

Existentialism is a response to nihilism. It is an attempt to find meaning in existence. Existentialism accepts a godless, meaningless universe, but it says that humans can create their own world. The nihilist has given up, while the existentialist says each person can find meaning in his or her own existence. What is valuable and true, for example, is each person's existence. People consider the nineteenth century Danish writer Søren Kierkegaard to have been a Christian existentialist. He was reacting against the lifeless Danish state Lutheran church and wanted to encourage believers to find meaning in the fact that God was in their lives since they could not likely find meaning in relation to their church.

The problems with these views are many, in addition to the ones mentioned above related to atheism in general. A nihilist cannot deny the reality of gravity, especially if a large rock is about to fall on him. If that person says, "This is not really happening" or "I'm not sure this is really happening," such an attitude renders everything meaningless, including those thoughts. An existentialist claims to find meaning in his existence, but the unborn person and the infirm or aged person have meaning whether or not those persons are aware of it.

Postmodernism

Postmodernism is a recent development that is in response to several threads of Western thought. The center of this development involves the changing nature of authority. In the premodern age, which we might also call the church age, most scholars and everyday people accepted the Bible as the authority and believed that objective truth existed and that they could know it. Then what scholars and philosophers came to call the modern period developed with the rise of human reason as the standard authority for thought. Reality for rationalists was not as the Scriptures taught it but as human reason identified it. If Scripture did not match up with human reason (or some human's reason), then so much the worse for Scripture. Evolution became generally seen in the West as the reasonable, rational explanation for the existence and operation of the world.

The rationalists' questioning of authority that dethroned God and objective, revealed truth did not stop with them. Further questioning, what philosopher Friedrich Nietzsche called "radical doubt," dethroned reason and declared that each person or each group, in his or her own story or their own story, is his or her own authority or their own authority. This is the essence of postmodernism.

The postmodernist says that we cannot know truth. All we have is "your truth," "my truth," and each group's "truth." Meaning is whatever people think it means. It is wrong, says the postmodernist, for one person or group to impose its concept of truth on another. The authority for each person or group is that person or group alone. There is no need to identify a rebel or reformer, for everyone is a rebel.

The postmodernist world is a world without a center. The locus of authority has moved from God, to mankind, to the individual.

Agnosticism

Agnosticism is a worldview developed by Englishman T. L. Huxley in 1869. He based this worldview on the belief that man cannot know the existence of anything beyond our physical experience. It is not a matter of not yet knowing anything of the metaphysical realm but that we cannot ever know it. Huxley proposed that we follow reason as far as it can take us but then acknowledge its limit.

If we can't know anything about the spiritual realm, the practical impact is that for agnostics, it doesn't matter.

Beijing, China

Unyielding Despair

British atheist Bertrand Russell (1872-1970) gave this summary of atheism:

> That man is the product of causes which had no prevision of the end they were achieving; that his origin, his growth, his hopes and fears, his loves and beliefs are but the outcome of accidental collocations of atoms; that no fire, no heroism, no intensity of thought and feeling, can preserve an individual life beyond the grave; that all the labors of the ages, all the devotion, all the inspiration, all the noonday brightness of human genius are destined to extinction. . . . that the whole temple of man's achievement must inevitably be buried—all these things, if not quite beyond dispute, are yet so nearly certain, that no philosophy which rejects them can hope to stand. Only within the scaffolding of these truths, only on the firm foundation of unyielding despair, can the soul's habitation henceforth be safely built.

What seems to you to be the best way to begin, live, and end each day?

For they exchanged the truth of God for a lie, and worshiped and served the creature rather than the Creator, who is blessed forever. Amen.
Romans 1:25

Assignments for Lesson 95

Worldview Recite or write the memory verse for this unit.

Project Finish your project for this unit.

Literature Finish reading *Revolution Is Not a Dinner Party.* Read the literary analysis and answer the questions in the *Student Review Book.*

Student Review Answer the questions for Lesson 95.
Take the quiz for Unit 19.

Floating Village on Tonle Sap Lake, Cambodia

20 Southeast Asia

In this unit on Southeast Asia, we begin by examining the South China Sea, a place of troubled waters because of competing claims regarding it. The country of the Philippines confronts several issues because of its geography. For many Americans, Vietnam is much more than just a geographic location. Indonesia, a country made up entirely of islands, is defined in great measure by water. The worldview lesson looks at how Jesus sees the world —in other words, Jesus' worldview.

Memory Verse Memorize John 1:1-5 by the end of the unit.

Books Used

The Bible
Exploring World Geography Gazetteer
Ann Judson: A Missionary Life for Burma

Project (Choose One)

1) Write a 250-300 word essay on one of the following topics:
   - In your own words, summarize the geopolitical issues surrounding the South China Sea and propose a solution to the situation. (See Lesson 96.)
   - Write a report on a city or geographic feature of Indonesia. (See Lesson 99.)
2) Interview a veteran who served in Vietnam. Prepare a list of questions to ask, including some about the geography he or she encountered and what it was like to be there. Be prompt in meeting him or her, respect his time, and express gratitude when you leave. (See Lesson 98.)
3) Prepare a presentation on the Philippines, including its history and current status. Make or reproduce a map and include pictures. (See Lesson 97.)

Literature

Ann Judson and her husband Adoniram Judson were among the first missionaries to leave the shores of the newly established United States of America in 1813. Setting out directly after their marriage, they served the people of Burma on the other side of the globe. Their years together were fraught with hardship and sacrifice. Their story has become a hallmark of Christian mission history.

In *Ann Judson: A Missionary Life for Burma*, author Sharon James quotes extensively from Ann Judson's own letters and writings as well as those of her contemporaries. These primary sources make for a vivid biography, capturing the genuine flavor of the subjects and their times.

Sharon James grew up in England where her parents had moved from South Africa to study at London Bible College. She grew up as a pastor's daughter and became a pastor's wife. Sharon studied at Cambridge University, Toronto Baptist Seminary, and the University of Wales. She has spoken widely at conferences and authored several books.

Plan to finish *Ann Judson: A Missionary Life for Burma* by the end of Unit 21.

Oil Drilling Rig, South China Sea

96 Troubled Waters: The South China Sea

Japanese troops storm the beach of the Pacific island. The troops carry out the assault in concert with soldiers from its two allied nations.

This scene took place not in 1942 but in 2018. Japan's allies in this military exercise were none other than the United States and the Philippines.

The joint maneuver involving the military of these three nations occurred about seventy-five years after Japan captured the Philippines and savagely fought the United States along the Pacific Rim of Asia during World War II.

The alliance of these three countries working together is not the only surprising strategic international move in the region. Vietnam and China are rivals. Their enmity goes back for centuries. Vietnam seeks an ally to counter China's growing economic and military power. To whom does Vietnam turn?

The United States.

Forty years after the Communist government of North Vietnam ended a long and bloody war against America, the U.S. and a now-unified Vietnam are drawing closer to each other out of a common concern about what China is doing and might do.

The focal point for these surprising cooperative efforts is the South China Sea. The geography of the region that includes this relatively small but strategically located body of water is largely the same as it always has been (with a significant exception we'll discuss later in this lesson). What is different are the actions of the nations that surround it, primarily China.

The South China Sea

The South China Sea is a western arm of the Pacific Ocean that lies against the Southeast Asian mainland. Its border consists of the Taiwan Strait to the north, Taiwan and the Philippines to the east, Borneo to the south, the Gulf of Thailand and the Malay Peninsula to the southwest, and Vietnam and China to the west. The sea encompasses about 1.4 million square miles.

The South China Sea is significant in today's world because of what happens on it and below it. The first issue is business. The Sea is part of the busiest and most important shipping lane in the world, primarily because of China's economic production. Ninety percent of all commercial goods that traders move between continents moves by sea (the other options are by air or, between Asia and Europe, by land). We know the trillions of dollars worth of goods that China produces. The seaborne goods that China ships out, as well as the raw materials and trading goods that come into China, arrive mostly through the South China Sea. At some point in these journeys, over half of the world's merchant fleet tonnage and a third of all the world's ocean traffic travels through the South China Sea.

The second issue is energy. China needs energy to produce the goods that it trades with other countries. China only has about one percent of the known world oil reserves, but it consumes ten percent of the world's oil and twenty percent of the world's energy. All that fuel for the Chinese economy has to come from somewhere, and much of it comes by tankers from the Middle East, through the Strait of Malacca by Singapore, and into (you guessed it) the South China Sea. Japan has a similar dependence on imported oil. Eighty percent of China's crude oil imports and sixty percent of Japan's energy come through the South China Sea.

But imports are not all of the energy story connected to the Sea. Exploratory drilling indicates that the South China Sea itself might hold significant oil and natural gas reserves that are as yet undiscovered. China would like to get its hands on as much of that as it can—but then, the other countries that border the Sea would like access to it also. On this issue conflict begins to emerge between China and the other countries that border the Sea.

During a 2014 dispute between Vietnam and China, these Vietnamese protesters gathered outside the Chinese embassy in London, England.

Which leads to the third issue, military security. China is involved in a huge military buildup. China has the world's largest army and the second-largest submarine fleet. It also has the world's second-largest military budget. This buildup includes the deployment of naval and air forces in the South China Sea. Part of this buildup involves China's activity in changing the geography of the South China Sea. By moving dirt and rocks around, China has transformed some of the tiny islands in the Sea into homes for airplane landing strips, docks for ships, and military outposts. This military presence protects Chinese shipping and the Chinese coastline, and it sends a message to other nations to stay away or tread lightly. Other nations see these moves as aggressive; China says it is merely defensive. The difference is how you understand the sovereignty of the South China Sea.

Background Perspective: The Law of the Sea

We can understand borders, treaties, and sovereignty issues involving the land areas of different countries, and even the rivers and lakes between countries; but what do we do with the wide open seas where no one lives but where ships from many nations travel?

The United Nations finalized its Convention on the Law of the Sea (UNCLOS) in 1982. As of October 2018, the United States had not ratified the UNCLOS but recognized it as codifying accepted international law.

Traditions from centuries ago created a generally accepted standard of maritime sovereignty. Nations agreed that a country could control the sea off its coast for as far as it could defend from shore. A cannon shot traveled about three miles, so most nations of the world accepted a three-mile sovereignty limit for each country (the "cannon shot rule"). As secretary of state, Thomas Jefferson wrote a letter to other foreign ministers in 1793 asserting this territorial limit and recognizing it for other countries.

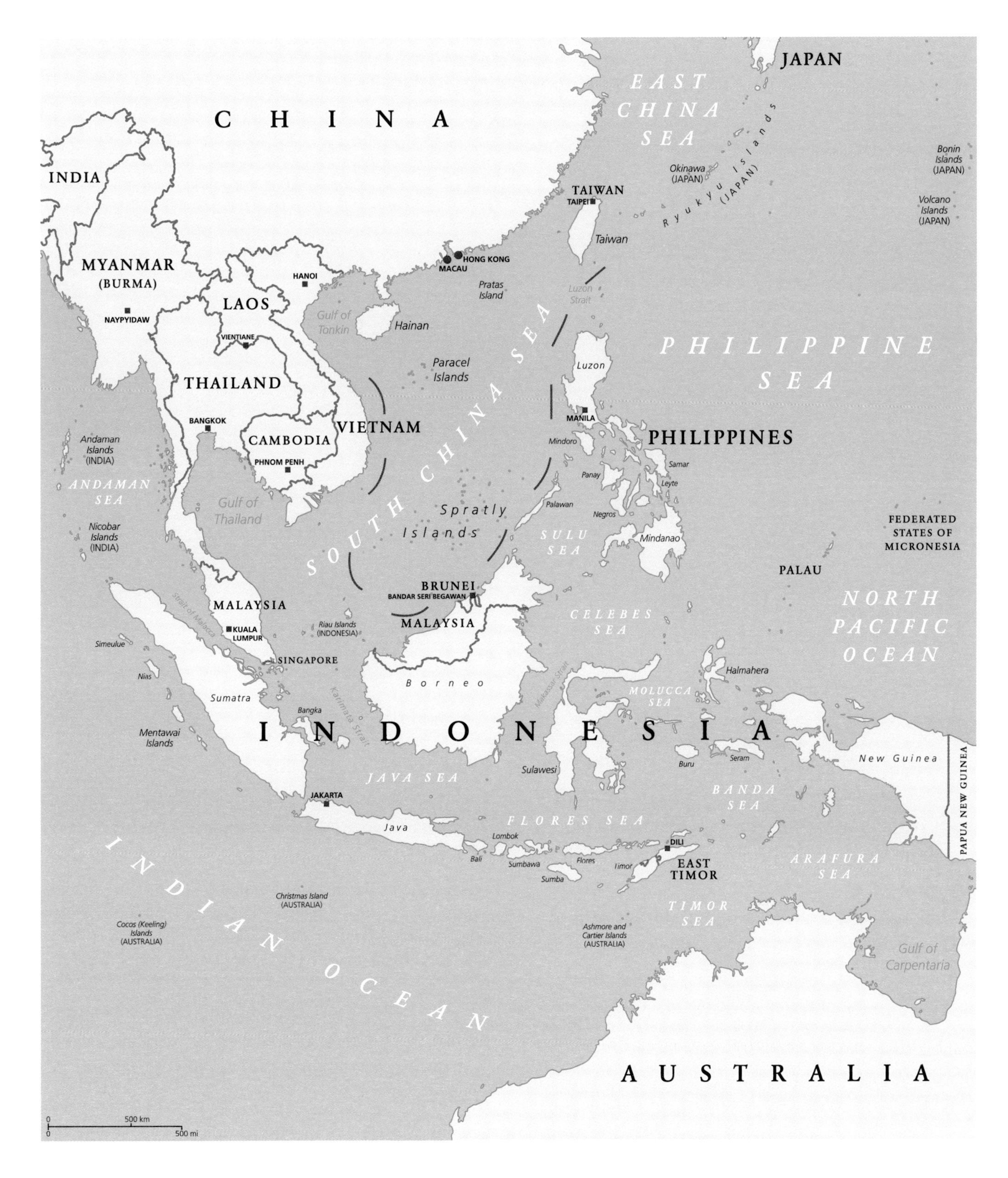

The baseline for a country is the low-water (low-tide) line along its coast. The UNCLOS has determined how countries deal with bays, mouths of rivers, reefs, and other unusual coastal features. Each country has complete sovereignty over internal waters. Ships from one country cannot claim the right of innocent passage through another country's internal waters.

The standard modern territorial limit that countries claim is now twelve nautical miles out from the baseline. A nautical mile equals one minute of latitude, or about 1.15 statute miles. This territorial sea includes the airspace above it and the seabed and minerals below it. Foreign ships passing through this territorial sea must obey the laws of the country that has sovereignty over it.

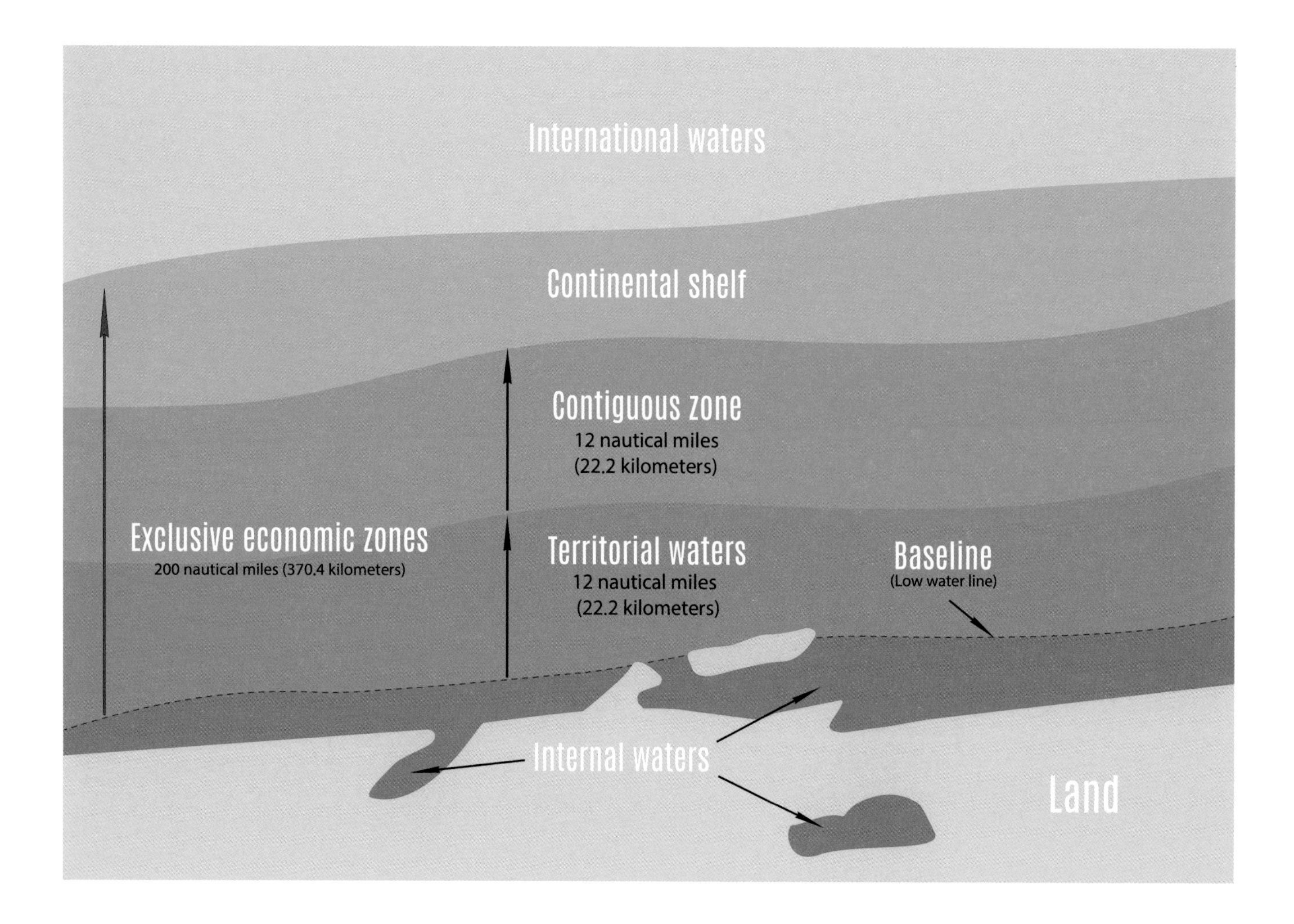

Countries may also claim a contiguous zone of up to 24 nautical miles from the baseline. A country may patrol this zone to oversee immigration, sanitation issues, and customs regulations that affect that country.

In addition, a country may claim an exclusive economic zone (EEZ) of up to 200 nautical miles from the baseline (or to the maritime boundary of another country), in which the country can exploit (including drill for) or protect natural resources, use the waters and winds for energy production, and establish artificial islands and have other activity within it. Each country also has sovereignty over the continental shelf that is within the EEZ below the sea.

The UNCLOS establishes rules for straits between countries that are more narrow than these maximum limits and for states that could make conflicting claims about sovereignty at sea (for instance, two countries that overlook the same bay).

This all sounds good sitting around a conference table, but at least two questions arise: (1) What happens when two or more countries claim sovereignty over the same area and don't want to play nice? and (2) What happens when the crews of ships from rival nations are staring at each other while sitting on disputed waters?

What Belongs to Whom?

The two most significant land formations within the South China Sea are the Spratly Islands and the Paracel Islands. The Spratlys are about one hundred tiny islands, reefs, atolls, and large rocks that lie mostly off the Philippines. British whaling captain Richard Spratly reported seeing what became known as Spratly Island in 1843, and the name eventually became associated with the entire archipelago. China, Taiwan, and Vietnam each claim the entire island chain, while Malaysia and the Philippines each claim some. About forty-five of the otherwise unoccupied islands have various small military outposts from China, Malaysia, Taiwan, and Vietnam. Brunei (a small country on the island of Borneo) claims a continental shelf and an EEZ in the area.

The Paracels are a similar island group that lies generally off of the coast of Vietnam. The name comes from a Portuguese word. French Indochina annexed the islands in 1932, and the successor state, Vietnam, claimed them also. Chinese forces expelled a South Vietnamese force in 1974 to lay claim to the islands. Vietnam and Taiwan also claim the islands.

The island groups offer excellent fishing, which provides food and jobs for many people. Control of the islands also enables access to any oil and gas that might be found near them.

In addition, China and the Philippines both claim the Scarborough Shoal (known as Huangyan Island in China), which lies about one hundred miles from the Philippines and five hundred miles from China.

So how should the six nations—China, Vietnam, the Philippines, Taiwan, Malaysia, and Brunei—divide up the Sea, the islands, and their resources?

China has a solution. It pretty much wants it all.

China has issued a map that shows a broken line, which is also called the nine-dash line or the cow's tongue line, which takes in 80-90% of the Sea (see the green line on the map on page 539). China says that the Spratly and Paracel Island groups have been considered part of China for centuries. Other countries dispute the claim, and many observers are unsure whether China claims just the islands or the entire water area. Vietnam claims that it has ruled over both island groups since the 1600s. The Philippines claim that their proximity to the Spratly Islands gives them a rightful claim to those islands.

China likes to negotiate with one country at a time. Critics charge that this gives China an unfair advantage in negotiations and want to see talks between China and ASEAN, the Association of Southeast Asian Nations, which consists of Thailand, Indonesia, Malaysia, the Philippines, Singapore, Brunei, Laos, Vietnam, Myanmar (Burma), and Cambodia. So far, China has refused.

Vietnamese Soldiers on Đao An Bang (Amboyna Cay), One of the Spratly Islands (2009)

Malaysian Fishing Boat in the South China Sea (2020)

In 2013, the Philippines took China to a tribunal under the UNCLOS over the sovereignty conflict. In 2016 the tribunal decided in favor of the Philippines. China boycotted the proceedings and says it will not abide by it.

China compares the role it wants in the South China Sea to the role that the United States played in the Caribbean Sea around 1900. The U.S. did not claim all of the Caribbean, but it exerted significant influence in what happened there. The Monroe Doctrine of 1823 and the Roosevelt Corollary of 1904 pretty much said that the United States would take care of things in the Western Hemisphere without any involvement from countries outside of the region. This is what China says now: "We'll take care of things in this region."

However, the United States has been a major player in Asia, Southeast Asia, and the South China Sea region for decades and does not want to give up its role. China wants to be the major player in that region. Other nations, such as Taiwan, Vietnam, and the Philippines, look to the United States for support and do not want China to have that primary role. This is why Vietnam is looking to the United States to balance the growing power of China.

The comparison does not do the situation justice. In 1900 the Caribbean was far from the world's main transportation routes; today the South China Sea is one of the most vital of such routes. In 1900 the United States wanted the nations to its south to be free, democratic, and engaged with the world economy and the family of nations. Today China has pretty much told other nations to stay out of their way except on China's terms.

What Happens Now?

The situation is at a diplomatic standoff. China's increasing military presence in the Sea and its economic presence throughout the region and the world put pressure on other nations to acquiesce to their claim of sovereignty. The United States and other nations occasionally send warships through the Sea or planes overhead to remind the Chinese that the Sea is not their lake. From time to time two ships from opposing countries get very close to each other, which makes for some tense moments. When Chinese warships warned two U.S. Navy ships away from the Paracel Islands in 2018, the Navy described the action as "safe but unprofessional."

Japan and Vietnam agreed to work together on security in the South China Sea. This agreement was the follow-up to the war exercises mentioned at the first of this lesson. Japan has no claim there, but it wanted to send its own message to Beijing. Chinese submarines occasionally shadow American Navy vessels as they go through the South China Sea. China has installed hi-tech undersea listening devices in the Pacific that can track submarine activity and perhaps even pick up transmissions between U.S. subs and their command bases. Neither side appears to be backing down.

China wants to control as much of the South China Sea as it can. The other nations with interests in the region want to have as large a presence as they can while minimizing China's. China hasn't stopped its buildup or abandoned its claims. We live in a world of virtual reality, where most "armies" and "naval vessels" do battle only in pixels; but the tension and uneasiness over this actual geographic region where much is at stake is real.

Psalm 107 speaks of those who do business on the sea in ships and who see the works of the Lord in the deep:

Those who go down to the sea in ships,
Who do business on great waters;
They have seen the works of the Lord,
And His wonders in the deep.
Psalm 107:23-24

Assignments for Lesson 96

Gazetteer Study the map of Southeast Asia and read the entry for the Paracel Islands (pages 165 and 163).

Worldview Copy this question in your notebook and write your answer: What do you think it means to have the mind of Christ? (See Philippians 2:5 and 1 Corinthians 2:16.)

Project Choose your project for this unit and start working on it. Plan to finish it by the end of this unit.

Literature Begin reading *Ann Judson: A Missionary Life for Burma*. Plan to finish it by the end of Unit 21.

Student Review Answer the questions for Lesson 96.

Palawan Island, Philippines

97 In the Middle: The Philippines

Homeschooling is a growing movement in the Philippines, just like it is in the United States. The Philippine government supports the right of parents who choose this route of educating their children, and many families have decided to take this approach. Active, dynamic leaders in the homeschool movement in the Philippines provide encouragement and suggest resources for families who have already begun a homeschooling journey; and they also provide information and guidance for the many families who are considering the possibility.

Being in the middle of a homeschool journey, and being in the middle of a growing movement are an appropriate place to be for homeschoolers who live in the Philippines. After all, the Republic of the Philippines is now and has been in the middle of many issues.

In the Middle of Southeast Asia

The Republic of the Philippines is strategically located, east of Vietnam and the South China Sea, near China, and on the western edge of the Pacific Ocean. Over seven thousand islands make up the Philippines archipelago, although the largest eleven islands account for 95% of its land area. The Philippine Trench, which runs north and south on the floor of the Pacific Ocean just to the east of the islands, extends to a depth of 34,578 feet, the second deepest spot in the world.

The Philippine Islands are in the middle of volcanoes and earthquakes. The mostly mountainous islands are located on the Ring of Fire, the rim of the Pacific where about 90% of the world's earthquakes and 75% of the world's volcanic eruptions occur.

Volcanoes erupt fairly often in the Philippines. The eruption of Mount Pinatubo in 1991 was the most severe eruption in the world in the twentieth century. It caused widespread damage and claimed over seven hundred lives during and after the eruption. This followed a magnitude 7.8 earthquake that struck in 1990.

In the Middle of Troublesome Weather

The islands are subject to severe weather. In a typical year about nine typhoons affect the islands and five or six cause damage. Because of the country's 22,000 miles of shoreline, tsunamis are also a threat.

In the Middle of Exploration and Trade

The Philippine Islands played a key role in the era of exploration. Ferdinand Magellan and his fleet, sailing on behalf of Spain, reached the Philippines in 1521. Magellan inserted himself into a conflict between indigenous groups and was killed. Spain continued to send conquistadors to the islands and claimed the archipelago as a colony in the mid-1500s. The Spanish named the islands for Prince (soon to be King) Philip II of Spain.

The Philippine Islands were an important port in early global trade. From 1565 until 1815, the Manila Galleon Trade Route was the route whereby one ship annually left Manila loaded with porcelain, spices, and silk from Asia, and followed the eastern tradewinds to the Spanish colony of Acapulco in Mexico. The Spanish then oversaw the transport of these goods overland to the east coast of Mexico, where another ship carried them to Spain. Another galleon sailed west from Acapulco to Manila each year, following the western tradewinds route, carrying silver and other cargo as well as government and Catholic Church officials. These ships often made calls at Guam and the Mariana Islands, which at the time were other Spanish possessions. Each trip across the Pacific took months to complete, and the ships docked at other ports in Asia and North America as well.

In the Middle of Colonial Empires

The Philippine Islands were part of two colonial empires. Spain ruled the islands as part of their world empire until the United States defeated Spain in the Spanish American War in 1898. At that time, Spain ceded the islands to the U.S. With the acquisition of the Philippines and Cuba as a result of that war, the United States suddenly became a colonial power.

Mayon Volcano, Luzon Island, Philippines

Almost immediately after the United States assumed oversight of the islands, Philippine rebels who wanted to be independent began fighting the Americans. This war continued until 1902, when the United States declared the conflict over.

The U.S. adopted a policy of easing out of colonial oversight of the islands. The United States granted the Philippines the status of self-governing commonwealth in 1935 with plans for independence in 1945. However, the islands were soon in the middle of the fighting in World War II. Japan seized the islands in 1942, and American forces and governing personnel left. U.S. and Filipino forces regained control in 1945. The U.S. finally granted independence to the Republic of the Philippines on July 4, 1946.

In the Middle of Ethnicity and Religion

The islands include dozens of people groups and language or dialect groups. The Tagalog people and language group is the largest single ethnic group in the islands. Several dialects of Tagalog exist, but a standard version of the language known as Filipino is one of the country's two official languages; English is the other. A dialect of Tagalog is the first or second language for more than half of the Philippine population. Despite centuries of Spanish rule, for the most part the people did not adopt the Spanish language. It remained primarily only the language of the government and the ruling class.

The Roman Catholic Church of San Agustin in Manila was completed in 1607. The bell tower on the left was damaged in an earthquake in 1880 and subsequently removed.

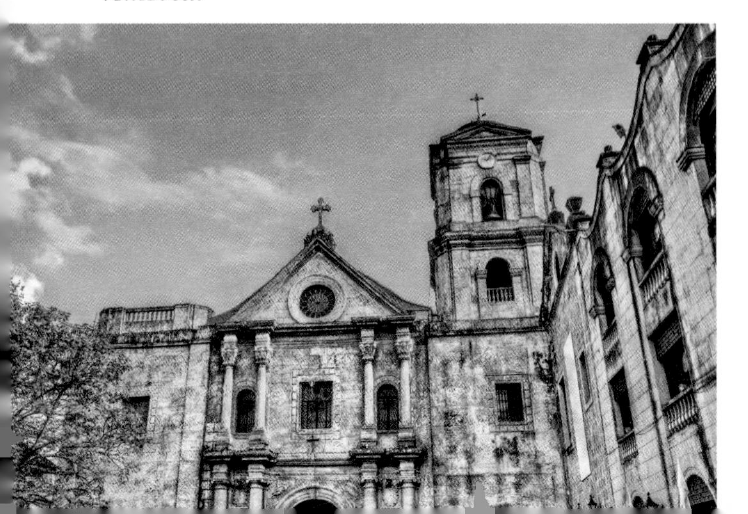

The presence of these various ethnic and linguistic groups meant that the islands actually did not constitute one entity in anyone's mind until the Spanish brought them under a single government. But even unified Spanish rule did not really make them a single people. The country is still fairly fragmented in terms of ethnicity, language, and religion.

About 80-85% of the people are Roman Catholic. In some northern provinces, Catholics make up over 90% of the population. The strength of the Catholic faith in the Philippines stems from their Spanish heritage, but for them it is not just a cultural religion. Generally the Philippine people hold their faith deeply and sincerely, although some people do hold to folk religion beliefs as well as their Catholic faith. The ninety percent of Filipinos who are Christian (Catholic, Protestant, and evangelical) make up the largest percentage of Christians of any country in Asia. This is in a part of the world that is predominantly Muslim, Buddhist, and (as in China) atheist.

The southern islands are another story. Arab and Malay traders came to that region in the 1300s and brought the Sunni Muslim religion. Today Muslims (called Moros, from the term Moors) make up a large part of the population in the south; in some provinces they are a majority. The people in the southern islands actually have more in common religiously and culturally with the people of Malaysia and Indonesia than with the Tagalogs to their north. The Moros never recognized the authority of the Catholic Spanish over them. The Catholic presence is much smaller in the southern islands. About 5-6 percent of the total Philippine population is Muslim. Islamic terrorist groups have operated out of the southern region for many years.

The Moros have tried to gain independence from the Philippines for years. This conflict has

People on the island of Panay have been celebrating the Festival of Ati-Atihan for hundreds of years. After the Spanish arrived, the festival acquired Catholic religious connections and the tradition spread to other parts of the Philippines.

caused over 120,000 deaths and two million displaced persons. In 2018 the Philippine Congress approved limited autonomy for the southern island of Mindanao. In addition to this Muslim agitation, a Maoist Communist terror group operates in other parts of the country and has caused considerable damage and loss of life.

Despite the religious and political differences, the country is not in constant turmoil. However, incidents of violence do occur from time to time. For instance, in January 2019 ISIS claimed responsibility for a double bombing at a Roman Catholic church in the southern Philippines that killed twenty and injured over one hundred others.

In the Middle of Economic Trial

The Philippines suffered economically under centuries of Spanish colonial rule. The Spanish were only interested in the wealth they could extract from the country and not in bringing about economic development there. After the United States took over the Philippines, the U.S. improved government operations and economic activity, but the islands were still a colony and depended to a great extent on the level of investment that the U.S. made in the country.

The Philippine economy has been growing in recent years, but it has a long way to go. It is one of the poorest national economies in Asia. The country has over 109 million people, the thirteenth largest national population in the world, but its gross domestic product in terms of purchasing power parity (how the average citizen compares to people in other countries in terms of real buying power) is 29th in the world. In terms of per capita GDP, it is 148th. By contrast the United States has the world's third largest population, the second highest purchasing power parity, and the 29th highest per capita GDP. Exports account for only one-fourth of Philippine economic activity, whereas for most Asian nations exports make up seventy-five percent of the economy.

One problem in the Philippines has been the extreme difference between rich and poor. Three-fourths of recent economic growth has gone to only forty families in the country. Corruption is widespread throughout the government.

Another problem has been the instability caused by Muslim and Maoist rebels. These conflicts prevent economic growth because people have to be concerned with self-preservation instead of being able to pursue productive economic activity. In addition, the conflict discourages foreign investment. Investors do not want to put their money where there is a strong possibility that it will be lost due to civil unrest.

In the Middle of International Conflict

Because of its strategic geographic location, the Republic of the Philippines has been caught in the middle of the tensions between China and the United States as it has dealt with its own internal problems. Even after independence, the country still greatly depended on the United States for military security and economic growth.

Ferdinand Marcos was president for twenty-one years (1965-1986), and ruled as dictator from 1972 to 1981 under martial law that he proclaimed. Marcos was an anti-Communist ally for the United States, but he was a corrupt ally. He and his wife Imelda, who also held several positions in the government, stole billions of dollars from the country. Marcos' rule kept the country from growing as it could have. After several years of unrest, Marcos resigned in 1986 (under pressure from the U.S. to do so) and he and his wife fled the country. The islands have seen repeated periods of political and economic instability since the last years of Marcos' tenure.

In a move that was intended to help the Philippines become less dependent on the United States, the U.S. closed Subic Bay Naval Base and Clark Air Base in the Philippines in the early 1990s. However, that was before China began developing

People at this vegetable market in La Trinidad on the island of Luzon wear face masks and use their cell phones in April of 2020 during the COVID-19 pandemic.

its military strength and began reaching into the South China Sea and claiming ownership of almost all of the sea and its mineral resources. These claims infringed on what the Republic of the Philippines understood to be its territorial waters. The Chinese claims have threatened the Philippines' security and economy.

Access to the South China Sea is an important part of the Philippines' possibilities for economic growth. The Philippines, as well as other countries in the region, send over one million fishing vessels into the South China Sea each year. In addition, the country imports all of its petroleum needs and obtains its natural gas from an offshore drilling field near Manila Bay. Filipinos have wanted to be able to engage in exploratory drilling in the South China Sea, but Chinese claims have thwarted this. On the other hand, China is the Philippines' third largest trading partner. China has also promised to put $24 billion in assistance and investment into the Philippines. This financial involvement can draw the two countries closer, but it can also make the Filipinos dependent on and indebted to China, which can be unhealthy. China's offer also potentially buys Philippine acquiescence to Chinese policies, such as its desire to take over Taiwan.

As you can see, the Republic of the Philippines has a complicated relationship with China. The Philippines do not want to be under the threat of Chinese military force, but the republic's own military is small and would be no match for the Chinese. Thus, the Republic of the Philippines has once again sought better military relations with the United States. China could easily control the Philippines with military force, but any move to do so would be a direct challenge to the United States' interests in the region.

In 2018 China eased tensions over issues in the South China Sea by signing agreements with the Philippines on fishing rights and on joint mineral exploration of part of the South China Sea. China appeared to be pursuing a cooperative and not confrontational relationship with the Philippines.

The Republic of the Philippines and its families are in the middle of longstanding historic trends and significant current issues. Homeschooling families in the Philippines, just like homeschooling families everywhere, want to find the best way through all of this for the future of their children.

Psalm 97 calls on the many islands of the world to be glad in the Lord:

The Lord reigns, let the earth rejoice;
Let the many islands be glad.
Psalm 97:1

Assignments for Lesson 97

Gazetteer Read the entry for the Philippines (page 172).

Worldview Copy this question in your notebook and write your answer: What do you think having the mind of Christ means concerning how you think about other people?

Project Continue working on your project.

Literature Continue reading *Ann Judson: A Missionary Life for Burma.*

Student Review Answer the questions for Lesson 97.

Dragon Bridge Over the Hàn River, Da Nang, Vietnam

98 More Than Just a Place: Vietnam

Vietnam.

For many Americans, it's more than a place. This simple geographic term carries great emotional weight. For your grandparents' generation, just saying the word conjures up thoughts of friends in the military who came home physically or emotionally scarred or who didn't come home at all. The word reminds that generation of angry protest marches and a long struggle to define our national purpose. Valley Forge, Gettysburg, Normandy, Korea, and Ground Zero of 9/11 are other geographic locations that have had this kind of deep impact.

Vietnam defined an era in our nation's history. It was the scene of a large and expensive American military involvement that cost the lives of over 58,000 Americans and perhaps three million Vietnamese but in which the United States did not achieve its military and political objectives. It was also the cause of bitter conflict among our citizens.

Before our military involvement there, most Americans knew little about Vietnam; it was just a name on a map. Since the war, America's relations with Vietnam have been transformed. The country has become a source for many consumer products that Americans buy and use. Vietnam still has geographic significance for the world, but in a way that is different from what it meant in the 1960s.

Geographic Background

Vietnam is located in Southeast Asia on the peninsula of Indochina. The term Indochina reflects the influence of India and China on the region. In the 1800s, as part of the race for colonies in which several European countries competed, France colonized the eastern part of the peninsula, which now includes Vietnam, Cambodia, and Laos. The region became known as French Indochina.

Japan seized part of French Indochina during World War II. The end of the war saw considerable turmoil as France tried to regain control while national groups arose and sought to assert their independence. One of these groups was the Communist Viet Minh led by Ho Chi Minh, who declared a Communist state with the capital in Hanoi. French forces and Ho's armies fought until 1954, when the Communists defeated the French at Dien Bien Phu.

A multinational conference in Geneva, Switzerland, declared a cease-fire line at the 17th parallel of latitude. This divided Vietnam into the Communist North led by Ho and an independent

South that was supported by the United States. Ho quickly declared his intention to unify all of Vietnam under his Communist regime. Communist rebels in South Vietnam who were dedicated to achieving an overthrow of the government of South Vietnam were called the Viet Cong.

American Involvement

The United States supported the government of South Vietnam, but that government was weak and corrupt. As a result, the U.S. faced a difficult geopolitical situation. Vietnam became an important proxy battlefield in the Cold War between the U.S. on one side and the Soviet Union and Communist China on the other (China was a lesser factor because the Vietnamese and the Chinese have not gotten along for centuries).

A stated goal of Communism was the domination of the world. The Soviet Union had taken over East Germany and the countries of Eastern Europe. Communists controlled China and North Korea. The stated intention of Communists in Vietnam was to take over all of that country. Part of the worldview of the American government and other world democracies at the time was the domino theory. This was the belief that if Vietnam fell to the Communists, other countries in Indochina and Southeast Asia would then be more likely to fall like dominoes to the emboldened Communists. Communist forces might take over that part of the world entirely like they did Eastern Europe and then keep going, perhaps even to Australia.

Richard Springman, serving in the U.S. Army, was captured in Cambodia in 1970. This photo shows him with a North Vietnamese officer before his release in 1973. He received the Bronze Star for attempting to escape during his time as a POW.

As a result of this geopolitical concern, the United States provided economic aid and military equipment and advisers to South Vietnam. Even though the government there was unstable, it was our best hope in the situation. Eventually the U.S. became directly involved militarily by sending fighting forces to South Vietnam and conducting extensive bombing campaigns against the North. As many as a half million American troops served in Vietnam at one time at the height of our involvement.

However, the United States and South Vietnam were never able to bring North Vietnam's army or the Viet Cong to their knees. American military and political leaders did not devise an effective strategy for the conduct of the war. U.S. and South Vietnamese forces never tried to invade the North; their goal was simply to defend the South. Meanwhile, the Communists executed quick attacks in the South and then disappeared into the jungle in a guerrilla ("little war") approach. Communist forces were fighting for the cause of uniting their country, whereas the American purpose was not so clear-cut or urgent to many Americans. Protests against American involvement in Vietnam weakened our effort and emboldened the enemy. Many Americans questioned why our military was helping to prop up a corrupt government in a location that did not seem to them to be vital to our national self-interest.

One geographic factor in the war were the heavy jungles that are common in Vietnam. The United States military attempted to clear out Communist hideouts in the jungles with a defoliant known as Agent Orange. It eventually became known that Agent Orange was a serious health hazard that affected many military personnel as well as Vietnamese civilians.

In the early 1970s, President Richard Nixon began withdrawing American troops and turning over more of the fighting to the South Vietnamese army. He also pursued negotiations with North Vietnam to end the fighting. In 1973 the Paris Peace Accords officially ended American involvement, but fighting between the North and South continued. The government and military of South Vietnam collapsed, and on April 30, 1975, North Vietnamese troops entered the South Vietnamese capital of Saigon and declared victory. Vietnam was once again united, but now it was under a Communist government. The Communist rulers renamed Saigon as Ho Chi Minh City. The Communist government killed thousands of those who had opposed it, and Communism did spread to other countries in Southeast Asia.

Since the War

The United States and Vietnam officially avoided each other until 1994, when President Bill Clinton ended the trade embargo that the U.S. had imposed on Vietnam. The restoration of diplomatic relations followed the next year, and trade relations grew from that point.

The Communist government of Vietnam pursued a course similar to that of China in opening the country to capitalist development. Vietnamese factory workers were willing to work for even less than Chinese workers were, so some companies shifted production from China to Vietnam. Vietnam went from being one of the poorest countries in the world to having a strong, growing economy. The percentage of people living in poverty in Vietnam decreased from 70% to 6%. The population there, 70% of whom are under 40, holds great potential for continued economic growth. In a recent survey, 95% of Vietnamese polled expressed support for market capitalism. They have seen it work in lifting their country out of poverty. With this rapid industrialization, however, have come the typical problems of pollution and unwise exploitation of such natural resources as sand, fish, and timber.

Communist Vietnam has developed closer economic and military ties with the United States, primarily because of Vietnam's uneasy relationship with China. China and Vietnam fought a serious month-long war in 1979 over Vietnam's involvement

Residents of Quang Phu Cau, Vietnam, make incense sticks for use in the celebration of Tet (Vietnamese New Year) and for export to other nearby countries.

Hang Son Doong Cave

In 1990 a Vietnamese farmer, Ho Khanh, was seeking shelter from a jungle storm when he entered a huge cave. He became disoriented and got lost leaving the cave area and so could not report its location. In 2008 Ho was hunting for food in the area when he found the cave again. This time he reported his discovery to British cave experts who were working nearby.

Hang Son Doong (Mountain River Cave) is the largest cave in the world. The main cavern is three miles long and reaches as much as 650 feet high. It could hold a block in New York City—skyscrapers and all—and a Boeing 747 jet could fly in it. Its largest stalagmite is 262 feet high. Cave-ins or dolines have allowed sunlight to enter through depressions in the ceiling, and this in turn has led to jungle growth inside the cave. Explorers have since discovered 57 other caves in the same region of what used to be southern North Vietnam, near the border with Laos. Three of the world's four largest caves have been found there.

In 2013 the government opened the cave to limited public excursions. The number of cavers and explorers coming to the area has increased significantly in the last few years, but that number is still relatively small because the area is so difficult to reach and tours are expensive. In the photo below, you can see tents in the entrance to the cave.

The area received heavy bombing during the Vietnam War. Unexploded shells remain in the jungle, and these shells have cost the lives of several people who have set them off when they were looking for metal they could sell. The increase in tourism has brought a financial windfall for the villages in the cave area, but this has brought the usual tradeoff of potential damage to the local ecosystem.

in Cambodia, an ally of China. As China has expanded its sphere of influence in Southeast Asia and the South China Sea, Vietnam has turned to the United States for military support to protect its interests. Once enemies, Vietnam and the U.S. have become allies against China. Few Vietnamese people carry any bitterness about American involvement there. In fact, if you ask most Vietnamese about "the war," they are more likely to think of the conflict with China than the one involving the United States.

Direct U.S. military involvement in Vietnam ended decades ago. American military personnel and millions of Vietnamese veterans and civilians still carry physical and emotional scars from that conflict. As time has moved on, however, the emotional toll in both countries has eased. The Vietnamese are growing in their economic well-being. Perhaps, if a trend develops toward political and religious freedom there, the people of Vietnam will have what the American troops fought and died for them to have; and the world will continue to turn from the failing practices of classic, hardline Communism.

The Lord can bring an end to war. It is by ceasing human striving that we really come to know God. In this way He is exalted among the nations.

Come, behold the works of the Lord,
Who has wrought desolations in the earth.
He makes wars to cease to the end of the earth;
He breaks the bow and cuts the spear in two;
He burns the chariots with fire.
"Cease striving and know that I am God;
I will be exalted among the nations, I will be exalted in the earth."
Psalm 46:8-10

Assignments for Lesson 98

Gazetteer	Read the entries for Cambodia, Laos, Myanmar, Thailand, and Vietnam (pages 167, 169, 171, 175, and 177). Read "My Duty to Serve" (pages 311-312) and watch the video interview using the link provided there.
Worldview	Copy this question in your notebook and write your answer: How does the video "My Duty to Serve" affect your worldview?
Project	Continue working on your project.
Literature	Continue reading *Ann Judson: A Missionary Life for Burma.*
Student Review	Answer the questions for Lesson 98.

Padar Island, Indonesia

99 A Nation Defined by Water: Indonesia

On the morning of December 26, 2004, residents and vacationers in the Indonesian city of Banda Aceh on the western coast of the island of Sumatra arose to enjoy the day after Christmas. About 8 a.m. local time, a strong 9.1 earthquake shook the bed of the Indian Ocean about one hundred miles to the west. Shortly thereafter and without warning, a huge wall of water estimated to be sixty to one hundred feet high struck the city beach area and swept inland.

Devastating waves moved in other directions away from the earthquake epicenter as well. The water struck Thailand, Sri Lanka, and other countries as far west as coastal areas of East Africa over 5,000 miles away. The death toll from this destruction reached over 167,000 in Indonesia alone, perhaps a quarter of a million total lives lost in all affected areas. Property damage was estimated at fifteen billion dollars. This wall of water is known as a tsunami. The 2004 event in Indonesia is considered the most costly tsunami in world history.

The Power of Water

Water gives life. People, animals, and plants need water. Water can also destroy. Storms and tsunamis can devastate property and take lives.

Water can separate. Especially in the past, it often left people on opposite sides of a river or sea with little or no contact. Water can also bring together. Travel in some cases is easier on water than on land; so water enables trade, travel, and exploration for those who are willing to embark.

A Nation Defined by Water

The country of Indonesia is defined to a great degree by water. Its name comes from the Greek words Indos (India) and nesos (island). Indonesia is the largest country in the world made up entirely of islands. It is the seventh largest country when combining land and sea area and the fourteenth largest land area of the countries of the world.

The official counts of the islands that make up Indonesia vary depending on how one defines an island, but the largest number is over 17,000, with more than 900 of them inhabited. The islands of Indonesia cover an area over 3,200 miles wide—one-eighth of the world's circumference—and over 1,100 miles from north to south. The three largest islands are Sumatra, Java, and Borneo.

The land area of Indonesia is about three times the size of Texas, but it is spread over a total area about the size of the continental United States. The country straddles the equator in Southeast Asia at the junction of the Indian and Pacific Oceans along some of the world's busiest and most strategic shipping routes. In this sense, the waters of Indonesia are a great benefit.

The People and the Land

The people of Indonesia are a fascinating mixture of local tribesmen who never leave their home area and urban dwellers who are connected to the world. Indonesia has the fourth largest population of the countries of the world, over 271 million, following only China, India, and the United States. The island of Java is one of the most densely populated areas on earth.

The people groups of Indonesia use over seven hundred languages. They are 87% Muslim. In fact, Indonesia has more Muslims than any other country in the world. This is true even though Islam did not come to the country in a significant way until the late 1200s and did not grow in significant numbers until even later than that. Hinduism and Buddhism predated Islam in Indonesia. As a result, the practice of Islam there has a different flavor than it does in the Middle East.

Despite its large population, Indonesia is the second most heavily forested country in the world, following only Brazil. The country has over 3,000 species of trees. Indonesia also has over 40,000 species of flowering plants, including 5,000 species of orchids. The country is home to two of the world's largest flowers, the *Rafflesia arnoldii* and the *Titan arum*. These flowers give off a terrible smell, which attracts flies, which help in pollination. Indonesia is also home to the Komodo dragon, the largest and heaviest lizard in the world.

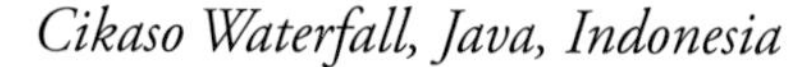

Cikaso Waterfall, Java, Indonesia

Great Mosque of Central Java, Semarang, Indonesia

The Making of a Tsunami

The surface of the earth rests on a layer of tectonic plates. Beneath these plates lies hot, molten (semi-liquid) rock, which moves and flows. This motion causes tectonic plates to shift and sometimes to collide. This action results in earthquakes and, when the molten rock breaks through the earth's surface, volcanoes.

Scientists believe that most of the Pacific Ocean rests on one large tectonic plate. The edges of this plate collide with other plates. The result of this activity is called the Ring of Fire. Like the Philippines, Indonesia is within the Ring of Fire. With over 70 active volcanoes, it has more than any other country.

When tectonic activity occurs beneath an ocean or extends into an ocean, it displaces a huge amount of water. This water forms a rise in sea level which moves in huge waves that hit coastal areas usually with little or no warning. The common term for this wave is tsunami, which is Japanese for harbor wave.

A tsunami can travel up to 450 miles per hour. At sea, the wave is a smaller tide that moves fast. When it gets closer to shore, where the seabed is more shallow, the tide slows down and increases in height, sometimes to as much as 100 feet. As the tide approaches land, it can briefly pull water away from the shore. When the tsunami strikes land, up to one thousand tons of water can hit a yard of coastline.

Tsunamis occur most often in the Pacific, and about 80% are the result of earthquakes. Volcanoes and landslides can also cause tsunamis, and they have been known to occur even on the usually calm Mediterranean Sea. The eruption of Mount St. Helens in Washington State caused a tsunami on nearby Spirit Lake. Earthquakes on land can sometimes cause tsunamis in the ocean. A tsunami can strike land thousands of miles from the event that caused it.

Damage from a tsunami can be extensive. It can destroy coral reefs and affect the flow of rivers. Destroyed buildings, trees, and other refuse can clog bays and harbors. The force of the water can alter coastlines. Pollution from damaged industrial sites can contaminate land and water. Even more significant is the loss of human life and the damage to human society. Homes, schools, hospitals, and businesses can be wiped out. Sanitation systems can become unusable. Roads can become cluttered or washed out, resulting in a delay in relief efforts and in the resumption of normal activity. With tens of thousands of people becoming victims within a matter of minutes in sometimes hard to reach places, sanitation issues that result from the bodies not being buried can be enormous. Effects from a tsunami can last years or even decades, although Indonesia has done a remarkable job in recovering from the 2004 disaster.

View of Banda Aceh, Indonesia, After the 2004 Tsunami

Tsunamis in History

A few tsunamis in history merit mention here. In 1883 on the Indonesian island of Krakatoa (modern spelling: Krakatau), a fissure in a volcano allowed seawater to come into contact with molten lava, causing the island to explode. The eruption changed the geographic makeup of the island and created a tsunami. Tsunami waves went around the earth four times. Significant weather and atmospheric changes also occurred around the world as a result of the huge amount of ash and dirt thrown into the atmosphere. Some 36,000 people died in the event.

In 1958 an earthquake in Alaska caused a landslide into the Gulf of Alaska, which caused the tallest tsunami on record to strike a thinly populated section of the Alaskan coast. Judging from the height of the vegetation damage, the wall of water is thought to have been about 1,700 feet tall.

A 1964 earthquake off the coast of Alaska caused a tsunami to strike Hawaii, 2,400 miles distant, about five hours later.

In British culture, Boxing Day is the day after Christmas. Traditionally, wealthy families made up gift boxes and presented them to people who served them throughout the year. The 2004 Boxing Day tsunami in Indonesia is thought to have had the energy of 23,000 atomic bombs. Over one million people became homeless. Scientists detected waves from it in Antarctica and South America. The tsunami was the first to be detected by a satellite that was monitoring tectonic plate and volcanic activity, but the data was not immediately available to scientists and relief workers. People around the world donated over seven billion dollars for relief in the first month following the tsunami.

An earthquake off the coast of Japan in 2011 caused tsunami waves around the Pacific and took almost 20,000 lives in Japan. In addition to the usual damage that a tsunami causes, one significant result was damage to the Fukushima Daiichi nuclear reactor in Japan.

People have offered various suggestions to try to limit the damage from tsunamis. Some have proposed building seawalls, but the tradeoff of a wall tall enough to provide significant protection would be the loss of coastal scenery. Remote warning systems such as wave gauges and seafloor pressure gauges are helpful but can provide false readings. A seafloor gauge has to transmit data to a buoy on the surface, which then transmits to a satellite, which then communicates the information to human technicians, who then have to get the warnings out to the population, all this when minutes can make the difference between life and death.

Another Kind of Disturbance

A political disturbance came to a head on an island in Indonesia in the last half of the 1900s. Timor is one of the southernmost islands in Indonesia. Portuguese traders came to the island in the early 1500s and established a colony. The Dutch came to Timor as part of their effort to control the entire region and fought with the Portuguese. In 1859 Portugal gave the Dutch the western part of Timor, which became part of the Dutch East Indies.

Japan occupied eastern or Portuguese Timor during World War II. Afterwards the Portuguese reasserted control, but the Timorese resisted this continued foreign domination. In 1975 Portuguese authorities withdrew to a small island. East Timor (Timor-Leste) declared its independence, but then Indonesian forces invaded a few days later to subdue the country and make it a province of Indonesia. Between 100,000 and 250,000 people died because of the Indonesian repression, and the Indonesian policy in the province did not end the people's desire to be independent.

In a 1999 referendum overseen by the United Nations, almost 80% of the people of East Timor voted for independence. However, anti-independence militia, supported by the Indonesian military, went on a campaign to subdue those who desired independence. Later in 1999, an Australian-led peacekeeping force under United Nations oversight ended the violence. Finally in 2002, the

The government of Indonesia gave this statue of Christ to the people of Timor in 1996, while Timor was still a province of Indonesia. It stands outside Dili, now the capital of Timor-Leste.

international community recognized Timor-Leste (East Timor) as an independent country. Instability continued for several years, but the country eventually developed a stable government and society.

Timor-Leste is slightly larger than Connecticut. It consists of the eastern part of the island of Timor, an enclave in the western half of the island, and two other islands. The population is about 1.3 million. The former province was an anomaly in Indonesia in that the population is about 98% Roman Catholic.

A New Capital City

Water and geography have contributed to an unusual policy decision by the government of Indonesia. In 2019 the president of Indonesia announced that the country would spend an amount equalling 33 billion U.S. dollars to move the national government from Jakarta on the island of Java to a new city to be built on a site in the province of East Kalimantan on the Indonesian portion of the island of Borneo (the country of Brunei and part of Malaysia are also on Borneo).

The main reason given for the move is that Jakarta is sinking at the rate of six or seven inches per year. The massive construction projects in the city, which has a metro population of about 32 million, put great weight on the ground. In addition, only about one-fourth of the residents of the city have access to piped water, which means that the many wells that have been dug have reduced the groundwater level. These factors have contributed to the sinking. Some experts cite rising sea levels as an issue also. About half of the city is below sea level and has been subject to several costly floods. A new location for the capital on Borneo might afford greater security from terrorist attacks as well.

On the other hand, the construction of a new city that might become the home for 1.5 million government workers and others could pose environmental risks to Borneo, such as deforestation. Moving a nation's capital is unusual but not unprecedented. Brazil built the new city of Brasilia to replace Rio de Janeiro. Nigeria moved its capital from Lagos to Abuja. Egypt is building a new city where it will move government operations from Cairo.

We should not think of Indonesia as a natural or political problem waiting to happen. Indonesia is a significant player on the world stage. It has one of the strongest economies in Asia. Its oil and natural gas reserves assist in the functioning of the world economy. It is located on vital world trade routes. At the same time, its commitment to Islam, which can sometimes take radical form, and its vulnerability to natural disasters are potential problems that Indonesia must confront.

The people of Indonesia, like the people of every nation, need to hear the greatness of God and the saving message of the gospel.

Sing to the Lord, all the earth;
Proclaim good tidings of His salvation from day to day.
Tell of His glory among the nations,
His wonderful deeds among all the peoples.
1 Chronicles 16:23-24

Assignments for Lesson 99

Gazetteer	Read the entries for Brunei Darussalam, Indonesia, Malaysia, Singapore, and Timor-Leste (pages 166, 168, 170, 173, and 176).
Geography	Complete the map skills assignment for Unit 20 in the *Student Review Book.*
Project	Continue working on your project.
Literature	Continue reading *Ann Judson: A Missionary Life for Burma.*
Student Review	Answer the questions for Lesson 99.

Putao, Myanmar

100 The Worldview of Jesus

Would you like to have the worldview of Jesus, to think the way Jesus thought? You can.

The entire New Testament is really about worldview: changing one's worldview to trust in Jesus as Lord, and what that trust means for all of a person's outlook, mindset, or worldview. For instance:

> *Do not be conformed to this world, but be transformed by the renewing of your mind, so that you may prove what the will of God is, that which is good and acceptable and perfect.*
> *(Romans 12:2)*

> *Set your mind on the things above, not on the things that are on earth. For you have died and your life is hidden with Christ in God.*
> *(Colossians 3:2-3)*

This lesson traces themes in Jesus' life and ministry that reveal His worldview: His purpose, priorities, and perspectives. Understanding how He thought while He was on the earth gives us something to aim for as we develop the worldview of Jesus.

1. Jesus' worldview is that God is working.

"My Father is working until now, and I Myself am working." (John 5:17)

Jesus said this on a Sabbath day, when the Jews were not supposed to work, commemorating when God rested from the days of Creation. But if God were not on the job on the Sabbath, sustaining our world, we would be in serious trouble. The commandment to remember the Sabbath and to refrain from work on that day must mean something besides just doing nothing because God is not doing nothing on the Sabbath.

This verse shows that Jesus was not a deist, and it also shows the fallacy of the worldview of Deism (see Lesson 60). Jesus' worldview is that God has the world in His hands and is always on the job.

2. Jesus' understanding about Himself is part of His worldview.

"The Father loves the Son and has given all things into His hand." (John 3:35)

[Jesus called] "God His own Father, making Himself equal with God." (John 5:18)

"I and the Father are one." (John 10:30)

These and many other verses teach that God exists and that Jesus believes in Him. Jesus understood

Himself to be God's unique, only-begotten Son. Jesus expressed His understanding of His purpose in many different ways. He came to make the Father known (John 1:14). Jesus said, "I have come down from heaven, not to do My own will, but the will of Him who sent Me" (John 6:38). The Christian worldview involves knowing who Jesus is in relation to God and also knowing who we are in relation to God. Coming to believe in the identity of Jesus changes a person's worldview.

3. Sin and Satan are real.

Jesus knew that Satan is real. Jesus confronted Satan in the wilderness during His temptations (Matthew 4:1-11). Satan entered Judas just before Judas began to plot with the Jewish leaders to betray Jesus (Luke 22:3). However, Satan is doomed. Jesus said that He could see Satan "fall from heaven like lightning" when the seventy cast out demons in Jesus' name (Luke 10:13). Satan continues to do evil in a limited way, but his destruction is sure (Revelation 20:10).

Jesus taught the reality of sin. He told the woman who had been caught in adultery to sin no more (John 8:11). Sin ruins the lives of those who commit sin as well as the lives of others who are affected by it.

However, people are not the enemy of God. Paul said, "Our struggle is not against flesh and blood, but against the rulers, against the powers, against the world forces of this darkness, against the spiritual forces of wickedness in the heavenly places" (Ephesians 6:12). Satan is the enemy, and He uses people to further his cause.

Festival of Light Boats, Laos

Jesus described His generation as "evil and adulterous" (Matthew 12:39), "unbelieving and perverted" (Matthew 17:17), "adulterous and sinful" (Mark 8:38), and "wicked" (Luke 11:29). That was Jesus' worldview about the generation during which He lived on earth. Can you imagine Jesus using those same words to describe the generation of today?

4. Jesus understood and loved people.

Jesus taught a great deal about people. He understood what is inside people.

> *Now when He was in Jerusalem at the Passover, during the feast, many believed in His name, observing His signs which He was doing. But Jesus, on His part, was not entrusting Himself to them, for He knew all men, and because He did not need anyone to testify concerning man, for He Himself knew what was in man.*
>
> *(John 2:23-25)*

Jesus understood what made people tick because (1) He created them and set them ticking and because (2) He was a person Himself. Although He was without sin, Jesus understood that "men loved the darkness rather than the Light, for their deeds were evil" (John 3:19).

Even with this knowledge, Jesus loved people and died for them. He did this because we needed Him to, but He did it even more because of His love for us, because of who He is.

Jesus looked past the stereotypes that people develop about each other and was able to see the real person who had spiritual needs. Jews had "no dealings with Samaritans" (John 4:9), yet Jesus had a conversation with a Samaritan woman that changed her life. The Pharisees dismissed tax collectors and sinners, but Jesus shared meals with them (Matthew 9:11). Jews kept their distance from Gentiles, but Jesus interacted with a Gentile Syrophoenician

Fishing on Nong Han Kumphawapi Lake, Thailand

woman (Mark 7:24-30). Bartimaeus was a blind beggar whom nobody wanted, but Jesus took time for him and healed him (Mark 9:46-52). This is Jesus' worldview concerning people.

Jesus understood what is truly important about people. He saw rich people putting their gifts into the temple treasury, but He took special note of the poor widow who put in two small coins, all that she had to live on (Luke 21:1-4).

Jesus described a group of Samaritans coming out to meet Him as fields that were white for harvest (John 4:35). The disciples probably did not share that opinion of Samaritans because of their worldview. Today, our opinion of whether people are hardened soil likely to reject the gospel or a field white for harvest depends on our worldview. We can at the very least sow and water, be salt and light, and let God give the increase. People are worth it; they were for Jesus.

Jesus believed in the potential of people—more accurately, the potential of what God can do in them. He called people to put away sin and to be different because He believed they could, for God is at work in the heart, mind, and life of the believer (Philippians 2:13). Yet with people who finally reject the gospel, there comes a time to shake the dust off our feet and move on to more fertile soil (Matthew 10:14).

Jesus even believed that those who would be our enemies are worth loving (Matthew 5:43-45) because, as Paul said, they are not really our enemies. We should see such people not as those we should crush and defeat but as people we should love, influence, and teach by word and example. They deserve for us to try to rescue them and not condemn them because that is what Jesus did for us.

5. Jesus opposed empty religious traditions and profiteering in the name of God.

Jesus cleared the temple of those who were making God's temple into a den of robbers (Matthew 21:12-13). He had strong words against the traditions of the Jewish elders which actually prevented people from obeying God (Mark 7:1-23).

On numerous occasions He directly challenged the traditional Jewish views of what was and was not "lawful" on the Sabbath.

Sometimes religion, even religion carried on in the name of God, is ungodly. How can we know? If it demeans people; if it promotes a system instead of the Savior; and if it exalts something or someone other than God and Christ, it is not doing God's will.

Jesus taught that it is possible to major in the minors, to be more concerned about tithing tiny garden seeds than about being people of justice, mercy, and faithfulness (Matthew 23:23). We must lovingly accept the conscience and scruples of those who honestly feel the need to maintain certain practices that Scripture does not directly command, but on such matters we need to ask, "Did Jesus die for this?" And even if we conclude that He didn't, we must remember that He died for those with whom we have differences over matters of opinion and faith, and treat them accordingly.

6. The way of Jesus requires a person's whole heart and life and involves selfless service.

Jesus said, "If anyone wishes to come after Me, he must deny himself, and take up his cross daily and follow Me. For whoever wishes to save his life will lose it, but whoever loses his life for My sake, he is the one who will save it. For what is a man profited if he gains the whole world, and loses or forfeits himself?" (Luke 9:23-25). Jesus calls on those who would be His followers to do nothing more than what He did Himself.

The way of Christ will not work if a person tries to follow it partially or part-time. We cannot serve two masters. In addition, it would be difficult to know accurately in each situation we face whether to be our Christian self or our worldly self. Instead, following Christ requires a radical, 180-degree change in a person's worldview.

Jesus said, "Unless a grain of wheat falls into the earth and dies, it remains alone; but if it dies, it bears much fruit. He who loves his life loses it, and he who hates his life in this world will keep it to life eternal" (John 12:24-25). Later that night, Jesus washed the feet of the disciples during the Last Supper to show them the full extent of His love.

Jesus also said that greatness in the kingdom of God comes by serving. He said, "If anyone wants to be first, he shall be last of all and servant of all" (Mark 9:35). In Philippians 2 Paul says "Have this mind [We might say, "Have this worldview"] among you, which was also in Christ Jesus" (Philippians 2:5). He then describes Jesus giving up His exalted status of equality with God to take on the lowly status of a human slave suffering death by means of a cross. This humble, faithful service was why God exalted Him to the highest position and gave Him the name that is above every name. It takes a remarkable, trusting attitude—a remarkable view of the world and other people—to wash feet and give your life in lowly service. Paul also had a high status as a Pharisee among Pharisees, but his worldview became one of counting everything he had as rubbish in order to know Christ (Philippians 3:7-8).

The way of Christ is where giving a cup of cold water to a little one in the name of a disciple does not lose its reward (Matthew 10:42). It is this kind of humble service that changes the world, not power, money, fame, or beauty. Few of us will have any significant quantity of those in our lives, but we can all be servants.

To be a servant requires a worldview of faith or trust that God will take care of you as you serve. Jesus teaches us that life is not to be filled with worrying. Instead, it is to be about trusting, serving, and glorifying God (Matthew 6:25-34). With faith in God, prayer will be an appropriate part of your worldview because you believe that God answers prayer (Philippians 4:6-7).

7. Jesus believed in the sovereignty of God.

God is in charge, He knows what He is doing, and His timing is perfect.

Jesus knew that God had a plan, even a timetable, for bringing about redemption. In a series of passages (John 2:4, 7:6-8, 7:30, and 8:20), either Jesus or John says that Jesus' time or hour has not yet come. Then in John 12:23-27, Jesus says that His hour has come. This idea is repeated in John 13:1 and 17:1.

The sovereign God can bring good out of what is evil. In Philippians 1, Paul describes how God brought much good out of Paul's imprisonment. The ultimate example of God's ability to bring good out of what is bad is the cross, when God used the unjust execution of the innocent Son of God to bring about redemption for all who would believe in Him.

Political power comes from God. When Jesus stood before Pilate, the Roman governor asked Him in some frustration, "Do You not know that I have authority to release You, and I have authority to crucify You?" Jesus answered, "You would have no authority over Me, unless it had been given you from above" (John 19:10-11). Political power comes not from financial clout, or military might, or bargaining skill, or even the voters. These all might have parts to play, but worldly authority comes from God (see also 2 Chronicles 10:15, Luke 1:52, and Romans 13:1-4).

Christians live and have lived in monarchies, empires, totalitarian regimes, and democracies. They have held positions of great authority, and they have been fed to lions by those in authority. Christians know that their primary citizenship is in heaven (Philippians 3:20). Because of the sovereignty of God, Christians can know that the world is not something they have to fear; instead it is where they can live with confidence, faith, and victory.

8. Resurrection and judgment are coming.

The bodily resurrection of Jesus vindicated His identity, status, and power and provided hope for our own resurrection from the dead. Jesus predicted the coming general resurrection of the dead and the last judgment (John 5:28-29, Matthew 25:31-46). In other words, we have a destiny before us; life has a

Christ Church (Anglican), Malacca, Malaysia

point, a goal. This reality affects what we value in our lives and what we should make our highest priority.

Paul wrote that at the last judgment, "at the name of Jesus every knee will bow, of those who are in heaven and on earth and under the earth, and that every tongue will confess that Jesus Christ is Lord, to the glory of God the Father" (Philippians 2:10-11). At that time, everyone's worldview will involve a recognition of Jesus as Lord. Any other worldview won't matter. As a matter of fact, in ultimate terms no other worldview matters now, either.

In the beginning was the Word, and the Word was with God,
and the Word was God. He was in the beginning with God.
All things came into being through Him, and apart from Him
nothing came into being that has come into being.
In Him was life, and the life was the Light of men.
The Light shines in the darkness, and the darkness did not comprehend it.
John 1:1-5

Assignments for Lesson 100

Worldview Recite or write the memory verse for this unit.

Project Finish your project for this unit.

Literature Continue reading *Ann Judson: A Missionary Life for Burma.*

Student Review Answer the questions for Lesson 100.
Take the quiz for Unit 20.
Take the fourth Geography, English, and Worldview exams.

Lake Wakatipu, New Zealand

21 Australia and New Zealand

This unit on Oceania begins with a survey of New Zealand. We then go to Sydney, Australia, which is well known for its harbor and for the bridge and Opera House there. We look at the fascinating Australian Outback and the land, plants, animals, and people associated with it. As the island continent, Australia has special connections with the Great Barrier Reef and with other islands surrounding it. The worldview lesson introduces John Mann, a research scientist and Christian who discovered a solution for a serious problem that Australia faced.

Memory Verse

Memorize Romans 1:20 by the end of the unit.

Books Used

The Bible
Exploring World Geography Gazetteer
Ann Judson: A Missionary Life for Burma

Project (Choose One)

1) Write a 250-300 word essay on one of the following topics:
   - Research and write about one element of the Australian Outback, such as the Aborigines, its desert land, the plants, unusual animals, Coober Pedy, or some other subject. (See Lesson 103.)
   - How do you hope to express your faith in your future career, family, or volunteer opportunities? (See Lesson 105.)
2) Create a drawing of the Sydney Harbor Bridge, the Sydney Opera House, or a portion of the Great Barrier Reef.
3) Create a presentation on the Maori people of New Zealand. You might focus on their acceptance of Christianity, their language, culture, dress, houses, or lifestyle. Include pictures in your report.

101 Kia Ora, Aotearoa (Greetings, New Zealand)

Why did they come?

What prompted people in Polynesia to embark in seaworthy voyageur canoes to come to what is now called New Zealand? Were they pushed out of the islands where they were living? Was there a spirit of adventure and discovery that pulled them onto the wide ocean? Did a small group discover the new lands, then return to lead others there? Did only a small group come and eventually populate the new islands?

These settlers had likely arrived in New Zealand by the 1200s. We will never know all the details, since this people did not leave a written record of their history; but in some context the people we know as the Maori (the word means normal or plain) became the first known inhabitants of the land they called Aotearoa, the land of the long white cloud.

The Europeans Come

We know more about the motivations of the first Europeans who came to the islands. At the time, explorers from many nations were going out to claim new lands as colonies for their home countries. The Dutch explorer Abel Tasman was the first European known to land in what became New Zealand, doing so in 1642.

Dutch cartographers named it Nova Zeeland after the Zeeland province in the Netherlands. The British explorer Sir James Cook came in 1769, anglicized the spelling of the name to New Zealand, and claimed it for Great Britain. British settlers began coming in significant numbers in the 1820s.

The Treaty of Waitangi

The different perspectives of the British and the Maori led to conflict. In 1840 British and Maori representatives signed the Treaty of Waitangi. In it, the Maori recognized British sovereignty over the land, and the British recognized Maori rights to ownership of the land they already possessed. The goal of the arrangement was to enable the two cultures to coexist in New Zealand.

Two trends created a new majority population there. First, the Maori population declined primarily because the Europeans brought diseases against which the Maori had no immunity. The British and the Maori occasionally had military conflicts, but

The photo above shows hongi, *a traditional Maori form of greeting. Two people touch noses and often foreheads while grasping the other person's hand or touching the other person's shoulder.*

conflicts between opposing groups of Maori were more common. By 1896, only about 42,000 Maori still lived in New Zealand.

Second, while the Maori population was declining, immigrants from Europe were coming in large numbers. By the late 1850s Europeans outnumbered the Maori. Thus the dynamics of life that the Maori had known for perhaps six hundred years changed dramatically and permanently.

The Land

New Zealand is part of the area known as Oceania, which consists of thousands of islands in the Central and South Pacific Ocean.

The primary geographic feature of New Zealand are the two large islands that make up the bulk of its landmass. The South Island is the 12th largest island in the world and has the more rugged terrain of the two. It is bisected north and south by a mountain range called the Southern Alps. Among these is Mount Cook (the Maori name is Aoraki), which is 12,316 feet high. The mountain used to be about 130 feet taller, but an avalanche at the summit in 1991 and the gradual erosion of the ice cap cost it some of its height. On Mount Cook is the Tasman Glacier, which is eighteen miles long and about a half mile wide. It is one of about 360 glaciers in the Southern Alps.

The North Island is smaller, the 14th largest island in the world, but about three-fourths of the people live there. This island also has mountains; and although they are generally not as dramatic as the ones on the South Island, volcanoes are more numerous here. New Zealand lies along the Pacific Ring of Fire, and earthquakes are fairly common. The North Island is home to Waitomo Caves, which are amazingly illuminated by glowworms seeking to attract their prey.

Kiwi Crossing Sign and Mt. Ruapehu

The land area of New Zealand is about the same as the state of Colorado. It is a relatively isolated country; its nearest neighbor, Australia, is about one thousand miles distant. New Zealand is about 2,800 miles south of the equator and 3,100 miles from Antarctica.

The Kiwis

A common nickname for the people of New Zealand is Kiwis, for the flightless bird that is native to the islands. As of 2019 the population of New Zealand was about 4.8 million (Colorado by comparison had 5.6 million). Some 86% of the population lives in urban areas. About 60% of Kiwis are of European descent, and about 15% are Maori. Most of the rest are Pacific Islander or Asian. As of 2017 about 37% of New Zealanders identified themselves as Christian, but almost half of the population indicated that they held to no religion.

Despite their geographic isolation, Kiwis are fully engaged in the world community. Many served with the Allies in the two world wars of the 1900s. The country conducts trade with many other countries. Tourism is a major part of the New Zealand economy. Consistent with the Kiwis' reputation as adventurous people, New Zealand is credited with being the home of bungee jumping and other adventure sports.

Te Reo Maori (The Maori Language)

The story of the language of the first settlers of New Zealand reflects the history of the people themselves. At the beginning of the 1800s *te reo* (or *te reo Maori*) was the predominant language spoken in Aotearoa. Regional dialects of the language reflected the geographic isolation of various parts of the islands and the likelihood that the original settlers came from different islands in Polynesia. There was no written te reo, but islanders generally

These signs in Wellington, New Zealand, are written in English, Maori, and Chinese. They encourage visitors to "Be a Tidy Kiwi."

understood the meanings of various carvings, knots, and weavings.

Early European settlers depended on the Maori for many of the items and skills they needed, so the Europeans had to learn te reo to obtain what they needed, especially in conducting trade. The need and desire for a written language grew as more Europeans came. Missionaries first started working on a written form of te reo in 1814. Cambridge professor Samuel Lee worked with Maori chief Hongi Hika and his relative Waikato to create a workable written language in 1820.

Missionaries reported that the Maori eagerly adopted the written form of their language. In the 1820s Maori around the country were teaching each other how to read and write. They used charcoal to write on leaves, wood, and animal skins when they could not get paper. For the next half century, many Pakeha (European New Zealanders) who were government officials, missionaries, and other prominent leaders learned to speak te reo. Pakeha children often became fluent in the language.

The Treaty of Waitangi reflects the significance of language. The treaty documents were presented to the signers in both English and te reo. The Maori signers understood what the te reo document said, but the meanings of some of those words probably didn't convey the same meaning that the English version did to the English signers (such as the word sovereignty). The question of whether the Maori leaders understood what they were agreeing to remains unresolved in the minds of some people today.

As the Pakeha became the majority of the population, English became the predominant language and te reo came to be used only in Maori communities. If a Maori wanted to get along in Pakeha society, by holding a job or participating in sports, for instance, he had to learn English. Official policy and informal practice was to discourage the use of te reo. Schools suppressed the language, and children were often punished when they used it. Since language reflects culture, much of what it meant to be Maori was in danger of being lost to the Maori people. However, through the period of World War II most Maori families spoke te reo at home. Maori people conducted public gatherings in te reo. Some newspapers published in te reo. As is always the case when speakers of different languages interact, English-speakers began using some Maori words and speakers of te reo began using some English terms.

During and after World War II, as New Zealand became more urbanized and industrialized, many Maori moved to the cities to work. Life there took place in English, and te reo became forgotten and was considered irrelevant. By the 1980s fewer than 20% of Maori knew te reo well enough to be considered native speakers.

However, beginning in the 1970s a growing number of Maori began to appreciate anew their cultural heritage. Emphasizing te reo was part of this movement. Maori groups encouraged schools and the New Zealand parliament to promote the language. Radio, television, and newspapers began using more te reo words. In 1987 the New Zealand parliament made te reo Maori an official language of the country, an identity it now holds with English and New Zealand Sign Language. With this greater emphasis, the percentage of Kiwis who can carry on a conversation in te reo is now slightly greater and is growing.

New Zealand demonstrates the blending of English and Maori language by the everyday vocabulary that people use and by having some place names in English (such as New Plymouth, Christchurch, and Wellington) and some in te reo (such as Pahiatua, Wanganui, and Kaitaia). People will use whatever language works in a given situation, so English will likely continue to dominate as New Zealand's national and international language; but the Maori people, as a major component of who New Zealanders are, can honor their ethnic heritage by also maintaining a working knowledge of te reo.

Waitangi Day is a national holiday, celebrated in New Zealand on February 6. Celebrations include Maori cultural demonstrations such as dancing and canoeing.

Britain, Empire, and Commonwealth

The island on which England sits is called Great Britain. England became Great Britain (commonly called Britain) when the government based in London proclaimed authority over Wales and later Scotland. During the 1600s, 1700s, and 1800s, Britain developed a worldwide empire by its citizens establishing overseas colonies and its explorers claiming lands for the monarch. This collection of colonies, protectorates, and possessions became known as the British Empire. By the end of the 1800s, Britain controlled nearly one-fourth of the land surface of the earth, from tiny islands to the subcontinent of India, and over one-fourth of the world's population. This 1910 map illustrates the truth of the saying that "The sun never sets on the British Empire."

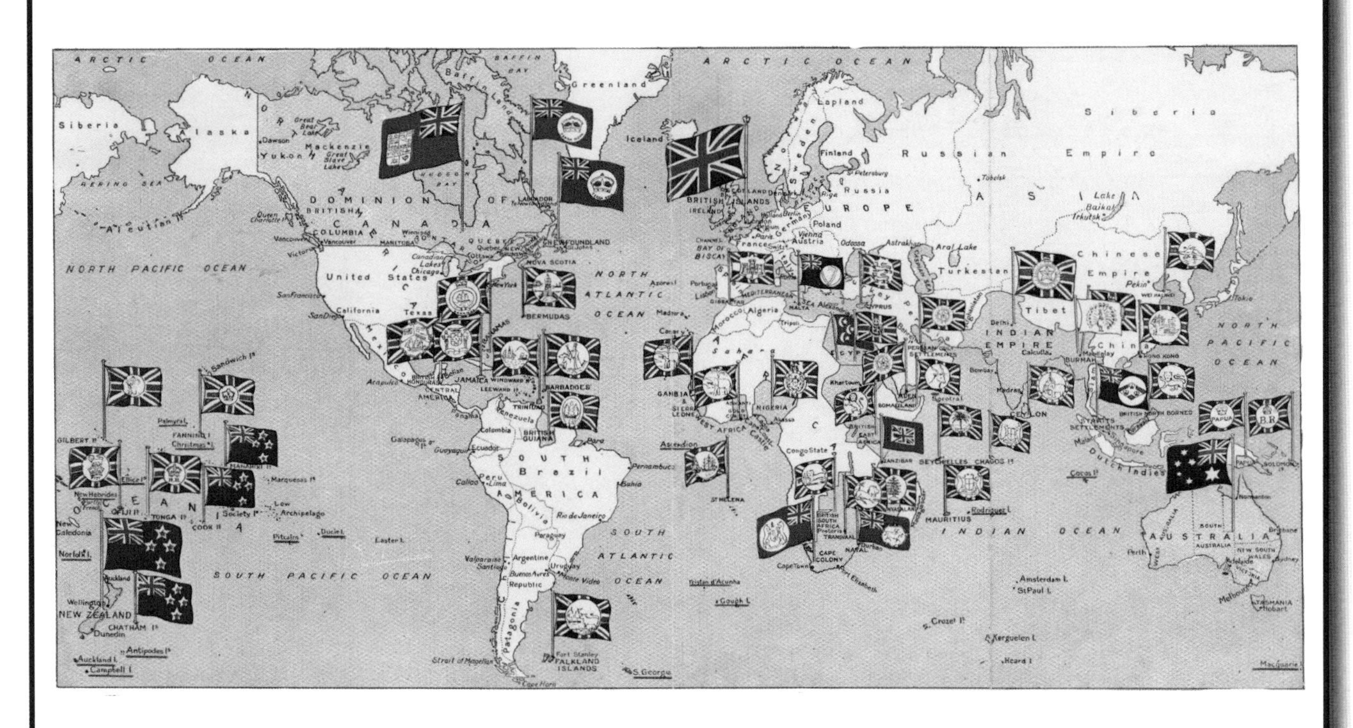

During the 1800s, some in the British government and the general British public promoted the idea of "responsible self-government" for their possessions. Under this principle a colony could receive the right to direct its own domestic affairs while the central government in London handled foreign affairs for the entire empire. One reason for the growing acceptance of this policy was the way that the central government mishandled relations with the colonies in America, which led to the American Revolution and the loss of those colonies from the empire.

A parallel development during the 1800s was the growing unpopularity of the whole idea of empire and colonies. The term that gradually replaced empire was commonwealth, which was less authoritarian and described a looser confederation of nations. The identity for several individual countries in the British Empire or Commonwealth that developed in the late 1800s and early 1900s was that of dominion, which described a self-governing nation within the commonwealth. Australia at first developed as several separate colonies, which united under one colonial government in the 1890s, and then received the status of dominion in 1901. New Zealand became a dominion in 1907.

CONTINUED

The change in the level of autonomy of the parts of the British Empire or Commonwealth is reflected in how those nations participated in the world wars and the events that followed. In 1914 when Great Britain declared war on Germany, it did so for all of its dominions, colonies, and possessions. After the war, the dominions signed the Treaty of Versailles separately and joined the League of Nations as independent members. When war came again in 1939, the dominions made their own decisions about whether to join the fight against the Axis powers.

A series of laws passed by the British Parliament in the 1900s defined and codified the relationships among the countries that had been part of the empire. Today the Commonwealth of Nations (the title no longer includes the term "British") consists of 53 independent countries, almost all of which are former British territories, plus the United Kingdom. The head of the Commonwealth is Queen Elizabeth II (pictured at right in 2019). The Commonwealth countries have selected her son, Prince Charles, as her successor; but the position is not officially hereditary.

Commonwealth nations share a common British heritage, enter into trade agreements with each other, and hold the Commonwealth Games every four years; but Britain (now officially the United Kingdom of Great Britain and Northern Ireland) does not rule the Commonwealth. Within the Commonwealth, sixteen nations are commonwealth realms, which recognize the British monarch as head of state. These include the United Kingdom, Australia, New Zealand, Canada, Belize, and several smaller countries. Thirty-two other Commonwealth member nations are republics, and five have their own monarchs.

Relations within the Empire or Commonwealth have not always been smooth. For instance, all of Ireland was once part of the British Empire/Commonwealth. The 26 southernmost counties established the Irish Free State in 1921-1922. It became simply Ireland in 1937 and a republic in 1949, withdrawing from the Commonwealth. The six northern counties in Ireland are now Northern Ireland and are part of the UK. India gained its independence in 1947, and Pakistan came into existence at the same time. Both countries are members of the Commonwealth. After 1960, several British colonies in Africa, Asia, and the Caribbean gained their independence but remained in the Commonwealth. In 1997 the UK gave its colony of Hong Kong to the People's Republic of China. Hong Kong is not a member of the Commonwealth of Nations.

God separated the peoples of the world by giving them different languages at the Tower of Babel. However, God provided the potential for the people of the world to come together on the Day of Pentecost by having the Christians there tell the story of the One Savior of mankind in multiple languages. Centuries earlier, God gave the prophet Daniel a vision of the coming of the Son of Man and of His dominion over people of all languages:

And to Him was given dominion,
Glory and a kingdom,
That all the peoples, nations and men of every language
Might serve Him.
His dominion is an everlasting dominion
Which will not pass away;
And His kingdom is one
Which will not be destroyed.
Daniel 7:14

Assignments for Lesson 101

Gazetteer Study the map of Oceania and read the entry for New Zealand (pages 178-179 and 181).
Read the Treaty of Waitangi (pages 313-314) and answer the questions in the *Student Review Book.*

Worldview Copy this question in your notebook and write your answer: Who is a prominent person (not in the ministry or a missionary) whom you respect who is open about his or her faith? Why do you respect this person?

Project Choose your project for this unit and start working on it. Plan to finish it by the end of this unit.

Literature Continue reading *Ann Judson: A Missionary Life for Burma.* Plan to finish it by the end of this unit.

Student Review Answer the questions for Lesson 101.

Sydney Harbor

102 A City Named Sydney

It's a big world in which we live. Our modern means of communication and travel enable us to know about and travel to places all over the globe, but this was not always true. In the distant past, few knew anything about the lands and people that were far away.

Was there land out there? Europeans speculated about it, but they didn't know for sure.

Who got to the largest landmass in the South Pacific first? We believe that the people we call the Aboriginal Australians came earliest. But when? From where? Over how long a time, and in how many waves? Since they did not keep written records, it's hard for us today to know.

Did Chinese explorers learn about that part of the world, or did Muslim traders from the Middle East or Indonesia? We don't know. Portuguese explorers heightened the interest of Europeans by repeating speculation of a "*terra australis incognita*"—an unknown southern land. A Spanish explorer sailing from Peru discovered the Solomon Islands in 1567, but we don't know what other lands he might have seen.

The Dutch were the first Westerners who had significant contact with that unknown southern land. In 1605 Willem Jansz sailed from the Dutch East Indies and saw the western side of Cape York Peninsula in the northeastern part of the terra australis incognita. More Dutch exploration took place in succeeding years. Best known of these was Abel Tasman, who in 1642 to 1644 sailed to the island later named for him (Tasmania), as well as New Zealand, and along the coasts of a much larger landmass. The Dutch called this landmass New Holland.

Then the British took an interest in developing trade possibilities in the region. William Dampier, who stopped briefly on the northwestern coast in 1688, published a book about his experience and his speculation regarding the land, then returned for more explorations. His book was critical of the new land, so interest lagged until the mid-1700s. Then, amid much speculation in England, Captain James Cook landed at Botany Bay on the eastern coast and named the land he claimed New South Wales. In 1801 the crown gave another British explorer, Matthew Flinders, the job of sailing around and charting the continent. In 1804 Flinders published his report that showed New Holland was one landmass and suggested that a new name be applied, one that stuck: Australia. Great Britain claimed all of Australia by 1827, including Tasmania and other small islands.

Transportation

Great Britain had a policy of transporting certain criminals to its colonies to serve out their punishment, a policy euphemistically called transportation. Britain sent criminals to its American colonies in the 1600s. When Britain lost those colonies in the Revolutionary War, it found another destination in Australia. The policy of transportation was attractive for many reasons: the destination was geographically distant from Great Britain, the policy relieved pressure on overcrowded prisons in Britain, and the practice provided colonists who could build up the British presence overseas. Transporting criminals was also defended as more merciful than capital punishment.

In 1787 Arthur Phillip led an expedition from Britain to New Holland that included about 730 prisoners. In January 1788 they landed in Botany Bay, but that area proved to be a poor location for a settlement. Phillip moved north, landed on the southern coast of Port Jackson harbor, and named the colony for Thomas Townshend, the British home secretary (similar to the American secretary of state), whose title of nobility was Lord Sydney. Thus, the city of Sydney was born.

William Bradley was a British naval officer and artist who made this drawing of "Taking of Colbee and Bennelong 25 November 1789."

Billy Blue arrived in Australia in 1801 to finish his seven-year sentence for stealing sugar. Upon his release, he married Elizabeth Williams, another former convict, and they had six children. Blue eventually operated a fleet of ferries across Sydney Harbor.

The Gadigal nation of aborigines was living in the location that the British settlers wanted, so there was a degree of conflict between the people living there and the incoming settlers. An outbreak of smallpox decimated the Gadigal people. The Gadigal were a clan of the Eora people. When the British asked the Eora where they came from, they replied, "Eora," meaning "here." Thus the British called them Eora. Descendants of the Eora still live in Sydney today.

The British captured two local men, Colbee and Bennelong, to attempt to learn their language and culture. Colbee soon escaped, while Bennelong spent more time with the British and eventually visited England. Because of his assistance, he asked the British to build a house for him on a peninsula that extended into Sydney Harbor.

Before the British government ended the policy of transportation in 1868, over 160,000 prisoners were taken to Australia. Among this number were some hardened criminals who continued their nefarious ways, but many of the prisoners had been convicted of minor crimes. A large number of those

transported took the opportunity to embark upon a new life. They became farmers or other productive citizens and made positive contributions to society. Many other people besides prisoners moved to Australia as well. According to research, about 20% of Australians today believe themselves to be descendants of transported convicts.

The Small, Amazing Continent

Australia is the world's smallest continent. It is about the size of the lower 48 United States (thus Matthew Flinders' job of sailing around it in the early 1800s was quite an endeavor). However, it is the world's sixth largest country.

Australia is the largest country in Oceania, the largest country that lies entirely within the Southern Hemisphere, and the largest country that does not share a land border with another country. It is the only continent that does not have any glaciers. The Eastern Highlands or Great Dividing Range of mountains in eastern Australia is the longest mountain range on the continent and the third-longest range on the earth's surface. The range is "dividing" in that all the rivers of eastern Australia flow from its crest. The range also divides the habitable areas along the eastern coast from the central Outback region of the continent. Australia is the world's largest net exporter of coal, producing about 29% of global coal exports.

The population of Australia is about 26 million people, less than one-tenth that of the United States. Most of the people live along the east coast and in the city of Perth on the west coast. Sydney, with its population of over five million, is home to over 20% of the country's people. Melbourne, on the southern coast, has almost five million. About half the country's population identify as Christian, and some 30% list their religious preference as "none."

Although central Australia is mostly desert, the continent does have geographic diversity and places of striking beauty. Queensland and New South Wales support a thriving wool industry, and significant mineral wealth lies beneath some of the desert areas. Australia has numerous plants and animals that are indigenous to it.

Sydney Harbor Bridge

Sydney Harbor and Bridge

A natural harbor is a body of water, surrounded by and protected by land, which is large enough for ships to enter and drop anchor. An artificial harbor is one that people have constructed. Harbor is often used interchangeably with port, but a port is technically a manmade facility for loading and unloading cargo and people.

Sydney Harbor (the British spelling is Harbour) is one of the largest natural harbors in the world. It is eleven miles long, covers about 21 square miles, has a perimeter of some 200 miles, and contains several islands. It is probably the deepest natural harbor in the world. The depth in some places exceeds one hundred feet, although it is quite shallow in other spots. Over five hundred species of fish have been found in Sydney Harbor. The water is surrounded by national parks and historic sites and provides opportunities for many water sports.

Sydney Harbor is an example of a ria, which is a coastal inlet formed by a drowned river valley that is open to the sea. A ria coast is a coastline that has several parallel inlets or rias, separated by ridges.

Proposals to build a bridge across Sydney Harbor emerged as early as 1815. Almost a century later, a civil engineer presented a plan for a bridge to the Australian parliament. World War I interrupted planning, and it was not until 1922 that funding became available.

Construction on the steel arch bridge began the next year and proceeded from the opposite shores. The two ends met in the middle (no small feat in itself!) in 1930, and the bridge was completed in 1932. A crowd of hundreds of thousands of people turned out for the event. Now Sydney on the south shore had a direct connection to the suburbs that had developed on the north shore.

The Sydney Harbor Bridge is 1,650 feet long, 160 feet wide, and 440 feet tall. It is the fifth longest steel arch bridge in the world. The bridge carries

Sydney Opera House During Vivid Sydney Festival (2017)

eight lanes of auto traffic, two rail lines, a pedestrian walkway, and a bike path. Walkways up the bridge's structure allow the daring to mount to the top of the bridge to enjoy breathtaking views or to take part in wedding ceremonies (fees apply for any activity at the top of the bridge). Like the Statue of Liberty in New York, the Gateway Arch in St. Louis, and the Golden Gate Bridge in San Francisco, the Sydney Harbor Bridge has become an iconic symbol of the city. Never ones to take themselves too seriously, Sydney residents have nicknamed the bridge "The Coathanger."

An even more widely known landmark of the city is the Sydney Opera House. People around the world recognize its white, sail-shaped shells that are part of its roof line. It stands on Bennelong Point, the peninsula where the Aboriginal Australian Bennelong lived when he worked to maintain good relations between his people and British settlers. After many years of discussion and planning, in 1957 the governing board chose a design submitted by Danish architect Jorn Utzon. The project cost more and took longer than original estimates. Queen Elizabeth II dedicated and opened the facility in 1973. It houses multiple performance and meeting halls.

Changing artistic and light displays have been projected onto the sails of the Sydney Opera House. In 2017 a regular display began featuring music and images celebrating Aboriginal Australian Dreamtime stories. Huge areas of coastal Australia were devastated by wildfires in late 2019 that continued into 2020. In January of 2020, to honor the firefighters, many of whom are volunteers, pictures of the firefighters were projected onto the sails of the Sydney Opera House.

People from many cultural backgrounds have interacted with the geography of Australia and Sydney Harbor. The Aboriginal Australians lived in many places across the continent, even in the rugged Outback. The British found a place (far away from Great Britain) for prisoners to get a new start in life. Modern Australians have created memorable structures that utilize and emphasize the geographic setting in which they are built. In this they have embodied the teaching of Ecclesiastes:

Whatever your hand finds to do, do it with all your might;
for there is no activity or planning or knowledge or wisdom
in Sheol where you are going.
Ecclesiastes 9:10

Assignments for Lesson 102

Gazetteer	Read the entry for Australia (page 180).
Worldview	Copy this question in your notebook and write your answer: How can and should science rightfully give glory to God?
Project	Continue working on your project.
Literature	Continue reading *Ann Judson: A Missionary Life for Burma.*
Student Review	Answer the questions for Lesson 102.

Camels in Queensland, Australia

103 Wonders of the Australian Outback

Australia is a geographic and natural wonder. There are some eleven million plant and animal species in the world. Of that number, 570,000 are native to Australia.

Over 80% of the continent's mammals, reptiles, frogs, and plants live only there. After all, what would Australia be without its kangaroos and koalas?

About 70% of the continent's insects live only in Australia.

The world's largest remaining intact tropical savannah is in Australia. It also holds the largest remaining temperate zone woodland habitat.

Only a few large natural areas remain in the world: the Amazon in Brazil; the boreal forests of Canada, Alaska, and Siberia; and the Sahara, for example. Another is the Outback of Australia.

The Australian Outback

The Outback is the vast heartland of Australia. The region covers about three-fourths of the continent, an area almost twice the size of India. The Outback covers all of the Northern Territory; most of Western Australia, South Australia, and Queensland; and even the northwestern corner of the relatively populous New South Wales.

The Outback is amazingly diverse, with rain forests, grasslands, large river basins, deserts, and salt pans. As dry as much of it is, eight wetlands have been designated as having international significance.

This central region of Australia is home to less than a million people, which is less than five percent of the country's population (imagine that number of people spread over an area almost the size of the continental U.S.). The Outback contains about 1,200 communities, but half of those communities have a population of less than one hundred. The Aboriginal people make up about one-fourth of the population of the Outback. People visit there to find adventure and to enjoy the beauty, but relatively few people live there.

Members of the Karjanarna Jaru people perform a traditional dance in Western Australia.

But then there are the camels.

Camels?

Europeans began wanting to explore Australia in the late 1700s. The Outback, however, limited exploration and development because of its widespread harsh conditions. Then someone thought of camels. Between 1870 and 1920, about 20,000 camels were imported to Australia from the Arabian Peninsula, India, and Afghanistan. About two thousand handlers or cameleers came as well. The camels proved to be a great idea. The use of camels to carry goods and passengers helped open supply lines, connected isolated settlements with each other, and enabled economic development.

By the 1930s, however, people had successfully introduced motorized forms of transport such as railroads, trucks, and automobiles into Australia. As a result, the camels were no longer needed or wanted. So what did Australians do with them? Why, they turned the camels loose into the wild, of course. Now there are about one million camels in the wilds of Australia, and the number continues to grow. A small number of entrepreneurs maintain herds of camels for dairy products and to produce such delicacies as camel milk chocolate.

Uluru and Kata-Tjuta

Rising from the plains of the Outback in the southwest corner of the Northern Territory—pretty much in the center of the Outback—is the geographic feature that the Aborigines call Uluru. This large, natural, roughly oval-shaped sandstone monolith is over 1,100 feet high and is almost six miles in circumference at its base.

About 25 miles to the west of Uluru is another outcropping, a collection of soaring rock domes called Kata Tjuta. The way that different groups look at these geographic features provides a good example of differences in worldview.

To the geographer, these formations are bornhardts, defined as bald, steep-sided, rock domes at least one hundred feet high. These formations get their name from Wilhelm Bornhardt, a German geologist and explorer of German East Africa. A bornhardt is a kind of inselberg, a rock formation that rises abruptly from the surrounding plain. Inselberg is a German word that means island mountain.

To the person who believes in God, these formations testify to the wonder and power of the

Uluru

Kata Tjuta

Creator. To the typical Aboriginal person, they are sacred, spiritual places that tell of the very earliest days of the world.

The first non-Aboriginal person known to have seen Kata Tjuta was Ernest Giles in 1872. He named the highest one Mount Olga after Queen Olga of Wurttemberg. The group came to be called the Olgas. The first European to see Uluru was William Gosse in 1873. He named it Ayers Rock after Sir Henry Ayers, the Chief Secretary of South Australia.

In 1894 a government-sponsored expedition into the area looked into possible mineral resources, studied the plants and animals living there, and researched Aboriginal culture. The main group of Aboriginals who lived near Uluru and Kata Tjuta were the Anangu. The expedition confirmed that farming was not possible in the area. Years later, in 1920, the government placed the two rock formations in the South West Reserve, one of a series of reservations created for Aboriginal people.

The result of this move was that few non-Aboriginal people came to the area. However, things changed in the 1940s. At that time the government reduced the size of Aboriginal reserves to enable mineral exploration. A dirt road was cut to Uluru in 1948. Miners and tourists began to come.

Ayers Rock National Park was created in 1950. Kata Tjuta was added in 1958. Tourism increased, and at the request of tourism business leaders the government encouraged the Anangu to move away from the site. In the late 1960s some Anangu and other Aboriginals began petitioning the government for the right to own their traditional homeland. At the end of a long process involving a new law and extended negotiations, in 1985 the Governor-General of Australia officially returned the park deeds to the Anangu people, who then immediately leased the land to the Australian National Parks and Wildlife Service (now called Parks Australia) for 99 years. A management board was created, the majority of which are Anangu.

In 2000, the Australian phase of the Olympic torch run heading to Sydney for the summer Olympic Games began with the runner making a lap around Uluru. The walkway around Uluru is a popular tourist activity, although the rock itself is closed to climbing out of deference to the Anangu, because of their view of Uluru as a sacred place.

Coober Pedy

In 1915 a teenager found an opal in the Outback of central South Australia. This discovery began a mining industry that now produces 70% of the world's opals. Individual miners who buy their own permits to dig, not large mining corporations, do most of the work.

But this is not the most remarkable thing about the town that grew up around the opal mining. The heat there is brutal, reaching 113 degrees in the summer. So the miners who came in the 1920s, many of whom had dug trenches in World War I, began digging caverns in which to live. Today about half of the town's 2,000 residents live underground. Life there is quiet, still, and a constant 75 degrees.

At least one underground resident has a swimming pool. The local hotel is partly underground, as is the Serbian Orthodox Church. Tourists can tent camp in an abandoned mine. Above ground, most people who play golf on the local course do so at night, using glow-in-the-dark golf balls. Course maintenance is fairly easy because there is no grass; golfers tee off from carpeted areas.

And by the way, the name of the town is Coober Pedy, a variation of Aboriginal words that mean "white man's hole."

At right is the underground interior of the Serbian Orthodox Church in Coober Pedy. Below is one of the "greens" on the golf course.

The ingenuity and determination of the people of Coober Pedy are remarkable. God can do even more amazing things, even in the desert and even in seemingly impossible situations. As the prophet Isaiah said in these words of the Lord:

Behold, I will do something new,
Now it will spring forth;
Will you not be aware of it?
I will even make a roadway in the wilderness,
Rivers in the desert.
Isaiah 43:19

Assignments for Lesson 103

Worldview Copy this question in your notebook and write your answer: What has been the influence of the theory of evolution on your generation?

Project Continue working on your project.

Literature Continue reading *Ann Judson: A Missionary Life for Burma.*

Student Review Answer the questions for Lesson 103.

Great Barrier Reef

104 In the Waters Near Australia

The great explorer Sir James Cook was marooned.

On June 11, 1770, during Cook's travels around Australia, Cook's ship H.M.S. *Endeavour* struck a coral reef. The crew made it to land, but the ship underwent six weeks of repair at a place that came to be called Cooktown. When the *Endeavour* was once again seaworthy, Cook could not find a way through the reef. He sailed north and landed on Lizard Island. Cook and the ship's botanist climbed to the highest point on the island and identified a space in the reef through which the *Endeavour* could sail. This is now known as Cook's Passage.

Cook had (pardon the expression) run upon one of the most significant geographic features of the earth, the Great Barrier Reef. Talk about human interaction with geography!

Coral Reefs

The coral is a tiny sea animal that forms an exoskeleton. Corals come together to form large ridges called reefs in the ocean. These reefs usually form in relatively shallow water, up to 130 feet in depth. Reef-building corals operate best in water between 72 and 82 degrees, so the South Pacific Ocean is a natural home for them. Some corals can live in colder conditions.

A coral reef is actually home to a variety of organisms. As the sea water washes over it, sand and mud collect on it. These become the habitat for seagrass and mangroves along with the algae that lives there. Storms and surf can build the natural structure into shoals and beaches, such that parts of the reef can become islands.

A reef can take many forms. One of these is an atoll, which is a coral reef surrounding a lagoon. Some scientists theorize that a volcano in the ocean might develop a reef around it, then the cone of the volcano collapses, creating a lagoon surrounded by the reef.

The Great Barrier Reef

The Great Barrier Reef lies on the continental shelf just off the northeastern coast of Australia. The distance from shore varies between ten and 100 miles. It consists of almost three thousand individual reefs and almost one thousand islands. The reef system stretches for over 1,400 miles and covers an area slightly smaller than Japan or California. The beautiful Reef is the world's longest and largest reef system.

Green Turtle and Coral,
Great Barrier Reef

Goose Barnacles on an Island in the Torres Strait

We understand that Aboriginal people were familiar with the Reef and fished the waters around it from canoes fitted with outriggers. The earliest documented sighting of the Great Barrier Reef by a European occurred on June 6, 1768, when the French explorer Louis-Antoine de Bougainville saw what became known as Bougainville Reef near the location of Cooktown.

Two years after Cook's encounter, the British captain William Bligh spent two weeks charting passages through Torres Strait, which lies between Australia and the island of New Guinea. This is the same Captain Bligh who commanded the H.M.S. *Bounty* in 1789 and against whom Fletcher Christian and eight others mutinied.

Matthew Flinders, who mapped the entire Australian coastline in the early 1800s (see Lesson 102), walked on the reefs. He discovered a safe passage through the reefs which today bears his name, Flinders Passage. In the 1800s and early 1900s, Chinese sea cucumber fishermen and Japanese pearl divers were active in the area of the Great Barrier Reef.

The Great Barrier Reef affords striking beauty above and below the surface of the water. It is one of the most complex ecosystems in the world. The Reef contains about 300 species of coral as well as lobsters, crabs, crayfish, worms, sponges, anemones, and many varieties of fish and birds. One can see sea turtles, over thirty species of whales and dolphins, and migrating butterflies there. The greatest natural threat is the crown-of-thorns starfish, which feeds on the living coral. The Reef is a popular tourist attraction, but this poses a threat to its preservation. Exploratory petroleum drilling has been limited around the Reef.

A recently noted phenomenon is called coral bleaching. Scientists believe that this occurs when the coral experience an environmental stress and expel the algae with which the coral has a symbiotic relationship. This removes the typical color from the coral and leaves it white. Coral can recover from bleaching.

People have listed what they think are the seven manmade wonders of the ancient world, the seven manmade wonders of the modern world, and the seven natural (we might say God-made) wonders of the world. The seven natural wonders are Mount Everest, the Grand Canyon, the Paricutin volcano, the Northern Lights, Victoria Falls, Rio de Janeiro Harbor, and the Great Barrier Reef. Impressive company, indeed. The Reef is visible from space. What is even more remarkable is that the corals in the reef are alive. Some have called the Great Barrier Reef the largest structure ever built by living creatures.

Islands of the Island Continent

Australia owns several of what it calls External Territories, which are islands in the Pacific, Indian, and Southern Oceans and the Coral and Timor Seas. Here are descriptions of some of these islands.

Fraser Island lies just off the southern coast of Queensland. It is the world's largest sand island, but on it are freshwater lakes, wetlands, and rainforests as well as sand dunes. Fraser Island is noted for its colored sands, which are the result of decaying vegetation leaching into the sand.

Fraser Island

Penguins and Seals on Macquarie Island

Macquarie Island is a distant territory in the Southern Ocean, closer to Antarctica than to Tasmania (which is itself an island). Macquarie Island is the crest of the undersea Macquarie Ridge. It is the only place on earth where rocks from the earth's mantle are exposed above sea level.

The uninhabited Ashmore and Cartier Islands are part of the Ashmore Reef that lies off the western coast of Australia. The waters near the West, Middle, and East Ashmore Islands have seen the work of Indonesian fishermen and American whaling men. Phosphate mining took place on West Island at one time. Nearby Cartier Island is an unvegetated sand cay that lies at the center of a surrounding reef. The waters around the island, which are part of the Territory of Ashmore and Cartier Islands, are home to 16% of all the fish species found in Australia.

Norfolk Island lies about 900 miles off the eastern coast of Australia. It was a penal colony for periods of time in the 1800s. The island has excellent farmland and was known as "Sydney's Food Bowl" for a number of years. In 1856 the island was given to 196 descendants of the mutineers from the H.M.S. *Bounty* and their Tahitian wives, who had been living on Pitcairn Island. At first they lived in what had been convict buildings but soon they received 50-acre land grants to build homes and farms. The population is now about 1,800, about half of whom are descendants of the Pitcairn Islanders. An airfield was built on the island during World War II to aid the Allied cause. After the war, the airfield encouraged growth in tourism on the island.

Lesson 104 - In the Waters Near Australia

One fun result of the human geography of tiny Norfolk Island, where many people have the same few last names, is a section in the telephone directory where customers are listed by their nicknames rather than their last names.

Mariners knew of the existence of a rocky island near Indonesia in the early 1600s, but it was only named Christmas Island in 1643 when the English captain William Myers of the British East India Company saw it on December 25. Britain's first landing took place in 1688, but the island generated little interest until phosphate was discovered there. Britain annexed it in 1888. Japanese forces seized the island during World War II but abandoned it at its surrender to the Allies. Britain placed the island under its colony of Singapore until 1958, when it was transferred to Australia. The island is 220 miles from Indonesia and 960 miles northwest of Australia. About two-thirds of the island is a national park.

The main annual event on Christmas Island is the red crab migration sometime between October and December, when millions of the crabs leave the forest and head for the coast to breed. They always start the trek before dawn after the first rainfall of the wet season during the last quarter moon as the tide is receding.

Norfolk Island

God has placed us in a fascinating, varied, colorful world with no end to the amazing things that demonstrate His power.

Red Crabs on Christmas Island

Behold, the nations are like a drop from a bucket,
And are regarded as a speck of dust on the scales;
Behold, He lifts up the islands like fine dust.
Isaiah 40:15

Assignments for Lesson 104

Project Continue working on your project.

Literature Continue reading *Ann Judson: A Missionary Life for Burma.*

Student Review Answer the questions for Lesson 104.

Prickly Pear, Queensland

105 John Mann, Scientist

When God created the world, He planned for living organisms to be in balance in given places. For instance, predators need certain growing things to live; at the same time, the predators keep those growing things from taking over an area. Growing things and their predators have to occupy the same geographic place to maintain the balance. Weather and other environmental conditions in a place can also prevent growing things from taking over.

No species of cactus is native to Australia; thus no predator that feeds on cactus would last very long there, unless something happened to change the balance. In 1788, something happened.

At that time, Captain Arthur Phillip of the English West India Company brought the prickly pear cactus from South America to Queensland, a state in northeast Australia. The cactus he brought was infested with a predator, the Cochineal beetle. Phillip wanted to start a dye-making industry there. The Cochineal beetle produces a maroon pigment in its body fluid and tissues that workers can gather to produce dye.

Unfortunately, the beetles died out fairly quickly. So much for the dye-making industry. However, this meant that the prickly pear cactus grew without any predators. And grow it did. By 1914, the prickly pear cactus had taken over about sixty million acres of land so densely that no farm crops could grow on it. This is about the size of the American state of Georgia.

Prickly Pear Forest, Queensland (c. 1930)

The land that the prickly pear cactus took over was land that farmers had used for grazing dairy cattle and growing grain. Families lost their farms because they could not support themselves. The government offered to give land to people if they could just maintain a six-foot-wide strip of land free of prickly pear as a border, but people weren't even able to do that. Nothing that farmers and government land managers did, whether with chemicals or machinery, stopped the cactus from growing and spreading. Neither weather conditions nor any other environmental factors stopped the cactus either.

The prickly pear infestation was a serious crisis in Australian society and in the Australian economy. The Queensland Prickly Pear Land Commission stated:

> It will never be known—not even remotely—what the pear pest has cost the State. Revenue, homes, and even lives, of all it has taken its toll. In place of well-kept farms, of prosperous homes, and of contented people, one sees all around the desolating blight of pear.

The Australian government sent researchers all over the world to try to find creatures that could control the cactus population. In 1926 scientists who were sent to South America brought back the *Cactoblastis cactorum* moth; but for some reason, which scientists could not determine at first, they did not breed well in Australia.

One of the team of scientists working on the project, John Mann, put light bulbs inside the insect cages to make the moths' environment warmer. He found that, when the moths were kept warm enough, they bred easily. Some three hundred million eggs were distributed throughout the affected area. The larvae from the eggs fed on prickly pear cactus and—thank the Lord—did not eat anything else.

By 1930 the prickly pear cactus crisis was solved. The land returned to productive use, and prickly pear was finally under control. Mann received many honors and wide recognition for his work. He became entomologist for the Government Department of Lands, a fellow of the Royal Zoological Society, a fellow of the Entomological Society of New South Wales, and served on many government committees. The Smithsonian Institution published his research. He made presentations on biological control to science conferences, grazing and growers organizations, and on the Australian Broadcasting Corporation.

The photo on the left below shows an abandoned property near Chinchilla, Queensland, in May 1928. Prickly pears had overtaken it. The photo at right below shows the same property in October 1929 after the introduction of the cactoblastis moth.

These Cactoblastis cactorum *larvae are eating prickly pear fruit.*

In 1969 Queen Elizabeth II made Mann a Member of the Order of the British Empire (M.B.E.). The M.B.E. is an honor awarded to people who make significant contributions to British society in the arts and sciences, in public service, and through charitable organizations.

John Mann was an accomplished scientist. He was also a Christian who had an unwavering faith in God. His beliefs guided his worldview.

Mann was born in 1905. In a 1982 interview in *Creation* magazine, Mann recalled that in the 1930s evolution was not as widely held a belief as it became in later years. He remembered reading an article in 1923 that proposed a theory of "the early humanoid in America" on the basis of a single molar tooth. The tooth had been worn down by water, and another scientist thought it might be the tooth of a bear. Mann remembered thinking that evolutionary theory appeared to be built on 99% imagination and 1% fossils. He decided that he would believe in the Bible until someone came up with definite proof that man had evolved from animals. No one ever did. Mann believed that the relationship between the cactus and the *Cactoblastis cactorum* moth was an example of a system created by God.

This May 16, 1929 issue of The Queenslander *magazine included an article on "Destruction of Prickly Pear". The cover shows a man who landed on the cactus after being thrown from his horse.*

Full index page 23

The Queenslander

ILLUSTRATED WEEKLY

6d

May 16, 1929.

"DESTRUCTION OF PRICKLY PEAR."—See Article on Page 62.

This monument was erected in Dalby, Australia, in 1965. After describing the invasion of prickly pear, it records "the indebtedness of the people of Queensland, and Dalby in particular, to the Cactoblastis Cactorum, and their gratitude for deliverance from that scourge."

He recalled having an astronomy instructor in college who described the movements of celestial bodies and then told his class, "Never let anyone tell you that these things happen of their own accord. There is a Supreme Being who guides these things in the way they go." That statement influenced Mann for the rest of his life. He often made his beliefs clear when he spoke to a group, regardless of the nature of the group.

Mann believed that the increasing acceptance of evolution was, in his words, "devastating! Young people in schools are being taught it as a straight-out fact. It is not a fact. It is a tragedy. People who are brought up as evolutionists cannot be brought up to believe in God."

John Mann passed away in 1994. He understood that even as a scientist—perhaps especially as a scientist—he could see a divinely-created plan that guided the universe. Mann was not alone in this belief. Accomplished and recognized scientists hold to the belief in God as Creator as the Bible teaches. With his Christian worldview, John Mann could see the balanced system that God had created in the world in which he lived, a system which helped his country in practical ways.

For since the creation of the world His invisible attributes,
His eternal power and divine nature, have been clearly seen,
being understood through what has been made,
so that they are without excuse.
Romans 1:20

Assignments for Lesson 105

Worldview	Recite or write the memory verse for this unit.
Project	Finish your project for this unit.
Literature	Finish reading *Ann Judson: A Missionary Life for Burma*. Read the literary analysis and answer the questions about the book in the *Student Review Book*.
Student Review	Answer the questions for Lesson 105. Take the quiz for Unit 21.

Yasawa Islands, Fiji

22 The Pacific Ocean and Its Islands

We begin this survey of Pacific islands by discussing navigation and an invention that revolutionized it. Oceania is usually divided into four subregions: Australasia, Melanesia, Micronesia, and Polynesia. We look first at Melanesia and specifically at the country of Papua New Guinea. We then survey Micronesia, focusing on the island of Guam; and we tell a remarkable story of survival after World War II. Finally, we look at Polynesia. The worldview lesson tells how the Psalms and prophets in the Old Testament used geographic features of the Creation to teach lessons about God.

Memory Verse Memorize Isaiah 55:12 by the end of the unit.

Books Used The Bible
Exploring World Geography Gazetteer

Project (Choose One) The time has come to write your research paper for this course. Read over the material in *Exploring World Geography Part 1*, pages xvi-xix, and begin working on it following the daily schedule suggested there. While you are working on your research paper, skip the weekly project assignments and focus on your paper.

Note that you do not have a literature assignment for this unit or for Unit 23 so that you can devote more time to the paper. Plan to finish your paper by the end of Unit 25.

If your parent decides that you will not write a research paper, choose one of the following projects to complete before the end of this unit.

1) Write a 250 to 300 word essay on one of the topics listed on page xvi of Part 1.
2) Write a poem or a song praising God for His Creation. See Lesson 110.
3) Write about an island in the Pacific you would like to visit and why you would like to go. See Lessons 107-109.

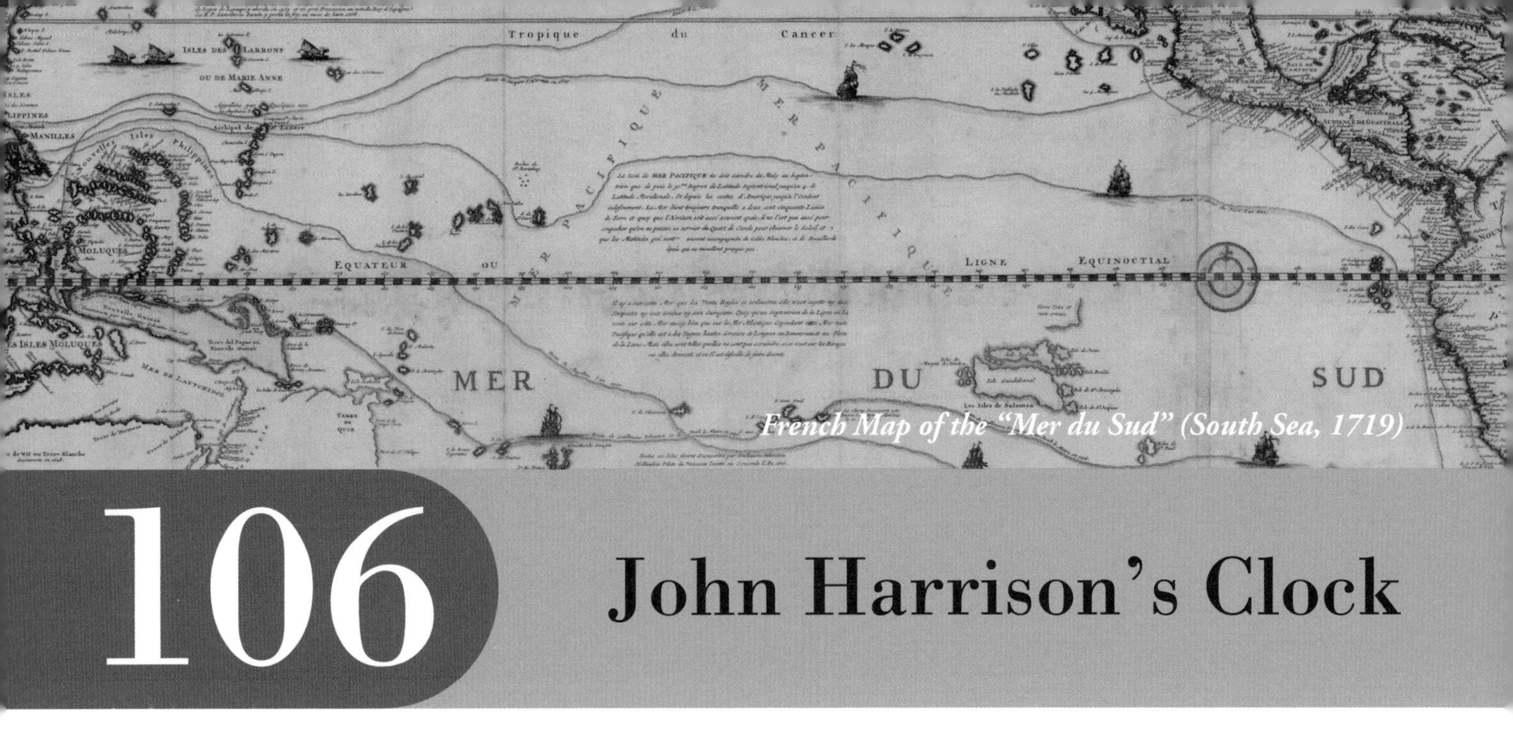

French Map of the "Mer du Sud" (South Sea, 1719)

106 John Harrison's Clock

Commodore George Anson didn't know where his ship was.

Anson left England in the H.M.S. *Centurion* with a squadron of five other ships in September of 1740, headed for the South Pacific. As he rounded Cape Horn on the southern tip of South America, a violent storm struck from the west. The storm raged for 58 days. Some of the ships in the squadron were not seen again, and several other sailors were lost at sea. Meanwhile, the trip was taking so long that the dreaded seamen's disease of scurvy was claiming six to ten victims every day.

Finally the skies cleared and Anson headed for Juan Fernandez Island, where he could find fresh water. After several days' sailing, he saw land. It turned out to be the western coast of South America. Anson had basically been sailing in circles. So he embarked once more. On May 24, 1741, he determined that the ship had reached the latitude of Juan Fernandez Island. He decided to sail west, but after four days he feared that he had once again been going in the wrong direction to reach the island. He turned the ship around, but soon he saw the coast of Chile. He had inadvertently come back to South America. Anson realized that he might have been only hours away from Juan Fernandez Island.

Once more, Anson headed west. The *Centurion* finally landed at Juan Fernandez Island on June 9, 1741. Anson lost over half his crew of five hundred men on this journey, mostly to scurvy.

The Problem

Anson's experience was extreme but far from unique. This was the issue. Sailors could determine their latitude north or south by taking readings of how far celestial bodies were above the horizon. They commonly did this with an astrolabe (starting in the late 1400s) or the more accurate sextant (developed about 1731). But once the ship was out of sight of land, determining a ship's location east or west (the longitude) was difficult. As a result, ships often ran aground on land they encountered sooner than expected, or they sailed seemingly forever hoping to reach their destination. Ships generally sailed near the coast or along well documented routes. This strategy had its own problems. Those shipping lanes could become crowded. And the sailors knew where they were, but so did pirates.

Navigators attempted many methods to determine longitude. One involved trying to anchor a ship at a given stationary point at sea, from which the ship would fire a cannon at given times. Then

Ships continued to carry sextants as a backup aid for navigation into the 20th century. In the 21st century, the U.S. Navy reintroduced training in celestial navigation out of concern about malfunctions of and deliberate attacks on GPS systems.

seamen would attempt to determine their distance from the sound based on the speed of sound. This proved to be impractical.

A more common way was dead reckoning, which involved throwing a log overboard and measuring how fast the ship moved away from the log. But this was notoriously inexact. Sailors and scientists knew that the inability to determine longitude accurately was the biggest navigational, mathematic, and geographic problem of the day.

Astronomers believed that they could devise a method to determine longitude using sightings of celestial bodies, but no attempt at doing so was successful. Rulers of Europe offered prizes to the person who could invent a method to determine longitude accurately. In 1714 the British Parliament offered the huge prize of 20,000 British pounds to the person who could do the job. The law creating the prize also created the Board of Longitude to judge any proposed solutions that people submitted. However, the board did not meet for several years because the ideas submitted were not worthy of consideration.

The key element in determining longitude was time. A navigator needed to know the time on his ship and the time at his home port (or another point of known longitude) at the same moment. The earth makes a complete revolution of 360 degrees in 24 hours, so one hour equals fifteen degrees of longitude. Each hour of time difference between the ship and its starting point meant fifteen degrees of longitude east or west. The navigator could set his onboard clock to noon when the sun was directly over him. When he knew the time at home at that same moment, he could determine the longitudinal difference and thus the distance they had sailed.

The challenge was developing a clock that would work on a ship. A pendulum-driven clock did not work on a rolling, unsteady ship. Variations in temperature and barometric pressure also affected an onboard clock's operation.

John Harrison

John Harrison was born in 1693 in Yorkshire, England. He became a skilled and respected carpenter. A quiet, thoughtful person, Harrison taught himself science, focusing especially on mechanics. He never apprenticed with a clockmaker, but in 1713 he completed his first pendulum clock. Almost all the parts were wood, as one might expect from an expert carpenter. The clock still exists, over three hundred years later. Harrison became known as the maker of exceptionally accurate clocks.

Harrison eventually turned his attention to the longitude problem. After working on a solution for four years, he presented his idea to individual members of the Board of Longitude in 1730. Receiving their encouragement, he then worked on the clock for five years. The completed clock weighed 75 pounds and sat in a cabinet four feet cubed. It was a complicated device of balances and springs that had to be wound daily. The clock successfully passed a trial run at sea in 1736. The next year the board met for the first time ever and gave its approval. The clock, called H1, still exists and still runs. But Harrison himself was not satisfied and wanted to pursue an improved version.

He delivered his second clock in 1741, but again pleaded for more time to improve the design. Harrison spent almost nineteen years developing his third clock, which had 753 separate parts. H2 and H3 also still exist and still run. His fourth device, delivered in 1760, was much smaller and more like a watch. On its trial run to Jamaica, H4 lost only five seconds in 81 days.

However, the Board of Longitude was dominated by astronomers, who were prejudiced toward their belief in an astronomical solution. They were skeptical of Harrison's clocks. The board kept requiring further trials for Harrison's devices and raising questions about their reliability, although the board did award Harrison small grants so that he could continue his work.

Meanwhile, some astronomers developed a method based on movements of the moon for calculating longitude. This approach required multiple sightings, detailed tables of lunar positions to be worked out years in advance, and logarithmic tables. Calculating a position at sea took four hours. In addition, the method did not work for several days during each month when the moon was not visible. Nevertheless, some members of the board were favorably disposed to this approach because of their background in astronomy.

Finally John Harrison began to receive justice. After an appeal to King George III, in 1773 Parliament (not the Board of Longitude) awarded him an amount of money that brought the total Harrison received over the years close to the original 20,000 pounds. A particularly obstinate member of the Board of Longitude insisted that the original H4 remain in England, so Sir James Cook used a device that was practically a replica of the H4 on a voyage and gave it high praise. H4 still exists, but because it was not handled and stored carefully for several years, it does not work.

John Harrison died in March of 1776. The lunar movements approach was popular for a time because it was cheaper, but as clockmakers developed Harrison's original ideas, their chronometers (the word that developed for clocks to measure time at sea) became more affordable. Chronometers became standard maritime equipment. Some observers believe that chronometers were a main reason why British ships came to rule the oceans of the world, and that this resulting nautical strength led to the development of the British Empire.

John Harrison's Final Chronometer: H5

The Geography of the Oceans

Our interest in the geography of the land areas of the earth is understandable. We live there. However, 71% of the earth's surface is water; and of that amount 96.5% is in the oceans. The oceans play a vital role in life on this planet.

Plants in the oceans produce over half of the oxygen we breathe. The seas hold fifty times more carbon dioxide than we find in the atmosphere. They are part of the water cycle that brings rain. Their currents and temperature levels moderate temperatures on land and contribute to other weather phenomena such as winds and storms. The oceans are an important source of food. The oceans are part of the "Goldilocks" nature of the planet: just right (see Lesson 1).

One specific way that humans interact with the geography of the oceans is by navigating over them to other places. We don't know when or how or why someone first made a boat of some kind and went somewhere by water, but ever since then sea travel has been a major part of human life. People have traveled by sea to discover, to explore, to emigrate, to trade, to fish and to hunt for whales, to conquer other peoples, and to have fun. The jobs that many people hold are directly related to the oceans.

Until recent times sea travel was in some ways easier than traveling overland (no climbing mountains or enduring deserts), but it was also potentially more dangerous. The surface on which seagoers traveled was not flat; it was constantly moving and undulating, sometimes casting up huge waves that threatened to swamp the vessels in which people were traveling. Until about 150 years ago, the idea of sea travel as recreation was unheard of. A ship being lost at sea was commonplace; today it is rare.

Getting There

People have traveled across the sea in tiny canoes, in sailing vessels, and in huge ocean liners. They have done it sometimes terrified, sometimes mystified, and sometimes confident. For thousands of years, they did it without the advanced technological devices and GPS systems that we have today. One big question that hovers above the story of sea navigation is: how? How did they do it? How did they know how to get where they were going?

A history of travel by water would include many topics. We could discuss the invention of different kinds of crafts, from hollowed-out canoes, bark canoes, and skin boats, to rafts, larger ships propelled by rowers or sails, or even modern nuclear-powered ships. We could describe the knowledge that seafarers needed, such as information about winds, tides, ice, currents, and large sea animals. We could survey the significance of naval activity, both military and commercial, for many civilizations of the world. For instance, when the Roman Empire encircled the Mediterranean, many Romans referred to that body of water by the Latin phrase Mare Nostrum—"our sea" (now there's a worldview for you!). We could discuss the knowledge of sea travel reflected in Homer's poetry from the 700s BC or the founding of sea-based empires by the Phoenicians and Greeks. For this lesson, however, we will focus on some aspects of the early development of navigation in the Pacific.

The RMS Tahiti *left Sydney, Australia, in 1930, heading for San Francisco, California. A propeller shaft broke and smashed the hull, leaving the ship disabled and sinking. Distress calls reached other ships, who were able to arrive two days later. All the passengers and crew were rescued safely.*

Navigation in the Pacific

We must confess at the outset that there is much we don't know. The Pacific Ocean stretches ten thousand miles from Ecuador to the Philippines. It contains some 21,000 islands and atolls (coral reefs), which have an average size of 23 square miles and most are much smaller. The landmass of the Pacific region is less than 1% of the surface area. Yet the great majority of these islands are inhabited, and those inhabitants speak an amazing variety of languages. How did that happen? The people who populated these islands apparently had a purpose; this kind of development does not happen by chance.

We don't know for sure how people first got to the Americas, although we are pretty sure that they came from Asia. We also don't know how they spread over the American continents and Caribbean islands and developed such widely different civilizations. We don't know if they were adventurers or if they had to leave their homes for some reason, such as war, disease, or volcanoes. We don't know why, if they came by sea, they didn't continue to explore to the east and "discover" Europe and Africa.

The oral traditions of the peoples of the Pacific and what little written information we have suggests that Pacific coastal residents first built rafts by lashing logs together and powered them by rowing. These people groups eventually built boats with sails, more suited to travel on the open ocean. They could add outriggers to canoes or build a deck between two small hulls to create a larger vessel.

Seagoers could travel significant distances by keeping known islands in sight. They apparently learned celestial navigation (determining their course by observing the sun, moon, stars, and planets). They also observed the patterns of wind and water currents and the migration and feeding patterns of fish, birds, and whales; specifically, they noted that the winds and creatures would go in one direction at one part of the year and return in another part of the year. They noticed how waves would look different as they approached land. In many places of the world, seamen learned not to travel at certain times of the year when weather was an issue.

Kaluniōpu`u, King of Owyhee Bringing Presents to Captain Cook by John Webber *(British, 1784). Webber was an artist who traveled with Captain James Cook on his final voyage to the Pacific Ocean, including this visit to Hawaii.*

The Chinese were apparently the first to develop a compass, around the time of Christ, although they did not use it in sailing for a thousand years. Europeans began using compasses a short time after that. A major development was the invention of the means of maneuvering, such as the rudder. Sailors used a rudder oar for some time. The first mounted rudder apparently developed in China in the first century AD, but it did not come into use in Europe until a thousand years later.

Navigation of the earth's waters is an important part of human interaction with geography. The peoples of the Pacific coastal areas, including the islands, contributed much to the progress, accomplishments, art and science, and mysteries of sea travel.

The psalmist praised the One who calms the storms and guides sailors to their destination.

He caused the storm to be still,
So that the waves of the sea were hushed.
Then they were glad because they were quiet,
So He guided them to their desired haven.
Let them give thanks to the Lord for His lovingkindness,
And for His wonders to the sons of men!
Psalm 107:29-31

Assignments for Lesson 106

Gazetteer Study again the map of Oceania (pages 178-179).

Worldview In your notebook, copy the lyrics of a hymn that praises God for the Creation, such as "This Is My Father's World" or "Hallelujah, Praise Jehovah."

Project Begin working on your research paper. Plan to finish it by the end of Unit 25. ***or*** Choose your project for this unit and start working on it. Plan to finish it by the end of this unit.

Student Review Answer the questions for Lesson 106.

Papua New Guinea

107 Papua New Guinea and Melanesia

Have you ever faced an uphill climb in dealing with something? Have you ever had to listen to someone go around the world to explain themselves? Can you ever let something in the past just be water under the bridge?

Geography influences our worldview (which we will discuss further in Lesson 125). Geography also influences the vocabulary we use. The world in which we live shapes the words we use because that world influences how we think. For instance, we have noted elsewhere that Scandinavian people have many words for fine distinctions regarding snow.

Linguistic researchers who have studied the language of the Yupno people of Papua New Guinea have found that they use terms that reflect the geography in which they live. About 8,000 Yupno live in the Finisterre mountain range. The village of Gua, the site of the linguistic study, sits in a valley. The valley has no roads; so the people carry food, water, and other things along sloped paths. Going uphill and downhill are everyday activities.

Some houses in Gua face uphill while others face downhill. Researchers found that when the Yupno speak of walking into a house, they talk of going uphill when describing going further into the house, regardless of the slope and orientation of the house. In another example, items close to the speaker are "up," while things farther away are "down." The Yupno even use geographic words when talking about time. They use uphill words when talking about the future and downhill words when discussing the past.

Researchers concluded that the Yupno create a mental geography that reflects the geographic setting in which they live, even when they are inside a building. The researchers wondered how the Yupno might communicate if they ever visit or move to a coastal area or a city, places with geography that is very different from what they have known in their village. The answer to this question is down the road a bit from where they are now—or is it up the road?

The Yupno are an example of the rich diversity of languages and cultures in the region of Melanesia and specifically in Papua New Guinea.

Melanesia, New Guinea, and Papua New Guinea

Geographers have tried to divide Oceania into four subregions based on linguistic and cultural factors: Australasia, Melanesia, Micronesia, and Polynesia. The specific islands grouped into each subregion vary depending on who is making the definition. One island group may sometimes be included in more than one subregion.

Dancers in Fiji

The Andesite Line, part of the Ring of Fire, is a line of volcanic and earthquake activity that separates Melanesia from Polynesia to the east and that separates Melanesia from Micronesia to the north. This division by subregion has a geographic background, but it also reflects differences in the ethnic and cultural backgrounds of the people in these subregions.

Australasia includes Australia and New Zealand and sometimes New Guinea and its nearby islands. We will discuss Micronesia and Polynesia in the next two lessons.

Melanesia includes the countries of Papua New Guinea, the Solomon Islands, Vanuatu (once called the New Hebrides), Fiji, and Caledonia. Melanesia consists of about 2,000 islands which have a total land area of about 385,000 square miles. The population of Melanesia is about eleven million people.

The term Melanesia comes from Greek words meaning black islands. The word black refers to the skin color of most of those who live in this region. The largest island in this region is New Guinea. New Guinea is the second largest island in the world after Greenland, and it is the largest and highest tropical island in the world with its many mountain ranges. Spanish explorer Ynigo Ortiz de Retez named the island New Guinea in 1545 because he perceived similarities between the external characteristics of the people who lived on this island and the people of the West African coastal area of Guinea.

The western half of the island of New Guinea is part of the country of Indonesia. The eastern half of the island is the main part of the independent country of Papua New Guinea (PNG). PNG also includes several smaller islands in the waters nearby. The island of New Guinea has several mountain ranges on it, and one of the world's largest swamps lies along its southern coast.

Various people groups from other areas settled in Oceania at some point in the ancient past. Living on the numerous islands and the rugged topography of islands such as New Guinea have tended to keep these people groups separate. The country of PNG has several thousand separate communities, most of which have only a few hundred people. They are divided by differences in language, custom, tradition, and in some cases longstanding conflict. A few of the tribes have been known to practice cannibalism. About 95% of Papua New Guinea is either Protestant (64%), Catholic (26%), or belong to some other Christian group (5%). Some 80% of the people live in rural or small-town areas.

Melanesia is also one of the most linguistically diverse regions in the world, with hundreds of languages spoken there, many of which have less than one thousand people who speak it. These languages are some of what are known as Austronesian tongues.

Minerals, especially copper, gold, and oil, account for almost two-thirds of PNG's export earnings. The country also has considerable natural gas reserves, and liquified natural gas has become a major export. The mountain terrain of most of the country (and most of the island) has hampered economic development.

Church in Palembe, Papua New Guinea

From Colonial Interests to Independence

European explorers began coming to this region in the 1500s. In 1660, the Dutch proclaimed sovereignty over the entire island of New Guinea, primarily to keep other European countries from claiming part of the Dutch East Indies (now Indonesia). Over the succeeding centuries Britain, Australia, Germany, and Japan established colonies in Melanesia. Colonization became more intense in the late 1700s.

In 1884 Germany occupied the northern part of the eastern half of the island of New Guinea, and Britain laid claim to the southern part. Britain gave its part to Australia in 1902, and Australia occupied the northern part during World War I as part of the Allied war effort against Germany. After the war, the League of Nations gave Australia the mandate to govern the entire island.

In 1962 the United Nations gave the western half of the island to Indonesia. In 1971 the eastern half of the island was renamed Papua New Guinea. Four years later, PNG attained full independence. It used to be part of the British Empire, and now PNG is a member of the Commonwealth of Nations.

Today we can see how distant colonizing nations could not provide effective oversight of lands that have hundreds of different cultures and languages on hundreds of islands. We can also understand the difficulty of those lands making the transition to independence and developing an effective central government. This is the process that PNG has endeavored to follow.

The Bougainville Issue

The French navigator Louis-Antoine de Bougainville visited the island that now bears his name in 1768. This island became part of PNG, although it is several hundred miles east of the island of New Guinea. Bougainville is the easternmost part of PNG and the largest of the Solomon Islands.

Wreckage of an Airplane from World War II on Bougainville

Along with several other nearby islands, it has constituted the autonomous region of Bougainville.

Bougainville and nearby islands experienced the same process as PNG itself: colonized by Britain, then occupied by Germany, and later overseen by Australia. Modern residents of Bougainville accuse Britain and Germany of enslaving their forebears, and this accusation is no doubt true. Japanese forces seized Bougainville Island in 1942 during World War II, and American troops liberated it in 1944. As part of PNG, it celebrated independence in 1975.

However, many Bougainvillers have for years noted the ethnic and cultural differences of those on their island group compared to the people of PNG on the island of New Guinea. They also say that, despite Bougainville being rich in copper, gold, and tuna, the people of Bougainville have not benefited appropriately from the export of these resources. However, the people of the province have had a high standard of living compared to the rest of PNG.

A full-scale revolt began in 1988. Estimates vary, but perhaps as many as 15,000 people lost their lives during the rebellion, both from fighting and from disease since the rebels did not have easy access to medical supplies. In 2001 the PNG government and Bougainville leaders signed a peace agreement. The agreement granted the Bougainville province autonomy and promised a referendum on complete independence.

That referendum took place over a three-week period in late 2019. Turnout of eligible voters was over 87%, and almost 98% of the voters preferred independence over greater autonomy. However, the referendum was nonbinding. Resolution of the issue is in the hands of the PNG and Bougainville governments. At the time of this writing, many issues such as borders, trade, and the makeup of security forces remained to be resolved. The PNG parliament must ratify any agreement, which might take months or even years to finalize.

The desire for independence by an island province of about 250,000 people in the South Pacific has international implications. It stands to become another point of conflict between China and the West. The United States helped finance the referendum and hopes to profit from trade with the potential new nation. Meanwhile China has representatives in Bougainville who are offering deals that they say will build the potential new country's economy. Both sides see potential economic gain for themselves, and neither side wants to allow the other side to have a new sphere of influence in that part of the world. In addition to this conflict of interest, other provinces and territories of PNG and other countries might decide to pursue independence. The entire scenario of finalizing separation agreements and world powers making competing offers might be played out several more times.

Boungainville Lagoon

You will likely never go to Bougainville, and you might well think that your life will be largely unaffected by what happens there. But every person matters and every place matters. To the people who danced and sang when the referendum results were announced and who live as descendants of those who suffered under colonialism and slavery, the issue is of great importance, as it would be to us if we lived in a similar situation. This is why a study of what is happening on the other side of the globe is important and enhances your understanding of human geography.

Jesus taught how important each person is to God:

Are not five sparrows sold for two cents? Yet not one of them is forgotten before God. Indeed, the very hairs of your head are all numbered. Do not fear; you are more valuable than many sparrows.
Luke 12:6-7

Assignments for Lesson 107

Gazetteer Read the entries for Papua New Guinea, Fiji, New Caledonia, Solomon Islands, and Vanuatu (pages 182, 184, 190, 193, and 195).

Worldview Copy this question in your notebook and write your answer: What is a geographic feature on the earth that reminds you of a characteristic of God? (Examples: "God is like a mountain because..." or "God's love is like a flowing stream because...")

Project Continue working on your research paper (or project).

Student Review Answer the questions for Lesson 107.

108 Guam and Micronesia

Sometimes the geography over which countries fight can become a new home for the soldiers who do the fighting. For instance, some Union soldiers who fought on Southern battlefields during the Civil War moved to the former Confederacy after the war to help their fellow countrymen recover from the devastation they had experienced. After World War II, some American soldiers attended Bible colleges in the U.S. and then returned to Europe and Japan as Christian missionaries.

For some Japanese soldiers in World War II, the jungles of Guam became a new home, but in a surprising way. Pacific islands offer great beauty and places for exotic getaways. For one man, this Pacific island provided a place to hide—for almost three decades.

The United States received the island of Guam as part of the terms of surrender by Spain in the Spanish American War in 1898. A few days after the Pearl Harbor attack in 1941, Japanese forces seized the island of Guam from the U.S. In July of 1944, American forces retook the island as part of the Allies' advance toward Japan.

When the Japanese command fell apart on Guam during the American invasion, about 5,000 Japanese soldiers were left to fend for themselves. By far most were killed or captured over the next few weeks, but at the end of the war in August 1945 about 130 holdouts continued to hide in the jungle individually or in small groups. They refused to believe that Japan had surrendered. For a time they captured local cattle for food; but then, fearing discovery, they lived off the jungle, eating rats, eels, and venomous toads. Over the next decades some died while most surrendered. When two of the soldiers died in flooding in 1964, that left only one: Sgt. Shoichi Yokoi.

Yokoi, born in 1915, was drafted into the Japanese army in 1941. In February of 1943 he was stationed on Guam. When the Americans invaded,

Sgt. Yokoi's original hideout was destroyed by a typhoon. This is a photograph of a reproduction on Guam.

Kiribati House of Assembly, Tarawa Atoll

Yokoi fled to the jungle. While he was hiding out, Yokoi constructed an underground shelter and fashioned an eel trap to help himself survive. For many years he believed that one day his fellow soldiers would rescue him. He did not want to die, but he would have preferred that to the shame of surrendering.

On January 24, 1972, 28 years after he went into hiding and after eight years of being by himself, some local hunters discovered Yokoi and brought him back to civilization. When he returned to Japan he was welcomed as a hero, but he never adapted to modern life. He visited Guam several times before his death in 1997. A small museum on Guam tells his story and displays the eel trap and other items related to this remarkable Japanese soldier.

Micronesia

To the north of Melanesia in Oceania is the subregion of Micronesia (from the Greek words for small and island). Micronesia contains about 2,100 islands, most of which are small coral reefs. The total land area is about 1,000 square miles, of which the largest, Guam, has 225 square miles. The total population of the subregion is about a half million people.

Micronesia contains five island chains—Caroline Islands, Line Islands, Gilbert Islands, Mariana Islands, and Marshall Islands—and many other individual islands. The region contains seven distinct political entities: the independent and sovereign countries of Palau, Kiribati, Nauru, the Republic of the Marshall Islands, and the Federated States of Micronesia (FSM); the U.S. Commonwealth of the Northern Mariana Islands, and the U.S. Territory of Guam. The United States previously oversaw Palau, the Marshall Islands, and the FSM. The United Kingdom oversaw Kiribati and Nauru.

The most prominent of the individual islands is Wake Island. Marshall Islands has a claim on record with the UN for Wake Island and has a connection with it dating back centuries, but in practical terms the U.S. Air Force administers the territory and the United States has an airfield there.

Each of the main political entities has fascinating geographic characteristics.

At 8.1 square miles (about one tenth the size of the District of Columbia), the single island of Nauru is the world's smallest independent republic. Only Vatican City and Monaco are smaller, and they are not republics. The population of the island is about 10,000. The origin of its people is unclear because its language does not resemble any other in the Pacific.

Because its island expanse straddles both the equator and the International Date Line, Kiribati is the only country that is in all four hemispheres: the northern and southern hemispheres, and the eastern and western hemispheres.

The Republic of the Marshall Islands is the easternmost country in Micronesia. It primarily consists of two parallel chains of coral atolls, the Ratak (Sunrise) Islands to the east and the Ralik (Sunset) Islands to the west. The testing of nuclear weapons took place on atolls here in the 1940s.

Yap Island is now part of the Federated States of Micronesia. For centuries, the people there have used limestone discs as a form of currency. The limestone was quarried on Palau and transported to Yap. Many of the discs were small and portable, but others were huge, as seen in the photo below. The larger discs are left in place and the community keeps track of who owns which ones through oral history.

Chamorro Residents of Guam During World War II

Palau lies in the southwestern part of Micronesia. This island chain has the most species of marine life of any area of similar size in the world. This marine variety has made Palau a popular scuba-diving destination. It also has a rich floral abundance. One frog species found on the island gives birth to live frogs instead of eggs.

The Federated States of Micronesia is sometimes called simply Micronesia, but this is confusing since the entire subregion also carries that name. The people of these islands are culturally and linguistically diverse. Eight languages as well as local dialects are spoken here, but English is the official language and is spoken widely. Just about the entire population claims adherence to a Christian religion. Just over half are Roman Catholics, and two-fifths are Protestants.

The fourteen islands of the Northern Mariana Islands lie about three-fourths of the way from Hawaii to the Philippines. In the 1970s, the islanders decided not to seek independence but instead sought closer ties to the U.S. The result was commonwealth status, which acknowledges dependence but allows for a degree of self-rule. About 90% of the population lives on one island, Saipan. Because of Spanish presence there centuries ago and current American interest, the culture is a mix of Spanish Roman Catholicism and contemporary Americana.

The southernmost of the Mariana Islands is the island of Guam, a U.S. territory.

Guam: The Home of the Chamorro

At some point in the past, the Chamorro people from Asia (some scholars believe they were from Taiwan) came to the Mariana Islands. They developed a complex, clan-based society. They carved latte stones, which are found throughout the Marianas. Each latte has two parts, the *haligi* (shaft) and the *tasa* (cap). They were used as supports for dwellings and community structures.

Our best understanding is that Ferdinand Magellan and his crew came to the Marianas, and probably Guam, on his around-the-world exploration in 1521 and claimed the islands for Spain. Spanish rule was harsh and the Chamorros resisted. For instance, the Spanish tried to move people from other parts of the Marianas to Guam, but many Chamorros escaped by outrunning the slower Spanish ships in their canoes. As a result, the Spanish banned the use of canoes.

By the late 1800s, Spanish rule was weakening. When the islands came into American possession as a result of the Spanish American War, administration of Guam was placed in the hands of the U.S. Navy. The Chamorros requested citizenship and civil rights, but Navy authorities on the island resisted this move.

Four hours after the Japanese bombed Pearl Harbor on December 7, 1941, the Japanese bombed Guam and invaded the island. In this fighting and in the three years of Japanese rule that followed, over 1,100 American soldiers died and 13,000 Americans were injured, imprisoned, put on forced marches, or required to perform forced labor.

Today Guam is an unincorporated, self-governing territory of the United States. Its citizens do not vote in American presidential elections, but the island does send a delegate to the U.S. House of Representatives. The delegate takes part in committee activities and votes in committee, but he or she does not vote on bills in the full House.

Guam is three times the size of the District of Columbia and has a population of about 168,000. The presence of the United States military in several installations on the island are a major source of the island's income and economic stability. About one million visitors come to Guam each year.

Why American Involvement in the Pacific?

What interest does the United States have in tiny islands in the middle of the Pacific Ocean, thousands of miles from the U.S. mainland? Didn't colonies, territories, and commonwealths go out with the British Empire?

At left below are latte stones on Guam. At right is a replica latte house on Saipan in the Northern Mariana Islands. Latte stones are prominent symbols of Chamorro history and culture.

The United States began acquiring these islands in the late 1800s, when one way to measure the strength of a country was to add up its overseas possessions. The United States took over some foreign lands, though never as many as such European powers as Britain and Germany. Over the years since, the U.S. has granted independence to some of these lands, including many in Micronesia; but it continues to control others.

As it turned out, the American presence on these Pacific islands helped the Allied cause in World War II. Our continued presence there has given us a strategic strength in recent years as China has sought to expand its influence in the Pacific and the world. Given the role of the U.S. in world affairs, our possession of these tiny islands has helped to preserve peace since the end of World War II.

Even little islands can play a big part in human history and geography. God's will is that the earth and all its inhabitants give Him praise.

Sing to the Lord a new song,
Sing His praise from the end of the earth!
You who go down to the sea, and all that is in it.
You islands, and those who dwell on them.
Isaiah 42:10

Assignments for Lesson 108

Gazetteer Read the entries for Guam, Kiribati, Marshall Islands, Federated States of Micronesia, Nauru, Northern Mariana Islands, and Palau (pages 185, 188, 189, 191).

Worldview Copy this instruction in your notebook and write your answer: Write down a passage in the Bible that really lifts you up and strengthens your faith.

Project Continue working on your research paper (or project).

Student Review Answer the questions for Lesson 108.

Bora Bora, French Polynesia

109 Not Quite Perfect: Polynesia

"One peculiarity that fixed my admiration was the perpetual hilarity reigning through the whole extent of the vale.

"There seemed to be no cares, griefs, troubles, or vexations, in all Typee. The hours tripped along as gaily as the laughing couples down a country dance.

"There were none of those thousand sources of irritation that the ingenuity of civilized man has created to mar his own felicity. There were no foreclosures of mortgages, no protested notes, no bills payable, no debts of honour in Typee; no unreasonable tailors and shoemakers perversely bent on being paid; no duns of any description and battery attorneys, to foment discord, backing their clients up to a quarrel, and then knocking their heads together; no poor relations, everlastingly occupying the spare bed-chamber, and diminishing the elbow room at the family table; no destitute widows with their children starving on the cold charities of the world; no beggars; no debtors' prisons; no proud and hard-hearted nabobs in Typee; or to sum up all in one word—no Money! 'That root of all evil' was not to be found in the valley.

"In this secluded abode of happiness there were no cross old women, no cruel step-dames, no withered spinsters, no lovesick maidens, no sour old bachelors, no inattentive husbands, no melancholy young men, no blubbering youngsters, and no squalling brats. All was mirth, fun and high good humour. Blue devils, hypochondria, and doleful dumps, went and hid themselves among the nooks and crannies of the rocks.

"Here you would see a parcel of children frolicking together the live-long day, and no quarrelling, no contention, among them. The same number in our own land could not have played together for the space of an hour without biting or scratching one another. There you might have seen a throng of young females, not filled with envyings of each other's charms, nor displaying the ridiculous affectations of gentility . . . but free, inartificially happy, and unconstrained."

—*Herman Melville,* Typee

This idyllic setting is novelist Herman Melville's description of life in the Polynesian islands. Sounds wonderful, doesn't it? Who wouldn't want to live there? Who would doubt that this little piece of geography was almost heaven on earth? Like other writers, Melville visited the South Pacific and found what he thought was a wonderful place.

There's just one problem with Melville's description. It wasn't completely true.

In the novel *Typee*, Melville goes on to describe a conflict his hosts had with a neighboring tribe. Later during Melville's actual visit on the island, he got the idea that his gentle island hosts were fattening him up to have him for dinner; so Melville escaped his island paradise. (By the way, I have not read *Typee*, nor would I recommend it.)

Over the years many writers and artists have traveled to Polynesia for several reasons. Herman Melville deserted his American whaling ship in Nuku Havi and found his way to Tahiti. Robert Louis Stevenson came for his health, hoping to find healing from tuberculosis. W. Somerset Maugham was working for British military intelligence during World War I. James Michener was assigned to the region as a naval historian during World War II. Each author wrote one or more books based on his time in Polynesia. The French artist Paul Gauguin lived there twice, including the last eight years of his life. These are only the best known of the many writers and artists who went to Polynesia for inspiration. What about Polynesia exerts this magnetic pull?

Breadfruit is an important food in Polynesia.

What Is Polynesia?

In 1756 the French scholar and politician Charles de Brosses gave the name Polynesia (which means many islands) to all of the islands of the Pacific Ocean. In the modern definition, Polynesia covers a triangular area across the Equator that includes about one thousand islands. Some islands are atolls while others are the peaks of undersea ridges or volcanoes. These islands and island chains include independent nations and territories of other countries. The points of the triangle of Polynesia are Hawaii (part of the United States) in the north, Easter Island (owned by Chile and lying 2,300 miles west of Chile) to the east, and New Zealand (an independent country) to the south. A small band of what is considered Polynesia extends west between Melanesia and Micronesia.

Among the islands of Polynesia are places that have been the ideal island getaways in the minds of many people around the world. For instance, French Polynesia is a largely autonomous territory of France. It includes the island of Tahiti. Samoa is an independent country made up of several islands, while American Samoa is a U.S. territory consisting

Tahitian Women on the Beach *by Paul Gauguin (French, 1891)*

of several islands just to the east of Samoa. American Samoa lies 2,200 miles southwest of Hawaii. Its territorial capital is Pago Pago.

Polynesia is a good study in ethnogeography, which is the study of the ethnic identity of a place and all that this identity involves. There is no single Polynesian culture. The various cultures found on the many islands are diverse and often complex.

Many aspects of life reflect social structure. For instance, on the Marquesas Islands, which are part of French Polynesia, the height and composition of the thatched homes reflect the residents' social standing. Families with a lower status might live in a house just a few inches off the ground while priests or chiefs might occupy a house built on a platform seven or eight feet high, the structure of which might include stones that weigh several tons each. Regardless of the geography that exists where people live, humans tend to devise ways to indicate their social status and hierarchy. Despite the peaceful reputation that the South Pacific has, tribes have frequently fought each other; and human sacrifice and cannibalism have taken place.

Agriculture varies depending on what is grown on given islands, for instance, coconut on some and breadfruit on others. Scientists believe that many of the islands did not have what the first settlers needed, so they brought plants with them from their homelands. One common activity in which all Polynesians share is fishing.

Rock Carvings on Easter Island

Influences on the Polynesian People

The complexity of Polynesian social life is matched by the numerous influences on them. Geography is a key influence. Such factors as the weather, the presence or absence of mountains, and the varying soil composition of the islands on which they live affect the people.

Western colonization has significantly affected life in Polynesia. At various times Great Britain, France, Germany, New Zealand, the United States, and Chile have claimed or continue to claim islands in Polynesia. One especially important presence has been the French military. Between 1962 and 1996, France exploded 192 nuclear weapons in French Polynesia. The first ones were exploded in the atmosphere, which resulted in significant nuclear fallout. The latter ones were exploded underground, but this caused the atoll where the tests took place to sink several yards. The French military presence had a huge economic impact on the islands. After testing ended, France helped the islands transition to an economy based more on tourism.

A positive influence from the West has been the work of Christian missionaries. The Christian faith has a widespread presence throughout much of Polynesia.

Another major influence has been urbanization. Many people who once lived in rural, farming areas moved to cities to work for the government or in related industries; and their descendants have stayed there or moved to other cities. Hawaii is home to about 70% of the entire population of Polynesia. More ethnic Samoans now live away from Samoa than live in Samoa. One effect of this population shift has been a decline in the practice of traditional cultures and knowledge of traditional languages, but efforts in recent years have helped interest in them grow. The image of the simple island-dweller tilling his small garden no longer applies to the majority of Polynesians, just as it no longer applies to much of the rest of the world.

Polynesian geography certainly includes everything you would expect: beautiful beaches, lush greenery, blue skies, and the opportunity to enjoy tropical weather. Beneath these attributes, however, lies a complexity that approaches (and in some ways imitates) Western society. The experience of writers and artists, explorers and colonizers, businessmen and military leaders, as well as tourists, tells us that if you go looking in Polynesia, you will probably find whatever it is you wanted to see.

Ekalesia Matavera is part of the Cook Islands Christian Church. This meeting place was established in 1857 on the island of Rarotonga.

The missionaries who ventured to the South Pacific found beautiful scenery and people who needed Jesus. They were ambassadors for Christ just as Paul described:

Now all these things are from God, who reconciled us to Himself through Christ and gave us the ministry of reconciliation, namely, that God was in Christ reconciling the world to Himself, not counting their trespasses against them, and He has committed to us the word of reconciliation.
2 Corinthians 5:18-19

Assignments for Lesson 109

Gazetteer Read the entries for American Samoa, Cook Islands, Easter Island, French Polynesia, Niue, Pitcairn, Samoa, Tokelau, Tonga, Tuvalu, and Wallis and Futuna Islands (pages 183, 184, 185, 190, 192, 193, 194, and 195).
Read the photo essay about the Pacific Remote Islands Marine National Monument (pages 315-317).

Project Continue working on your research paper (or project).

Student Review Answer the questions for Lesson 109.

Samoa

110 "Let the Many Islands Be Glad"

What does geography say to you? Does it make you sing?

The meaning of geography is more than the bare facts of the earth's physical makeup. The Bible says that to the heart of faith, geography speaks loudly and clearly of the power and majesty of God. This is definitely a distinctive worldview!

The book of Psalms and the words of the Old Testament prophets bear compelling witness to the meaning that the people of Israel drew from the geography they knew. In addition to describing the geography of Bible lands, the writers of the Bible, inspired by God, used geographic features in their teachings. The Psalms and the books of the prophets contain most of the poetic literature of the Bible. Many passages in these books use geography to make their points.

The Homeland

Any people who has a homeland will cherish that place. In a very real sense, a homeland sings to its people. We see this in the way that the land of America sings, as given voice in such compositions as "My Country 'Tis of Thee," "America the Beautiful," and the more recent "God Bless the U.S.A." Each of these songs praises elements of geography in patriotic pride.

The Israelites were no exception. They cherished the land that God had given them. In some of the psalms of David, he rejoices in their covenant homeland. Psalm 65:9-13 is one example:

You visit the earth and cause it to overflow;
You greatly enrich it;
The stream of God is full of water;
You prepare their grain, for thus You prepare
the earth.
You water its furrows abundantly,
You settle its ridges,
You soften it with showers,
You bless its growth.
You have crowned the year with Your bounty,
And Your paths drip with fatness.
The pastures of the wilderness drip,
And the hills gird themselves with rejoicing.
The meadows are clothed with flocks
And the valleys are covered with grain;
They shout for joy, yes, they sing.

After the Jews' captivity in Babylon, they were able to return to the land God had given them. Another psalmist uses geographic terms to express his joy at the return of the Southern Kingdom from captivity:

When the Lord brought back the captive ones
of Zion,
We were like those who dream.
Then our mouth was filled with laughter
And our tongue with joyful shouting;
Then they said among the nations,
"The Lord has done great things for them."
The Lord has done great things for us;
We are glad.
Restore our captivity, O Lord,
As the streams in the South.
Those who sow in tears shall reap with joyful
shouting.
He who goes to and fro weeping, carrying his
bag of seed,
Shall indeed come again with a shout of joy,
bringing his sheaves with him.
Psalm 126

The Jews knew their history. Now they could sing their song of redemption.

God's Creative Power and Continuing Control

The ancient Israelites probably had no idea how big the earth is, but their praise of God's power as evidenced in the created world went as far as they did know. Isaiah uses the impressive geographic features of the earth to illustrate how far above them their Creator is:

Who has measured the waters in the hollow of
His hand,
And marked off the heavens by the span,
And calculated the dust of the earth by the
measure,
And weighed the mountains in a balance
And the hills in a pair of scales? (Isaiah
40:12)

The psalmist praises God as the One who "has founded [the earth] upon the seas and established it upon the rivers" (Psalm 24:2). Psalm 65:6 reminds us that God "establishes the mountains by His strength." Amos says that God "forms mountains and creates the wind" (Amos 4:13).

Nauru

Psalm 95 tells us that it is God,

In whose hands are the depths of the earth,
The peaks of the mountains are His also.
The sea is His, for it was He who made it,
And His hands formed the dry land.
(verses 4-5)

Psalm 147 urges us to sing praises to God,

Who covers the heavens with clouds,
Who provides rain for the earth,
Who makes grass to grow on the mountains.
He gives to the beast its food,
And to the young ravens which cry.
(verses 7-9)

Significantly, despite this emphasis on physical strength and majesty, the Lord does not take special notice of people who are physically strong or who exercise power over others. Instead, "the Lord favors those who fear Him, those who wait for His lovingkindness" (Psalm 147:11).

Jeremiah rebukes Judah because they did not properly fear the Lord, who created the world in which we live:

"Do you not fear Me?" declares the Lord.
"Do you not tremble in My presence?
For I have placed the sand as a boundary for the sea,
An eternal decree, so it cannot cross over it.
Though the waves toss, yet they cannot prevail;
Though they roar, yet they cannot cross over it."
(Jeremiah 5:22)

A psalm of Moses expresses God's identity this way:

Before the mountains were born
Or You gave birth to the earth and the world,
Even from everlasting to everlasting,
You are God.
(Psalm 90:2)

The Geography of Earth Portrays God's Characteristics

How might you try to describe God? The writers of Scripture sometimes did so by comparing His traits to geographic features.

Your lovingkindness, O Lord, extends to the heavens,
Your faithfulness reaches to the skies.
Your righteousness is like the mountains of God;
Your judgments are like a great deep.
O Lord, You preserve man and beast.
(Psalm 36:6)

The psalmist tells us that we can find reliable strength in God, even if we go through cataclysmic, earth-shattering experiences:

God is our refuge and strength,
A very present help in trouble.
Therefore we will not fear, though the earth should change
And though the mountains slip into the heart of the sea.
(Psalm 46:1-2)

Tuvalu

Isle of Pines, New Caledonia

Isaiah offers a similar idea in the words of God:

"For the mountains may be removed and the
hills may shake,
But my lovingkindness will not be removed
from you,
And my covenant of peace will not be shaken,"
Says the Lord who has compassion on you.
(Isaiah 54:10)

God offers blessings to His people that come as a result of their faithfulness. He describes those blessings by using illustrations from geography that highlight features with which the people could identify:

If only you had paid attention to
My commandments!
Then your well-being would have been
like a river,
And your righteousness like the waves
of the sea.
(Isaiah 48:18)

The Physical World Praises God

The Psalms and the prophets tell us that Creation praises God.

Let the rivers clap their hands,
Let the mountains sing together for joy
Before the Lord, for He is coming
to judge the earth;
He will judge the world with righteousness
And the peoples with equity.
(Psalm 98:8-9)

They who dwell in the ends of the earth
stand in awe of Your signs;
You make the dawn and the sunset
shout for joy.
(Psalm 65:8)

Shout for joy, O heavens, for the Lord
has done it!
Shout joyfully, you lower parts of the earth;
Break forth into a shout of joy,
you mountains,

O forest, and every tree in it;
For the Lord has redeemed Jacob
And in Israel He shows forth His glory.
(Isaiah 44:23)

For you will go out with joy
And be led forth with peace;
The mountains and the hills will break forth
into shouts of joy before you,
And all the trees of the field
will clap their hands.
(Isaiah 55:12)

Praise the Lord from the earth,
Sea monsters and all deeps;
Fire and hail, snow and clouds;
Stormy wind, fulfilling His word;
Mountains and all hills;
Fruit trees and all cedars;
Beasts and all cattle;
Creeping things and winged fowl;
Kings of the earth and all peoples;
Princes and all judges of the earth;
Both young men and virgins;
Old men and children.
Let them praise the name of the Lord,
For His name alone is exalted;
His glory is above earth and heaven.
(Psalm 148:7-13)

As you live in and think about our world, listen to how it praises the Creator and learn what it can teach you about Him. This is not the same as the pantheist view, which holds that God is in and equivalent to the created world. God is separate from and above Creation, but as His handiwork it bears eloquent testimony to who He is. As you view the world, let it show you the hand, mind, and heart of God.

Let heaven and earth praise Him,
The seas and everything that moves in them.
Psalm 69:34

Assignments for Lesson 110

Worldview Recite or write the memory verse for this unit.

Project Continue working on your research paper. ***or***
Finish your project for this unit.

Student Review Answer the questions for Lesson 110.
Take the quiz for Unit 22.

Lake Winnipeg, Manitoba, Canada

23 North America Part 1

This unit is the first of two on the geography of North America. In the first lesson, we describe how a Northwest Passage has become a reality after many centuries when explorers searched for one. As a backdrop to the geographic accomplishment of the St. Lawrence Seaway project, we introduce Pierre Boucher, an early pioneer and hero in New France. The city of New Orleans has a fascinating and sometimes challenging geographic setting. The fourth lesson discusses some of the waterways in the United States that have contributed to the country's economic and cultural success. The worldview lesson tells the story of Dr. Joseph Murray, surgeon, Nobel Prize winner, and Christian.

Lesson 111 - There Is One: The Northwest Passage
Lesson 112 - Pierre Boucher's Canada
Lesson 113 - Not Always Easy: New Orleans
Lesson 114 - America's Amazing Waterways
Lesson 115 - Personal Application: Joseph Murray, Surgeon and Researcher

Memory Verse

Memorize Psalm 139:13-14 by the end of the unit.

Books Used

The Bible
Exploring World Geography Gazetteer

Project (Choose One)

If you are writing a research paper, continue to work on it following the schedule outlined on pages xviii-xix of *Part 1* of this curriculum. Remember that you do not have a literature assignment for this unit.

If you are not writing a research paper, choose one of the following projects and plan to complete it by the end of this unit.

1) Write a 250 to 300 word essay on one of the topics listed on page xvi of *Part 1*.
2) Describe an experience you and/or your family has had on water. Include what you learned about water from the experience. (See Lesson 114.)
3) Would you like to be a pioneer, exploring and settling new lands? Tell why or why not. (See Lesson 112.)

Beaufort Sea, Canada

111 There Is One: The Northwest Passage

People thought it was there, but then they decided it wasn't. Now it is, but sometimes it isn't. It's one of the most intriguing aspects of geography on the planet.

To Find a Water Route to Asia

For many years, the primary goal of European explorers was to find a way to sail to Asia because they knew—or thought they knew—about the trade and treasures they could find there. This quest drove Europeans to sail around Africa. It is what motivated Columbus to go west again and again; he just didn't expect to find a continent in the way.

The quest for Asia spurred many mariners who sailed west from Europe, often with tragic consequences. In 1497 the Italian navigator John Cabot sailed west from England and landed on the coast of Canada, but he thought he had arrived in Asia. The next year, Cabot set sail west once more with a fleet of five ships. He was never heard from again.

The English navigator Sir Humphrey Gilbert wrote a treatise extolling the westward quest and encouraging many others to try it. In 1583 Gilbert, heading west, landed at Saint John's, Newfoundland. He then headed back to England and was lost at sea.

The Dutch navigator Henry Hudson at first tried to find a Northeast Passage to Asia by sailing to the north of Europe and Russia. Abandoning that route in 1609, Hudson headed west to find a river reported by an earlier explorer to North America. He found it and headed inland, but he soon realized that it didn't go all the way through the continent. That river became known as the Hudson River.

Two years later Hudson led another expedition further north and entered the large bay that came to bear his name. However, his crew mutinied and set Hudson and a few others adrift. They were never heard from again.

Jacques Cartier of France and Francisco de Ulloa (who started from west of North America and headed east) also tried, but they too failed. It took time for most Europeans to latch on to the idea that there might be a way to obtain wealth from America and not just see it as an obstacle. Old ideas die hard.

Centuries later, the hope of finding a westward water route lingered. One of the motivations for the Lewis and Clark expedition from 1804 to 1806 was to find out if there was a water route through the American continent to Asia. They didn't find one. The British Royal Navy officer Sir John Franklin set out in 1845 with two ships and 128 men to complete the quest by sailing further north. All

The Sea of Ice *by Caspar David Friedrich (German, 1842) depicts a shipwreck in the Arctic Ocean.*

were lost. Diving expeditions in 2014 and 2016 discovered the wreckage of the ships to the north of Canada. The Irish explorer Robert McClure and his crew completed the transit in 1854, but not entirely by ship. They made part of the journey by sled over the ice.

Completing the Journey

People gradually gave up on the quest of a feasible Northwest Passage by sea, but one man did not give up the idea. Roald Amundsen grew up in Norway dreaming of navigating the seemingly mythical Northwest Passage. In 1903 he sailed west in the waters north of Canada in a single 70-foot-long ship with a six-man crew. They progressed slowly through the ice-choked waters. During two winters while they were stuck in the ice they conducted scientific observations. Amundsen hoped to reach the magnetic North Pole. He did not get there, but he established that the magnetic pole had moved since it was first discovered in 1831. This was a major scientific discovery. Amundsen pressed on, and in August 1906 he arrived at Nome, Alaska.

In the Bering Strait Amundsen met a whaling ship from San Francisco, and he knew that he could complete the Northwest Passage route, the quest that had motivated men for centuries. Amundsen had done it, but the drawback was that what he had done was not commercially feasible. He had made it through, but taking three years to do it would not make a profit in trade. Making a profit would require less ice in the Arctic waters, or a ship that could break up the ice, or both. In addition, Amundsen's route included some shallow waters, at times only 3-4 feet deep. This would not work for heavy freighters. Sgt. Henry Larsen of the Royal Canadian Mounted Police accomplished the route in one season on a schooner in 1944, but again this did not open up commercial possibilities.

The Ships

The Northwest Passage as it is currently defined is a series of deep channels that extends about 900 miles. It runs between the North Atlantic north of Canada's Baffin Island near Greenland and the Beaufort Sea north of Alaska. The Passage actually includes several alternative routes through and around the islands of northern Canada. The Passage holds thousands of icebergs (some of them 300 feet in height with much more below the water's surface), as well as other frozen areas and huge chunks of ice.

To get through the ice-bound Northwest Passage obviously requires icebreaker ships. An icebreaker does not try to cut through the ice, which can be several feet thick, with a pointed bow. Instead, the ship typically has a rounded, spoon-shaped bow. The captain pushes the vessel up onto the ice and breaks through the ice with the ship's weight. Shipbuilders

Russian Icebreakers

The Nordic Orion *in the Netherlands (2019)*

constantly pursue the design and construction of heavier icebreakers with more powerful engines.

Another issue is that the typical freighter is wider than most icebreakers, so new icebreaker designs have to cut a wider swath through the ice to enable commercially successful shipping through frozen waters. Russia is the most aggressive builder of icebreakers. Russians built the first one in 1898. The Russian icebreaker fleet numbers over forty, and more are constantly under construction. The United States, by contrast, has less than a half-dozen.

Yet another issue facing commercial ships that use the Northwest Passage is that they must have reinforced hulls. Even following an icebreaker, the risk of damage from the ice requires a stronger hull than is typical.

Other Factors

Better icebreakers and stronger cargo ships make the Passage more accessible, but these are not the only developments that have increased interest in the waterway. A significant change that has opened the Northwest Passage to more shipping is that, since 2000, the ice has been receding in the Arctic Ocean. In 2007 the route was ice-free for the first time in recorded history. This makes the route accessible for several months, but not year-round.

The first successful trip through the Northwest Passage by a large bulk carrier took place in 2013. The *Nordic Orion* carried a load of coal from Vancouver, Canada, east to Finland. It had an escort of icebreakers. The next year another cargo ship made the journey without an icebreaker escort.

The Northwest Passage offers an economic advantage to shippers. All other factors being equal, sailing north of Canada requires a week less time and is one thousand miles shorter than going through the Panama Canal. In addition, the Canal has weight and size restrictions that limit what ships can pass through it. These limitations do not apply in the Northwest Passage.

People have begun taking other kinds of watercraft through the Northwest Passage, such as adventure boats that carry only a few people. The first cruise ship to make it through did so in 1984, and such voyages are becoming more and more common.

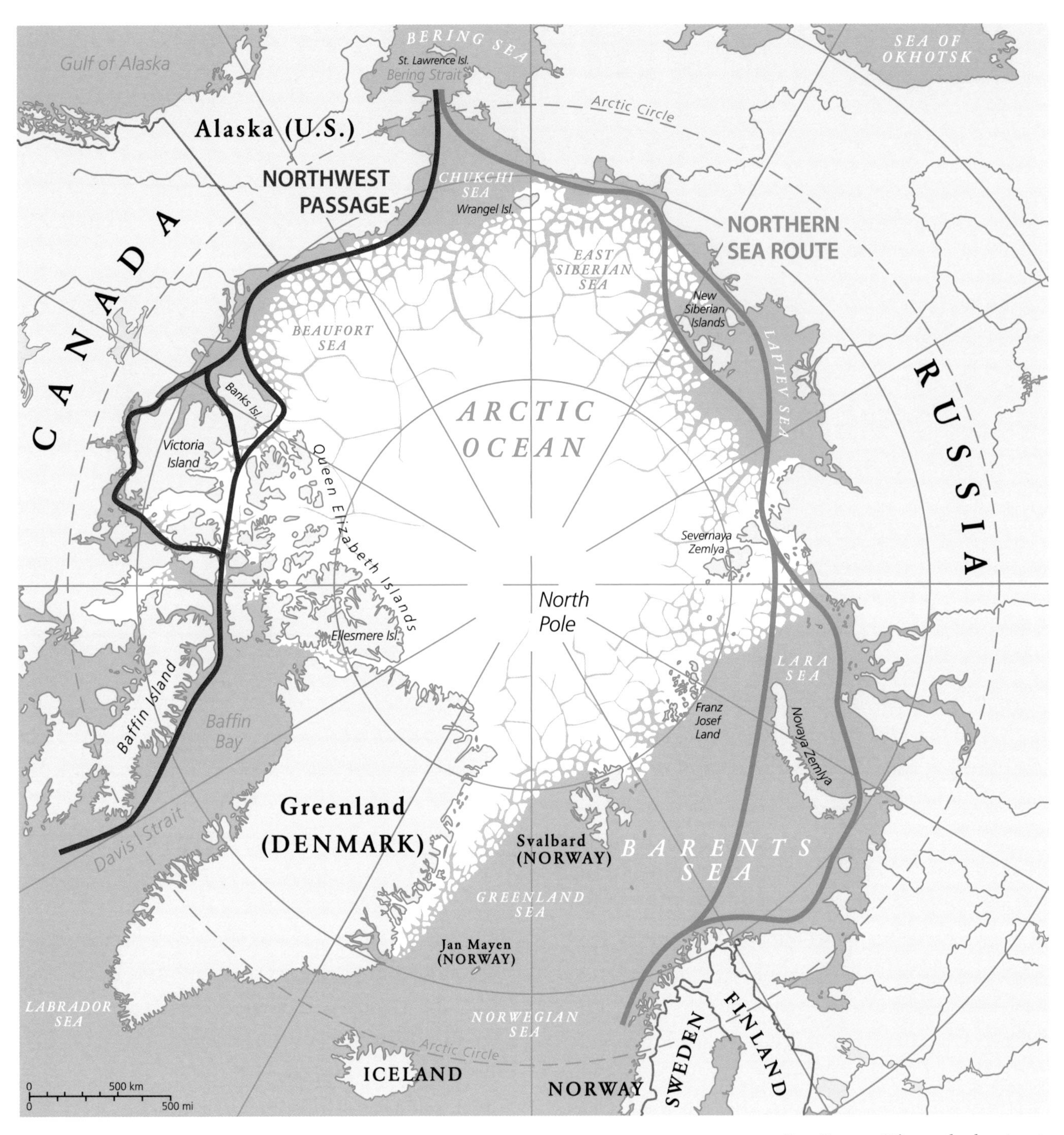

Sea Routes Through the Arctic

The greater traffic has made relevant an old question that was once thought to be merely theoretical: who owns the Northwest Passage? Canada believes that at least portions of it pass through its territorial waters and thus it is subject to Canadian jurisdiction and control. The United States sees it as an international waterway open to all. In the 1980s the two countries agreed that American ships could pass through but they had to request permission first. Canada fears that, as traffic increases, other countries will take the American position, infringe on Canadian sovereignty, and increase the risk of environmental impact that Canada would have to pay for. The countries involved have not reached a resolution that is satisfactory to all parties. The Russian government is also pursuing the opening of the Northern Sea Route through the Arctic Ocean north of Russia. Russia believes that it can control and defend this route.

In spite of this controversy, a Northwest Passage does exist. At some times in the year, following some route, now at last a ship can go between Europe and Asia without encountering a continent.

"Tears in My Eyes"

When Roald Amundsen realized that he was going to complete the transit of the Northwest Passage, he wrote in his diary:

> The North-West Passage was done. My boyhood dream—at that moment it was accomplished. A strange feeling welled up in my throat; I was somewhat over-strained and worn—it was weakness in me—but I felt tears in my eyes.

Amundsen's reaction reminds us of the power of a dream and of the role that geography plays in many of the dreams that people have.

Paul knew the experience of completing an arduous journey when he survived a shipwreck and arrived at Rome.

At the end of three months we set sail on an Alexandrian ship which had wintered at the island, and which had the Twin Brothers for its figurehead. After we put in at Syracuse, we stayed there for three days. From there we sailed around and arrived at Rhegium, and a day later a south wind sprang up, and on the second day we came to Puteoli. There we found some brethren, and were invited to stay with them for seven days; and thus we came to Rome.
Acts 28:11-14

Assignments for Lesson 111

Geography Study the maps of North and Central America and North America (pages 196 and 197).

Worldview Copy this question in your notebook and write your answer: What do you know about the human body that amazes you and convinces you that God created you?

Project Continue working on your research paper. Plan to finish it by the end of Unit 25. ***or*** Choose your project for this unit and start working on it. Plan to finish it by the end of this unit.

Student Review Answer the questions for Lesson 111.

Ships on the St. Lawrence Seaway

112 Pierre Boucher's Canada

I would like to thank Charlene Notgrass for writing this lesson about her ancestor. —RN

Five Great Lakes—Lake Superior, Lake Michigan, Lake Huron, Lake Erie, and Lake Ontario—dominate the map of North America. Lake Superior is the largest freshwater lake on earth by surface area. Together the lakes contain almost 20 percent of all the earth's fresh water. Rivers connect the Great Lakes to each other. Water flows out of the easternmost lake, Lake Ontario, into the St. Lawrence River.

The St. Lawrence has numerous islands, including the Thousand Island chain that begins at its headwaters. The "Thousand Islands" chain actually has more than 1,400 islands. The city of Montreal is built on the largest island in the Hochelaga Archipelago. The Hochelaga has more than 200 islands. The St. Lawrence River flows east for almost 744 miles until it enters the Gulf of St. Lawrence. For the last 155 miles, salt water from the Atlantic Ocean mixes with the fresh water of the river. The mouth of the St. Lawrence is one of the widest and deepest estuaries in the world.

Queen Elizabeth II and President Eisenhower Aboard the Britannia

The Great Lakes-St. Lawrence Seaway System

In the summer of 1954, Canada and the United States began the joint St. Lawrence Seaway project. In a striking example of human interaction with geography, this massive engineering and construction project improved shipping between the Great Lakes and the Atlantic Ocean and increased the generation of hydroelectricity. The project involved building locks and canals and deepening navigation channels. It was one of the largest-scale engineering feats of the 1950s.

On June 26, 1959, Queen Elizabeth II of the United Kingdom and U.S. President Dwight D. Eisenhower formally opened the St. Lawrence Seaway at Montreal, Quebec, Canada. In a speech, the Queen called it "a great joint enterprise between our two countries." Afterwards, she, her husband

Prince Philip, and President and Mrs. Eisenhower boarded the Royal Yacht *Britannia* and sailed away from Montreal Harbor. When the *Britannia* passed through the lock at St. Lambert, the St. Lawrence Seaway was officially open.

The St. Lawrence Seaway extends from Montreal to the middle of Lake Erie. The seaway is part of the Great Lakes-St. Lawrence Seaway System, which extends from the Port of Duluth-Superior at the western edge of Lake Superior to the Gulf of St. Lawrence and the Atlantic Ocean. The Port of Duluth-Superior includes twin ports in Duluth, Minnesota, and Superior, Wisconsin. It is the largest freshwater port in the world. The distance from Duluth to the Atlantic Ocean is 2,038 nautical miles. The trip takes eight and a half days.

The Great Lakes-St. Lawrence Seaway System allows ships to carry export goods from the central region of North America to the world. Ships from more than 50 countries travel the seaway. These ships make a total of 3,000-4,000 trips per year. They transport a wide variety of products, but the majority carry bulk goods, such as iron ore for the steel industry; coal which fuels power plants and steel production; limestone used in construction projects and steel manufacturing; general cargo, including heavy machinery and products made of iron and steel; cement, salt, and stone; and grain grown in the great North American breadbasket region in the U.S. and Canada.

The two heads of state who dedicated the St. Lawrence Seaway spoke English, but the ceremony took place in Quebec, Canada's only province where French is the primary language. Why?

Cartier, Champlain, and New France in the 17th Century

Jacques Cartier was born in France the year before Christopher Columbus first sailed to the New World. In 1534 King Francis I sent Cartier across the Atlantic Ocean to search for a northwest passage to the Pacific. Cartier entered the Gulf of St. Lawrence and sailed by what are now Prince Edward Island and New Brunswick. He met members of the Iroquois people, placed a cross in the region of the Gaspé peninsula, and claimed the area for France. Cartier conducted two other expeditions to the St. Lawrence, the last in 1542. He explored and mapped the St. Lawrence River as far as Montreal. Cartier heard two indigenous people say the Iroquoian word *kanata*, which means village. This is the origin of the name of Canada.

In 1608 French explorer Samuel de Champlain built a fortress beside the St. Lawrence River at what is now Quebec City. This would become the capital of New France.

The Boucher Family Arrives in New France

In 1634 Gaspard Boucher sold his farm in Mortagne-au-Perche in France. In 1635 he and his wife Nicole sailed to New France with their children. Their oldest son Pierre was 13 years old.

In the early 1600s, Jesuits were actively working as missionaries among the native nations who lived near the St. Lawrence River. Gaspard Boucher was a carpenter. Jesuits who lived on a farm near Quebec City hired him to work for them. The Jesuits educated the Boucher children, paying particular attention to Pierre. In 1637 young Pierre went to live in a Jesuit mission near the shore of Lake Huron. Pierre served as an assistant at the Sainte-Marie Among the Hurons mission for four years. While there Pierre lived among members of the Huron Nation, learning their customs and dialects.

Soldier, Interpreter, and Government Official

Pierre Boucher returned to Quebec City in 1641. He became a soldier, an interpreter, and an Indian agent. He also became an assistant to Hualt de Montmagny, the first governor of New France. Boucher rose quickly from private to sergeant.

The three major cities along the St. Lawrence River in Quebec today are from east to west Quebec City, Trois-Rivières, and Montreal. In 1634 Samuel de Champlain arranged for a fort to be constructed at Trois-Rivières. Missionaries founded Montreal in 1642.

In 1644 Pierre Boucher, now 22 years old, became the official interpreter and clerk of Trois-Rivières. His parents joined him there. Pierre soon became the captain of Trois-Rivières. Settlers living there were in constant danger of attack by Iroquois. Boucher required all residents to contribute labor to the town and to move inside the stockade.

Samuel de Champlain had a dream of French men marrying women of native nations, thus beginning a new race of people. Boucher agreed with Champlain. In 1649 he married Marie-Madeleine Ouebadinoukoue of the Huron Nation. She had been a student at the first school for girls in North America. The Ursuline order of nuns had established the school in Quebec City in 1639. Marie died in childbirth in December of 1649. Three years later Boucher married Jeanne Crevier. Jeanne had also been born in France and had become a pioneer in New France with her family.

In 1653 most of the settlers of New France were so fearful of the Iroquois that many talked of abandoning the colony altogether and returning to France. That August 600 Iroquois surrounded the fortifications at Trois-Rivières. Inside were Pierre Boucher and about 40 men, plus women and children. Most of the men were either teenagers or elderly. For nine days, the Iroquois laid siege to the fort. Every time the Iroquois attacked, the citizens of Trois-Rivières returned fire from inside the fort. Finally the Iroquois asked for peace. Boucher went out of the fort alone so that the Iroquois would not know that such a small number was inside the fort. Surprisingly the Iroquois agreed to return all prisoners they had previously captured and to send their leaders to Quebec to make peace with the governor of New France.

This is a reconstruction of the Jesuit mission Sainte-Marie Among the Hurons.

This stained glass window is in the Church of Notre-Dame in Mortange-au-Perche, France. It depicts Pierre Boucher leaving France to return to Quebec in 1662.

Emissary to France

When a new governor arrived in Quebec in 1661, he decided to send someone to the court of King Louis XIV in France to ask the king to send help to the struggling colony. The governor chose Pierre Boucher. Less than two months after the new governor arrived in New France, Boucher was on a ship bound for Paris. Boucher met with Louis XIV and received his promise of aid. A member of the king's court asked Boucher to write a report about the natural resources of New France and the reasons why the king should keep the colony. The king gave Boucher a title of nobility, making him a seigneur (similar to a lord in England). Boucher was the first citizen of New France to receive this honor.

When Boucher returned from France, the king sent with him two ships, 100 soldiers, and ammunition. Boucher himself borrowed money to pay for the passage of 100 more male settlers. Boucher's mission had met with great success. It was a major turning point in the history of New France. From then on, the French government took great interest in its colony in North America. Boucher wrote the report that a member of the court had asked him to write, which was published in Paris in 1667 as *Histoire veritable et naturelle des moeurs & productions du pays de la Nouvelle France, vulgairement dite le Canada*. It describes the flora and fauna of New France and the people who lived there. The book had a great impact. The king sent even more troops to help the colony.

Seigneur of Boucherville

At age 45, Boucher left public life and moved to his seigneury (manor) on the Percées Islands, which lay along the southern shore of the St. Lawrence River. The government of New France was organized in a similar fashion to medieval Europe. It gave certain seigneurs tracts of land. Other settlers lived and worked on the land but did not own it. The seigneuries along the St. Lawrence River were narrow and deep so that each seigneur would have access to the river.

Boucher had a dream of establishing an exemplary seigneury. Settlers living on a seigneury paid dues to live there. Boucher made his dues low and farmed his own lands himself, rather than requiring other settlers to work for him. By 1673 Boucher had built a manor house. He invited 37 settlers who had shown themselves worthy to join him, and he gave them titles to their land. In 1686 the governor of New France wrote that the Boucher family had worked the hardest for the good of the colony. He said that Boucher's seigneury, which came to be called Boucherville, was one of the finest in New France.

One of Pierre Boucher's great-granddaughters was Marie-Marguerite d'Youville (1701-1771). She founded the Sisters of Charity and operated a hospital in Montreal that cared for orphans, the elderly, and the disabled. She was the first person born in Canada to be canonized by the Roman Catholic Church.

Pierre lived to be 95 years old. He died in Boucherville in 1717. He had lived in New France for 82 years. Jeanne lived to be 91. She died in 1727. During their long lives together, Jeanne Crevier Boucher gave birth to fifteen children. She and Pierre reared them all to adulthood.

From New France to the Commonwealth of Nations

The Seven Years War began in Europe in 1756. The conflict spread to the New World. It is called the French and Indian War in the United States. In 1759 the British defeated New France at Quebec City during the Battle of the Plains of Abraham. The era of New France came to an end, and Great Britain took control of Quebec and all of what became Canada.

Many descendants of Pierre and Jeanne Boucher still live in Canada. However, after the fall of New France, a few of their children and grandchildren moved south to Illinois country. One of those was my ancestor, Jacques Timothe Boucher sieur de Montbreun, who later moved further south to Tennessee. Some descendants moved to Louisiana, others to the island of Mauritius in the Indian Ocean, some to the West Indies, and others back to France. When Ray and I attended the 350th anniversary of the founding of Boucherville in 2017, we met my relatives from Canada, France, and the United States.

One of the activities at the 350th anniversary of the founding of Boucherville was an All White Ball (where everyone dressed in white). Here are Ray and Charlene Notgrass having fun!

Monuments to Jeanne Crevier and Pierre Boucher in Boucherville with the St. Lawrence River in the Background

When Queen Elizabeth II and President Dwight Eisenhower opened the St. Lawrence Seaway in 1959, it had been 200 years since New France fell to the British. Though Canada remains a member of the Commonwealth of Nations, the heritage of New France continues in the citizens of Quebec who cherish their French language and culture.

In his old age, Pierre Boucher wrote down his last wishes. He told his children that he had done what he could to live without reproach. He encouraged them to try to do the same.

Likewise urge the young men to be sensible; in all things show yourself to be an example of good deeds, with purity in doctrine, dignified, sound in speech which is beyond reproach, so that the opponent will be put to shame, having nothing bad to say about us.
Titus 2:6-8

Assignments for Lesson 112

Gazetteer Read the entry for Canada (page 199).
Read the excerpt from *Canada in the Seventeenth Century* by Pierre Boucher (pages 318-320) and answer the questions in the *Student Review Book*.

Worldview Copy this question in your notebook and write your answer: How have you or your family been helped by a physician or surgeon who was a person of faith?

Project Continue working on your research paper (or project).

Student Review Answer the questions for Lesson 112.

Satellite Image of New Orleans

113 Not Always Easy: New Orleans

New Orleans, Louisiana, is not on the Gulf Coast.

The city actually lies about 110 miles north of the mouth of the Mississippi River. The river's southernmost extremity flows through delta swampland and bayous on a curving, twisted course to the Gulf of Mexico.

The city's location north of the Gulf is only one of the geographic surprises that New Orleans offers. It lies in a large, crescent-shaped bend in the Mississippi River. This is why one nickname for the city is the Crescent City. Below the city is the English Turn, a bend in the river so sharp that sailing ships had to stop and wait for the wind to change to continue on their journey. Time and again, the geographic setting of the city has made it the scene of dramatic events.

A Center for Trade and Culture

The Company of the West chose the site for New Orleans, at the head of the Mississippi delta, in 1717. The company was centered in Paris, France; but its director was the Scottish entrepreneur John Law. Law believed that, because of its location, New Orleans could become a major trading center. The city's port could receive goods that came from the interior of the North American continent by means of the Mississippi, Missouri, and Ohio Rivers. Those goods would be bound for the American East Coast

View & Perspective of New Orleans *by Jean-Pierre Lassus (French, 1726)*

and for other continents. The city could also send goods from other countries inland by the same routes. Law's vision proved accurate. Trade passing through New Orleans grew consistently over the years until, by the end of World War II, New Orleans was the second busiest port in the United States.

The building of the city, which began in 1718, took place on the east side of the river. The project endured many trials, including two major hurricanes in 1721 and 1722 and diseases spread by mosquitoes that inhabited the nearby swamps.

Because of its role in trade, New Orleans grew into an international city, with influences from France, French-speaking Canada, Spain, Ireland, Germany, Italy, Americans descended from Europeans, and African Americans. It was a busy city but also a fun city; this earned it the nickname of the Big Easy. The many cultural influences were the result of its location, and those influences continue today.

The Battle of New Orleans

After its loss to Great Britain in the French and Indian War, France ceded New Orleans and the Louisiana territory to Spain in 1762. In 1800 French leader Napoleon re-acquired the city and territory as part of his dream to build a world empire. But his plans fizzled, and in 1803 France sold them to the United States. Now the U.S. controlled this major western port city.

During the War of 1812, Great Britain and the United States each enjoyed some victories in battle. The British military decided to focus on capturing New Orleans in what it hoped would be a decisive victory. The plan called for British troops to capture New Orleans and then move up the Mississippi to join other British troops invading from Canada. This would cut the United States in two, and it would also strangle the trade that passed through New Orleans on which the U.S. depended.

In the fall of 1814, British forces and American forces (led by Andrew Jackson) fought a series of smaller battles along the Gulf Coast starting in Florida and moving west toward New Orleans. Jackson understood the strategic importance of the city and knew he had to make a decisive stand against the British there. He oversaw the positioning of lines of troops and the building of a series of defensive barriers along the eastern side of the river south of the city. He also put a smaller part of his forces on the western side since it was easier to defend.

The British hoped to sail their troop ships through the Rigolets, a passage that leads into Lake Pontchartrain to the north of New Orleans, land them on shore a few miles from the city, and then overwhelm American forces and take control of the city. However, the Americans ably defended the Rigolets, so the British had to land at Bayou Bienvenue and slog through marshlands to approach the city. This took place on the eastern side of the Mississippi River in late December and early January, so the British troops were cold, wet, and miserable.

On the morning of January 8, 1815, the Americans unleashed a vicious attack on the approaching British and inflicted serious casualties while suffering relatively few themselves. British forces on the west side of the river pushed the Americans on that side back and advanced toward the city, but the British commander on the western side realized that he could not take the city with the

This is a preserved portion of the Chalmette Battlefield, site of the Battle of New Orleans on January 8, 1815.

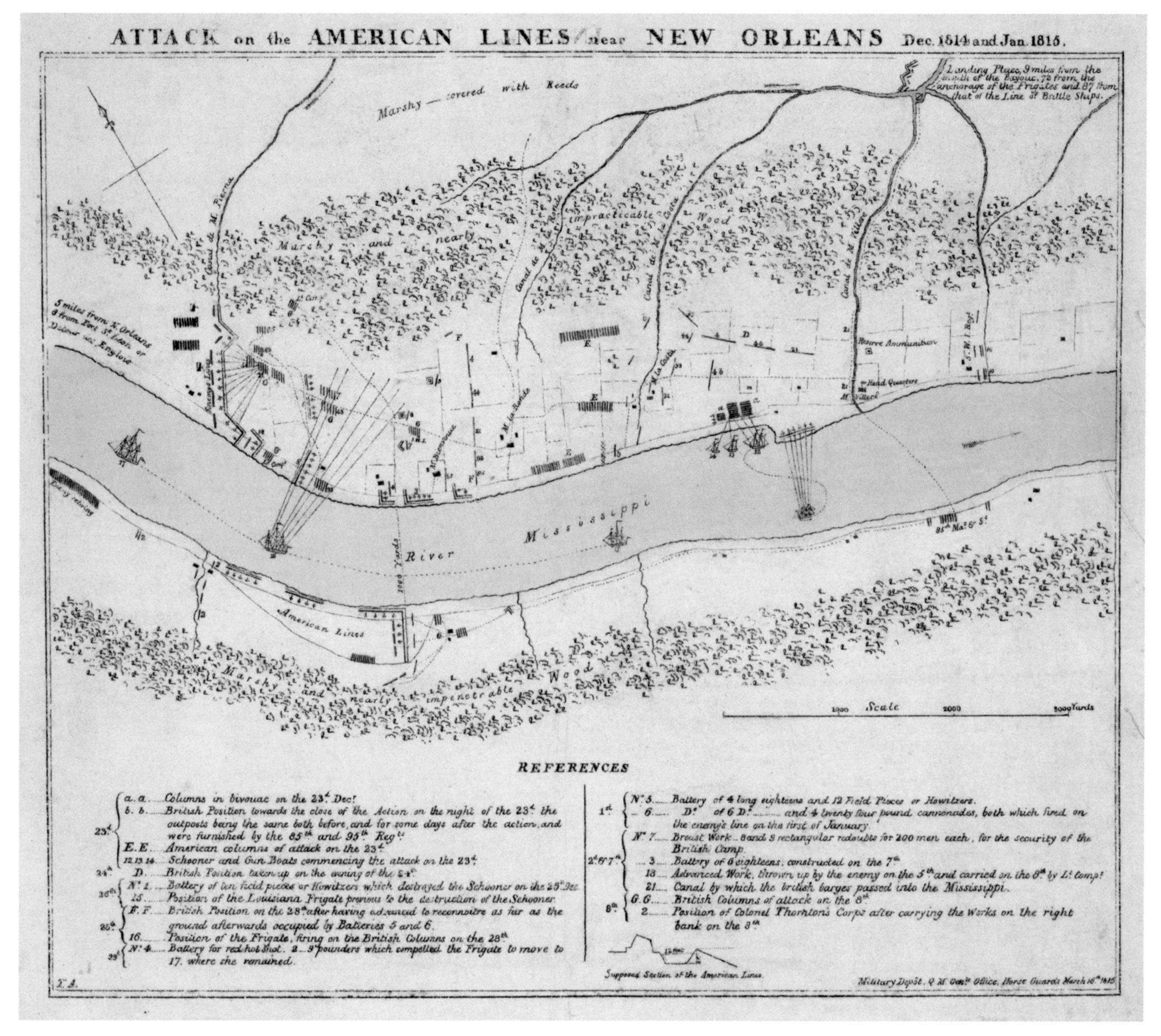

This is a British map of the battles near New Orleans in December 1814 and January 1815. The British troops are shown in red and the American in blue.

contingent he had and withdrew. The battle lasted about two hours. The surviving British troops sailed away in defeat the same way they had come, while the Americans celebrated their victory.

The fact that the battle took place after the two countries had signed a peace treaty in Europe a few weeks earlier did not diminish the Americans' rejoicing nor the perception in European capitals that the United States was indeed a nation to be reckoned with. Andrew Jackson became a national hero and eventually won election to the presidency in 1828.

Hurricane Katrina

Fast forward almost two centuries later, to August of 2005. New Orleans was attacked again. This time it was not a foreign army but a terrible natural disaster—a hurricane.

The metropolitan population of New Orleans at the time was over one million people. The urban area of New Orleans had spread widely, on both sides of the Mississippi. However, much of the city lay at or even a few feet below sea level. The government had constructed a long series of levees along the river and

Hurricane Katrina
August 28, 2005
Mobile, AL
Biloxi, MS
Pensacola, FL
New Orleans, LA
Houston, TX
Gulf of Mexico

along Lake Pontchartrain to keep floodwaters out of inhabited areas, although inspectors had warned that the levees might not hold during a major storm.

Katrina was a major storm.

After crossing southern Florida, Hurricane Katrina entered the Gulf of Mexico, strengthened, and headed for the Gulf Coast. With winds estimated at 170 miles per hour, Katrina made landfall on the Mississippi coast on August 29, devastating Gulfport and Biloxi. New Orleans did not take a direct hit, but ten inches of rain and a storm surge overwhelmed the levees and flooded the city. The city government had ordered a mandatory evacuation and over one million people left, but some people did not or could not leave.

By the next day, an estimated 80% of New Orleans was flooded, in some places to a depth of twenty feet. The water destroyed many houses and businesses. Rescue workers plucked many people from the roofs of their homes. Thousands of people crowded into the Louisiana Superdome and the New Orleans Convention Center. The breakdown of the city's sewer system and the inability of the Superdome and other facilities to provide adequately for the refugees created a public health emergency. Unfortunately, the city, state, and federal governments did not put an organized rescue and relief plan in place for several days. Eventually buses took some New Orleans residents to Houston, Texas, and other distant cities. By September 6 fewer than ten thousand people remained in the city.

The hurricane and the resulting flooding and other disruptions cost the lives of more than 1,800 persons. Damage estimates exceeded $160 billion. The storm eroded seventy-three square miles of coastland. New Orleans has recovered to a significant degree, but it may never recover completely. Some neighborhoods have not been completely rebuilt. Even more than a decade after Katrina, the city's population was still below what it was before the storm. Many people who left New Orleans and Louisiana to escape the storm never went back.

These U.S. Coast Guard images show aerial views of flooding in New Orleans after Katrina.

Does Geography Matter?

John Law would say that it does. Andrew Jackson would say that it does. The victims of Hurricane Katrina would say that it does. Sometimes we don't realize this truth until a major development occurs or a major disaster strikes, but even in our modern, everyday lives geography plays an important part.

Lamentations tells of the great city of Jerusalem that had been destroyed:

How lonely sits the city
That was full of people!
She has become like a widow
Who was once great among the nations!
She who was a princess among the provinces
Has become a forced laborer!
Lamentations 1:1

Assignments for Lesson 113

Gazetteer Read the entry for the United States (page 201).

Worldview Copy this question in your notebook and write your answer: Do you know a scientist who is a person of faith? What have you learned from this person that has helped you in your faith?

Project Continue working on your research paper (or project).

Student Review Answer the questions for Lesson 113.

Barge on the Mississippi River, Iowa

114 America's Amazing Waterways

You can get on a boat in Minneapolis, Minnesota, and go . . . practically all over the world by water: Chicago, New York, New Orleans, Los Angeles, London, cities on the continent of Europe, the great cities of Asia, you name it. You would eventually have to go overland to get to many places on the globe, but you can get much closer to them by water.

The simple facts about the waters of the United States are astounding:

- Counting its rivers, canals, other inland bodies, and coastal waterways, the United States has more navigable waterways than the rest of the world—combined. The total is 25,481 miles, of which 12,000 miles are used for commerce.

- The American Midwest, which is laced by the Greater Mississippi River watershed, is the largest single area of farmland in the world. The waterways there have tremendous significance in transporting American farm products to the rest of the world.

- The Atlantic coast of the U.S. has more major ports than the rest of the Western Hemisphere combined.

- The Mississippi-Missouri Rivers system is the third longest (or fourth longest, depending on who is doing the counting) river system in the world. The Mississippi is 2,350 miles long, and the Missouri is actually about 100 miles longer.

It is hard to overemphasize the importance of rivers, lakes, coastal waterways, and man-made canals to human life. The United States began as colonial settlements along the Atlantic coast. Most major cities were founded along a coastline or on a river. American rivers significantly aided the settlement of the continent west of the Appalachian Mountains. Rivers were the original interstate transportation routes in our country, before railroads and highways. When you drive by a river or cross one on a bridge and see a towboat pushing a 15-barge tow, you are seeing the equivalent of 870 tractor-trailer trucks carrying coal, petroleum, raw materials, or other products.

The completion of the Erie Canal, which connected New York City with the Midwest via the Hudson River and the Great Lakes, was a major economic and social accomplishment. The American-built Panama Canal, following by a few decades the completion of the Suez Canal, created an around-the-world waterway.

The story of the rest of the world reflects the same importance of coastal areas, rivers, and canals. The transport of people and goods by water is several times cheaper than the same movement by land routes. Waterways not only have economic importance; but they also play a significant role in the environmental health of the world by providing water, serving as habitats for marine life, and lessening the use of petroleum fuel.

A Providential Land

In *The Federalist* Number 2, John Jay discussed what he saw as the way God's hand had guided the formation of the United States.

> Providence has in a particular manner blessed it with a variety of soils and productions, and watered it with innumerable streams, for the delight and accommodation of its inhabitants. A succession of navigable waters forms a kind of chain round its borders, as if to bind it together; while the most noble rivers in the world, running at convenient distances, present them with highways for the easy communication of friendly aids, and the mutual transportation and exchange of their various commodities.

The chain of "navigable waters" around the country's borders that Jay referred to consisted of, in his day, the Atlantic Ocean, the Great Lakes, the Gulf of Mexico, and the Mississippi River. Since then the United States has expanded to the Pacific Ocean, but his point remains just as valid.

Half of Canada is taken up by the Canadian Shield, the largest area of exposed rock of the earth's crust in the world. This geological feature and the cold temperatures in much of Canada have limited development in that direction. To our south lie the desert regions of northern Mexico, although, to be sure, much of the U.S. Southwest is desert also. As John Jay described, the United States is blessed with varied geography and abundant natural resources.

On the Ohio River *(Unknown American Artist, c. 1840)*

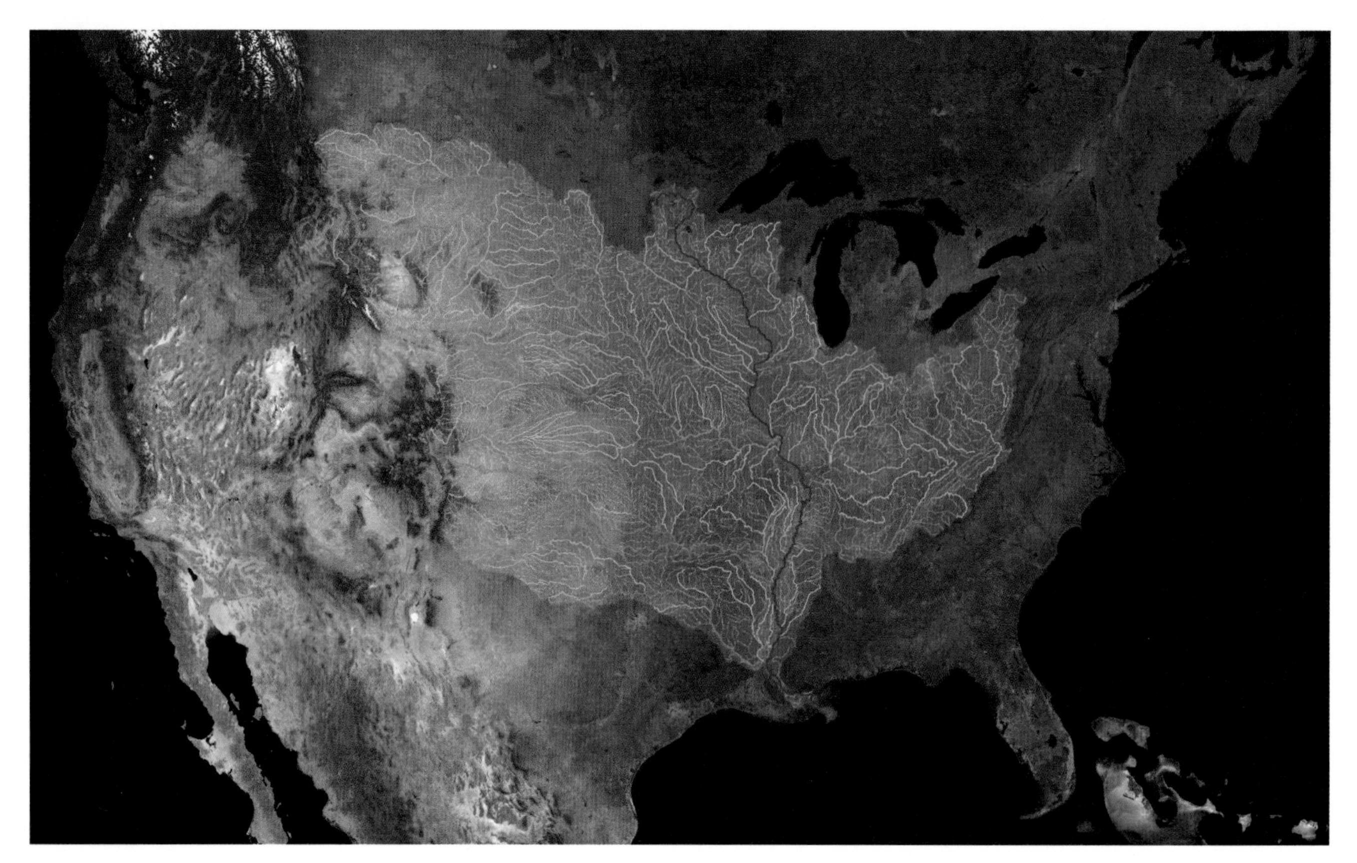

This satellite image has the Mississippi River highlighted in dark blue with its tributaries in lighter shades.

In this lesson we will look at three rivers in the United States as examples of how important rivers are. A study of the waterways where you live would reveal similar importance as well as features unique to your setting.

The Mississippi

For centuries the Mississippi River has helped millions of people have a better life. Members of native nations used it for food and to travel where they wanted to go. European explorers hoped to find aspects of America they had not seen before. Fur traders paddled along it to take their pelts to market. Pioneers floated down it to reach the places where they hoped to start new lives.

Today the river is a vital highway for business as individuals, families, and companies depend on it as part of the way they support themselves. The Mississippi watershed region covers all or parts of 32 states (40% of the continental United States) and two Canadian provinces. In addition to transporting the goods mentioned above, 92% of U.S. agricultural exports, 78% of American feed grains and soybeans, and a majority of livestock and hogs get to market by the Mississippi River and its tributaries. Millions of Americans use it as their source for drinking water. Vacationers refresh their lives in its waters by fishing and boating. A series of 29 locks and dams on the upper Mississippi enhances the river's impact.

The Mississippi blesses the physical life of more creatures than humans. It is home to over 260 species of fish, one-fourth of the fish species found in North America. Sixty percent of all North American birds fly through the Mississippi basin during their fall and spring migrations.

The Colorado

The Mississippi River flows through a part of the country that is usually lush and green. The Colorado River, by contrast, flows through lands that are usually dry and brown. A major problem with the Mississippi is its occasional flooding. A major problem with the Colorado is the frequent scarcity of water.

Headwaters of the Colorado River rise in Colorado and Wyoming. The river is about 1,450 miles long and flows through the Grand Canyon. It used to flow across the border with Mexico, but the river hasn't had enough water to do that since 1998. The watershed covers all of Arizona and parts of six other states. The river and its tributaries supply municipal water for thirty million people, including Denver, Salt Lake City, Albuquerque, Los Angeles, and San Diego, and irrigate almost four million acres of farmland.

Population growth in the major cities and an increase in farming in the region have put severe demands on the water that is available from the Colorado. This water is so precious that interstate agreements, Supreme Court decisions, and even international treaties govern its use. Water management policy regarding the use of reservoirs behind Glen Canyon Dam (Lake Powell) and Hoover Dam (Lake Mead) is a major political, social, and economic issue.

The Duck

The Duck River is an example of the many small rivers that grace the American landscape. It flows for 284 miles from east to west across mostly rural southern Middle Tennessee and empties into the Tennessee River. Archaeological evidence indicates that native nations lived along the river thousands of years ago.

You might think that there is nothing special about this little river, but there is. The Duck River is home for more species of fish than all the rivers in Europe combined. It has the most varieties of fish per mile of any river in North America. The Duck supports 151 species of fish, 60 species of freshwater mussels (which have disappeared across much of the U.S.), and 22 species of aquatic snails. In addition, one can find river otters, beavers, mink, hawks, osprey, and herons in and alongside its waters. Conservationists have called it one of three hotspots for fish and mussel diversity in the world.

The Duck is an illustration of the potential conflict between the need to protect natural resources and the need to supply a growing human population. As indicated above, the river is a natural treasure.

Colorado River in Arizona

Duck River Near Manchester, Tennessee

However, it is also the sole water source for a quarter of a million people who live south of Nashville in an area where the population is booming.

This conflict came to a head in the 1970s. The Tennessee Valley Authority proposed building two dams on the Duck River for flood control, water supply, and recreation. They were never intended to produce electricity. One dam was completed in 1976. The other dam further downstream was started in 1973, but lawsuits by environmentalist groups to protect endangered species in the river delayed and then halted construction in 1983. TVA had spent $83 million on the project by the time it was abandoned, and the partially completed dam was torn down in 1999.

God provided the waters of the earth to help sustain plant, animal, and human life on our planet. The prophet Amos used the image of flowing streams to illustrate his call to the people of Israel for them to practice justice and righteousness.

[L]et justice roll down like waters
And righteousness like an ever-flowing stream.
Amos 5:24

Assignments for Lesson 114

Geography	Complete the map skills assignment for Unit 23 in the *Student Review Book*.
Project	Continue working on your research paper (or project).
Student Review	Answer the questions for Lesson 114.

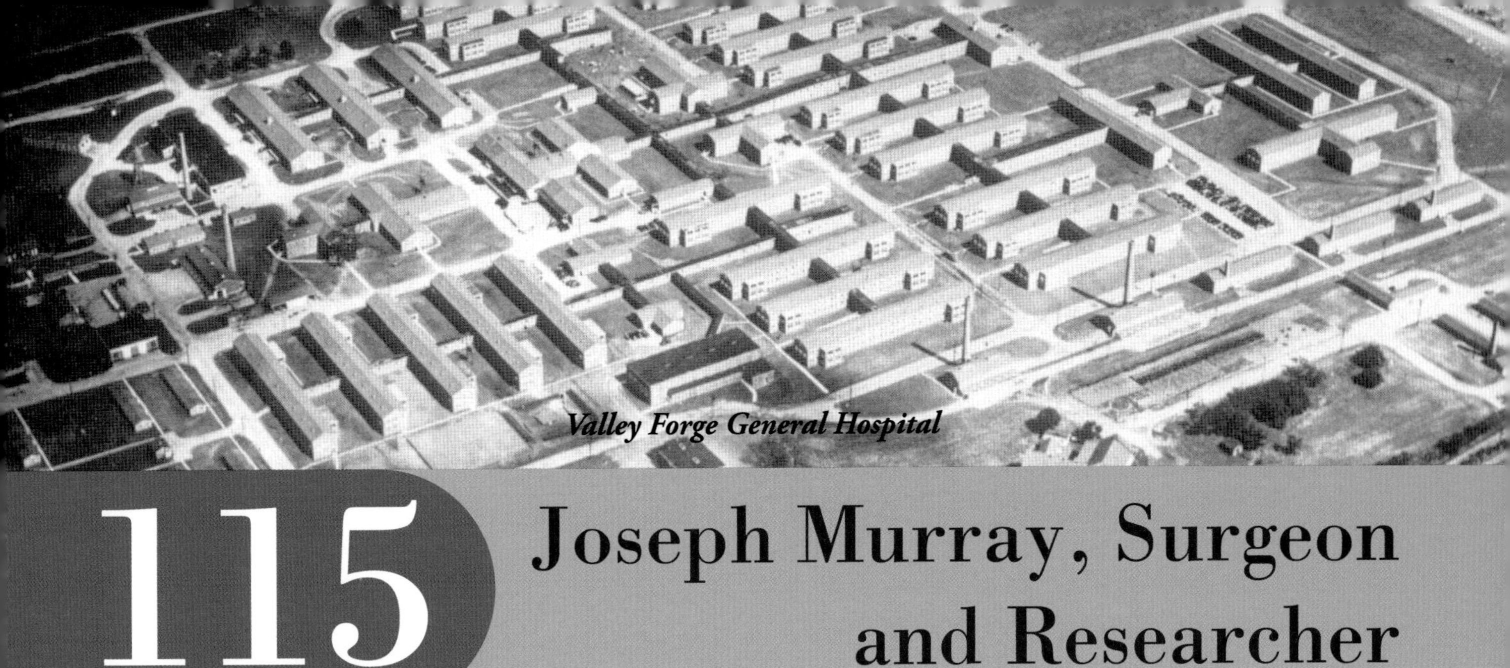
Valley Forge General Hospital

115 Joseph Murray, Surgeon and Researcher

> The more we learn about creation, it just adds to the glory of God. I've never seen a conflict [between religion and science].

> We're just working with the tools God gave us. There's no reason that science and religion have to operate in an adversarial relationship. Both come from the same source, the only source of truth.

Physician Joseph Murray was a man of science and a man of Christian faith. He achieved world-changing accomplishments, but he was also a humble man who cared deeply about the patients he served and who never lost his sense of wonder about the world in which God had placed him.

Joseph E. Murray was born April 1, 1919, in Milford, Massachusetts, to his attorney/judge father and schoolteacher mother. His parents dedicated themselves to giving their children the best education in the world. In his Nobel autobiography, Murray recalled:

> As a second year high school chemistry student, I still have a vivid memory of my excitement when I first saw a chart of the periodic table of elements. The order in the universe seemed miraculous, and I wanted to study and learn as much as possible about the natural sciences.

Joseph attended Holy Cross College. He wanted to become a doctor because of the great respect he had for his family's physician, so in 1940 he enrolled in Harvard Medical School. Drafted into the Army, Murray and other medical students were rushed through training so they could join the war effort.

Murray was assigned to Valley Forge General Hospital in Pennsylvania, where he helped soldiers who had been burned and disfigured in the war. He learned about the intricacies of the human immune system when dealing with skin transplants. He also learned about the human spirit from soldiers who endured multiple surgeries, intense and prolonged pain, and permanent disfigurement, but who persevered.

One patient who had a profound impact on Murray's life and career was Charles Woods, an aviator who had been burned over 70% of his body, including his entire head and hands. The medical team of which Murray was a member performed 24 operations on Woods to give him a functional face.

Charles Woods went on to be a successful businessman in real estate, radio, and TV. He also ran for political office several times. Woods became friends with evangelist Billy Graham in the 1950s. Graham is second from left and Woods in second from right in the photo above.

Murray learned important biological principles regarding transplants from his work with Woods, but even more, Murray learned about human courage. Murray later wrote in his autobiography, *Surgery of the Soul*:

> If I had been Charles Woods, would I have had the will to survive for six weeks with 70 percent of my body burned, and the ability to endure two years of unimaginable pain? Would I have had the dignity and confidence to walk out into the world with a disfigurement that would cause people to stare at me for the rest of my life? I honestly don't know. I know only that Charles taught me what a human being can achieve, and should strive for, in dealing with adversity.

Murray went on to do groundbreaking work in plastic and reconstructive surgery, helping burn victims and those with congenital deformities, as well as becoming an expert in head and neck surgery. What he learned helped him achieve breakthroughs in the field of organ transplants.

On December 23, 1954, Murray performed the first successful kidney transplant, between identical twins Ron and Richard Herrick. Richard had advanced kidney failure, and his only hope was to receive one of his brother Ron's kidneys. Murray gave serious, prayerful thought to the ethical question of whether it was right to take a kidney from a live donor. No doctor had ever done that before. The procedure was successful and made medical

The surgical team at Brigham and Women's Hospital in Boston performed the successful kidney transplant in 1954.

The World Transplant Games (started in 1978) and the U.S. Transplant Games (started in 1990) have provided opportunities for donor recipients and donors to participate in athletic competitions together to raise awareness of organ transplantation. These photos are from the 2008 U.S Games in Pittsburgh, Pennsylvania.

history. It led to the now-widespread practice of organ transplants. Richard Herrick lived eight more years. He got married, and he and his wife had two children. At the time he performed the operation, Joseph Murray was 35 years old.

Two days later, Murray performed another procedure that demonstrated his humility and dedication to healing. On Christmas Day he sewed up a simple cut on a young boy's forehead in a hospital emergency room. "I would never not respond in an emergency," Murray later said. "That's what being a doctor is."

In 1959 Murray performed the first kidney transplant from a non-identical donor. Then three years later, he performed the first successful kidney transplant that used the organ of a deceased donor. Murray helped develop the protocols that surgeons have used around the world for this kind of surgery. He is considered the father of organ transplantation.

Murray maintained a lifelong connection with and compassion for his patients. In 2004 Murray and original kidney donor Ron Herrick attended the U.S. Transplant Games in Minneapolis and together lit the Olympic Torch. When Ron Herrick died 58 years after the surgery, Dr. Murray had no doubt that he wanted to attend the funeral in rural Maine. Multiple generations of the Herrick family attended the service. They all looked like the twins. Murray realized that Richard's children and grandchildren would not have existed if Murray had not performed the surgery.

Murray became one of the leading experts in his field. He taught numerous medical students at Harvard Medical School and shared the Nobel Prize for Medicine in 1990. One comment about Murray's work was that it "proved to a doubting world that it was possible to transplant organs to save the lives of dying patients." Yet he always remained humble about his own and mankind's understanding of the world:

> I think the important thing to realize is how little we know about anything—how flowers unfold, how butterflies migrate. We have to avoid the arrogance of persons on either side of the science-religion divide who feel that they have all the answers. We have to try to use our intellect with humility.

Physician Joseph Murray had a worldview that honored God in his heart and in his work. This is how Murray expressed his faith regarding his work:

> Work is a prayer, and I start off every morning dedicating it to our Creator. . . . Every day is a prayer—I feel that, and I feel that very strongly.

Dr. Joseph Murray was married to the wife of his youth for 67 years. When Murray died in 2012, he was survived by his wife, six children, 18 grandchildren, and nine great-grandchildren. As Murray had requested, a song played at his memorial service was one made famous by Louis Armstrong: "What a Wonderful World."

The psalmist expressed this same wonder at how God created human beings:

For You formed my inward parts;
You wove me in my mother's womb.
I will give thanks to You, for I am fearfully and wonderfully made;
Wonderful are Your works, and my soul knows it very well.
Psalm 139:13-14

Assignments for Lesson 115

Worldview Recite or write the memory verse for this unit.

Project Continue working on your research paper. ***or***
Finish your project for this unit.

Student Review Answer the questions for Lesson 115.
Take the quiz for Unit 23.

North America at Night (2012)

24 North America Part 2

American landscape artists have created beautiful representations of our country's geography. Capitalizing on its geographic setting, Babcock Ranch, Florida, was the first community in America powered completely by solar energy. The Diomede Islands stand sentinel off of Alaska on either side of the International Date Line. The beloved books by Laura Ingalls Wilder reflect the geographic settings in which she lived. The worldview lesson focuses on the importance of place in the Biblical narrative.

Memory Verse Memorize John 14:2-3 by the end of the unit.

Books Used The Bible
Exploring World Geography Gazetteer
The Country of the Pointed Firs and Other Stories

Project (Choose One)

If you are writing a research paper, continue to work on it following the schedule outlined on pages xviii-xix of *Part 1* of this curriculum.

If you are not writing a research paper, choose one of the following projects and plan to complete it by the end of this unit.

1) Write a 250-300 word essay on one of the following topics:
   - Discuss the pluses and minuses of petroleum and of alternative sources of energy. See Lesson 117.
   - Choose one of Laura Ingalls Wilder's books and discuss how the geography there played a part in the story. See Lesson 119.
2) Draw or paint a landscape of the setting where you live. See Lesson 116.
3) Research and write about one of these Biblical settings: Canaan (or the Promised Land), Egypt, Mt. Sinai, places in Jesus' life and ministry, places in the book of Acts. See Lesson 120.

Literature

Sarah Orne Jewett created the fictional seaside town of Dunnet Landing, Maine, for *The Country of the Pointed Firs and Other Stories* to showcase the personalities, desires, eccentricities, affections, and humors of everyday people. Jewett had a special talent for the art of short stories. This unusual novel reads like short stories focused on characters rather than on a driving plot line. Maine's stunning, sometimes harsh landscape shapes their lives and livelihood and connects them all.

Sarah Orne Jewett was born in South Berwick, Maine, in 1849. Her writings sit firmly in the geography of Maine where she lived most of her life. She published her first short story at the age of eighteen. Jewett traveled widely and knew many great contemporary authors, including Mark Twain, Alfred Lord Tennyson, Christina Rossetti, Rudyard Kipling, and Harriet Beecher Stowe. After sustaining injuries in a carriage accident in 1902, she was unable to continue writing. Jewett died in 1909, leaving behind an impressive body of significant American fiction.

We do not find the introduction or afterword of the Signet Classics edition to be helpful, so they are not assigned as part of the literature reading.

Plan to finish *The Country of the Pointed Firs and Other Stories* by the end of Unit 25.

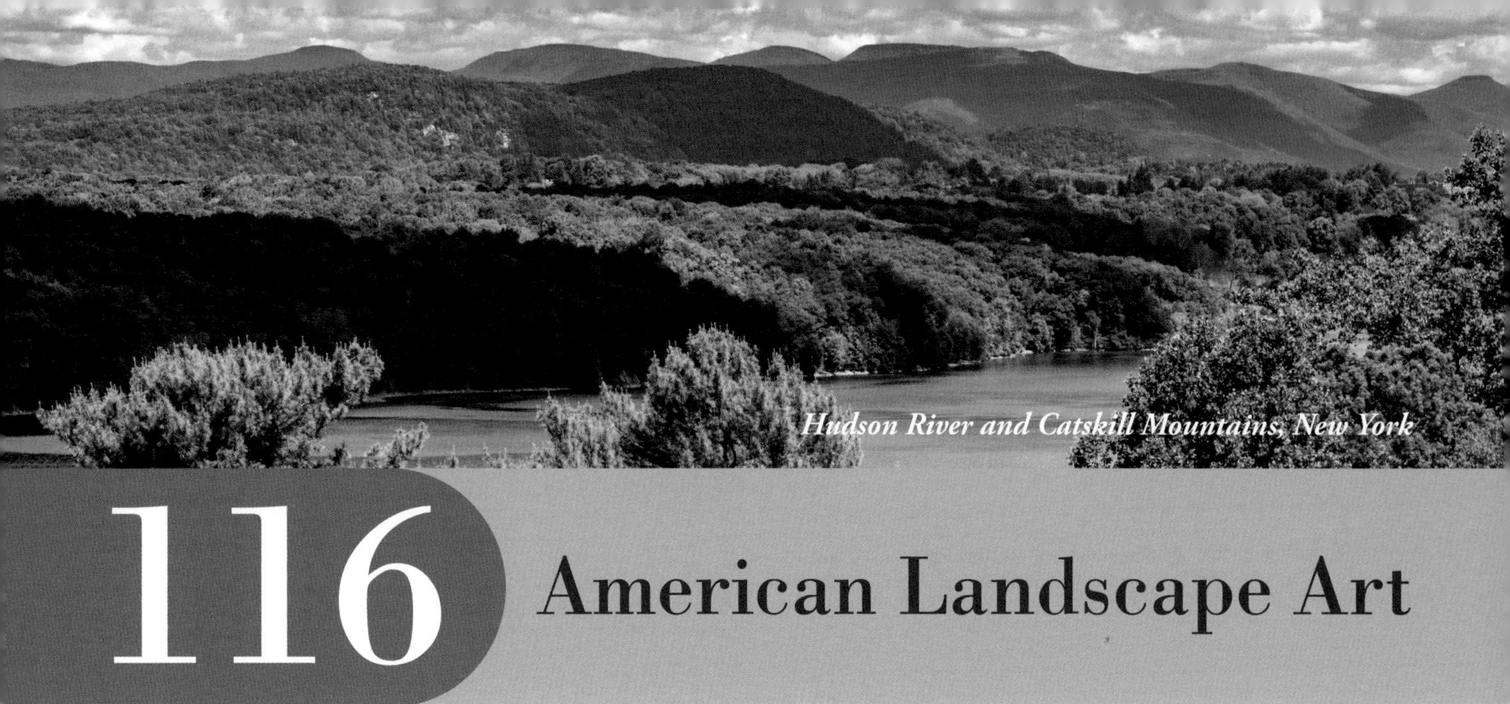
Hudson River and Catskill Mountains, New York

116 American Landscape Art

When you stand in awe of a beautiful scene, whether it is a tumbling waterfall, majestic mountains, endless prairie, or sweeping valley, you are admiring geography. Geography is politically strategic, economically important, and scientifically fascinating; but in many instances it is also beautiful. Since human geography examines how people have interacted with geography, this lesson focuses on how a group of artists interacted with the geography of America.

America had a self-image problem. The nation was dealing with some insecurities.

In the early 1800s, the United States had defeated Great Britain in war—twice. America's land policies had taken its borders all the way to the Pacific Ocean. The country had begun developing as the economic powerhouse it would become later in the century.

And yet, many countries in Europe still did not take the United States seriously. To nations that had been developing their national identities since the 1200s and before, a nation that had begun in 1776 seemed little more than an upstart. America's army and navy were tiny. The nations of Europe, having dealt with the threat of Napoleon twice, considered America's claim to authority in the Western Hemisphere as expressed in the Monroe Doctrine, and smiled faintly. Europeans doubted whether the U.S. could back up its bold words with bold actions, at least without help from the British navy.

As far as intellectual culture is concerned, to many Europeans Americans still seemed to be babes in the woods. Only James Fenimore Cooper had achieved any level of literary success in Europe. Aspiring American artists studied in France and Germany because the United States had nowhere that provided what was considered to be world-class training.

An informal group of American painters began to change that, starting in the 1820s. The United States had given rise to several accomplished portrait painters, such as Gilbert Stuart and Charles Willson Peale. Americans had produced renderings of historic scenes, such as the Boston Massacre and the signing of the Declaration of Independence. While American artists had painted famous and influential Americans and some coastal scenes, no one had yet really captured the grandeur, beauty, and possibility of America as a place, a nation. The artists in this group knew the profound beauty that was there, waiting to be revealed to the world and even to most Americans.

The Hudson River School

A Pic-Nic Party
by Thomas Cole (American, 1845)

During the period from approximately 1825 to 1870, a group of artists began to capture the beauty of American geography. Their works got the attention of art critics and the general public. The usual term for their work was landscape art. An art critic writing in 1870 dubbed the group the Hudson River School because much of their work centered on the Hudson River Valley and the Catskill Mountains of New York State.

Out of their love for and awe of the beauty of nature, these artists attempted to show the splendor, majesty, and remoteness of the American interior. The scenes they painted were serene, quiet, grand vistas of woodlands, rivers, mountains, and cliffs. They portrayed untamed wilderness and hinted at the limitless resources and the promise of prosperity that this country, blessed by God, contained. There was even a hint of manifest destiny as some paintings showed settlers moving west into the amazing beauty. Those who saw the paintings could imagine the greatness that lay before America, perhaps even culminating in the coming of the millennium of Christ, which was an idea that gained popularity at the time.

These works were part of the Romantic movement. The late 1700s had seen the rise of the Enlightenment and an emphasis on reason. While reason is necessary and has a proper place in God's order, it does not always satisfy the human soul's God-given desire for beauty and wonder. Romanticism in literature and art centered on human feelings and the emotional impact of a work. These paintings were intended to be inspirational.

Harbor Landscape
by Thomas Doughty (American, 1834)

Romanticism was also a response to the increased industrialization taking place in America at the time, and these artists emphasized the natural environment. The scenes portrayed in landscape art were not the towns and cities where businessmen built factories. There was also a degree of idealism in these works. The scenes that artists portrayed were sometimes even more beautiful than what people actually saw in those places. In addition to painting small farming villages, these artists often showed people on outings or passing through a place, not the cities where many people lived. The artists did not usually portray indigenous Americans or African American slaves, groups about whom the U.S. was having serious policy conflicts at the time.

In terms of technique, these artists skillfully used lighting to compose their works. Contrasts of light and dark were common; a thunderstorm in one part of a painting might play off sunshine in blue sky in another part. In portrait painting, the figure of the subject dominates the canvas. By contrast, in landscape art the grand scene is the subject and the people are a small part of it. Their size provides perspective on the size of the setting.

The Hudson River artists were the first school of painting that developed in the United States. Although some of them studied in Europe, they celebrated their independence from the traditions of European art. There were, after all, no castles in America to paint; but these artists found worthwhile settings to portray what America had to offer.

The later generation of this school moved further west and portrayed scenes west of the Mississippi River. These latter works are known as Western American Art (the term Western Art usually refers to art of Western Civilization). Some artists are called the Rocky Mountain School. Some of the locales featured in these works were even more untouched than the Hudson River Valley. Many of these paintings portray everyday life for cowboys and ranch hands. Ranch animals, especially horses, are prominent. Some paintings and sculptures convey stories in which action is taking place.

American Landscape Artists

Many of these artists knew each other or studied under each other. Many bought homes in the Hudson River Valley.

Thomas Doughty (1793-1856) was one of the first American artists to do landscapes. He took some art classes and saw some European landscape paintings, but he was largely self-taught. One of his best-known works is "Harbor Landscape."

Thomas Cole was one of the most influential of the Hudson River artists. He was born in England in 1801, and his family emigrated to the U.S. in his teen years. Cole studied in Europe. His works include "Schroon Mountain" (1838), "A View of the Two Lakes and Mountain House, Catskill Mountains, Morning" (1844), and "The Oxbow" and "A Pic-Nic Party" (both 1846). Cole became a close friend of Romantic poet William Cullen Bryant. He died in 1848 in his home in the Catskill Mountains of New York.

Asher Brown Durand (1796-1886) studied in Europe. Thomas Cole was a major influence on Durand's work. One of Durand's most famous works is "Kindred Spirits." Durand completed this work in 1848 as a tribute to Thomas Cole, who had died earlier that year. In the painting Cole and his friend William Cullen Bryant are standing together in an idealized setting in the Catskills. Bryant had delivered the eulogy at Cole's funeral. One of Durand's students was Mary Josephine Walters (1837-1883), one of a few female artists of this school. One of her best-known works is "Landscape with Three Ladies Seated Under a Tree."

Frederic Edwin Church (1826-1900) was one of the most commercially successful painters of this school. He studied under Thomas Cole. Church's works were large and dramatic, such as his "Niagara" of 1857. Church branched out to portray other landscapes besides the United States. His "Heart of the Andes" (1859) and "The Icebergs" (1861) are two examples. Church sometimes presented single

Kindred Spirits
by Asher Brown Durand (American, 1849)

Heart of the Andes
by Frederic Edwin Church (American, 1859)

Chasm of the Colorado
by Thomas Moran (American, 1873-74)

Looking Down Yosemite Valley, California
by Albert Bierstadt (American, 1865)

painting exhibits, as did other artists. For "Heart of the Andes," he arranged a series of public showings. Thousands of patrons paid twenty-five cents each to view that one work. Church died in his house on the Hudson River.

Thomas Moran was born in England in 1837. His family emigrated to Pennsylvania when he was a small child. Moran's work had a significant impact on American policy and interest in the Yellowstone region. In 1871 Moran accompanied an exploratory party to Yellowstone and made several sketches. Before then, many Americans considered Yellowstone as a dangerous place with its geysers, hot springs, and bubbling mudpots. Moran showed Congress his pictures, which proved that Yellowstone was really a natural wonderland. Congress established Yellowstone as the first national park in 1872. Moran completed "Grand Canyon of the Yellowstone" also in 1872, and the 7' x 12' work hung in the U.S. Capitol. Moran painted "Bridalveil Falls" in Yosemite Valley in California, the "Chasm of the Colorado" (what we know as the Grand Canyon) in Arizona, and the "Mountain of the Holy Cross" in Colorado. In this latter geological formation, crevices in the side of a mountain retain snow after it has melted or blown off from the rest of the surface, leaving the shape of a cross.

Albert Bierstadt was born in Germany in 1830. He died in New York in 1902. Bierstadt was one of the last of the Hudson River School artists. He took many trips to the western U.S. but did much of his work in his New York studio. His use of light in his paintings became known as luminism. Among his many large works are "Storm in the Rocky Mountains" and "Looking Down Yosemite Valley, California."

Another popular American artist of this time period was Frederic Remington (1861-1909). Educated at Yale, Remington traveled in the West, like Bierstadt, sketching and taking photographs; but he did much of his work in New York City. His paintings usually tell a dramatic story and convey much motion and energy. He did considerable work as an illustrator for popular magazines of the day. Remington was also an accomplished bronze sculptor. He completed his first sculpture, "The Bronco Buster," in 1895.

People interact with geography in their work, recreation, and everyday lives. Some people record this interaction through art. These works convey meaning and emotion to observers, who through them can develop a greater appreciation of the wonders of God's Creation.

From the end of the earth I call to You when my heart is faint;
Lead me to the rock that is higher than I.
Psalm 61:2

Assignments for Lesson 116

Gazetteer Read the entries for Bermuda and Greenland (pages 198 and 200).

Worldview Copy this question in your notebook and write your answer: The Israelites who came out of Egypt passed miraculously through the Red Sea. The Israelites who entered Canaan miraculously passed through the Jordan River. What significance do you see in God giving each generation a similar geographic experience?

Project Continue working on your research paper. Plan to finish it by the end of Unit 25. ***or*** Choose your project for this unit and start working on it. Plan to finish it by the end of this unit.

Literature Begin reading *The Country of the Pointed Firs and Other Stories*. Plan to finish by the end of Unit 25.

Student Review Answer the questions for Lesson 116.

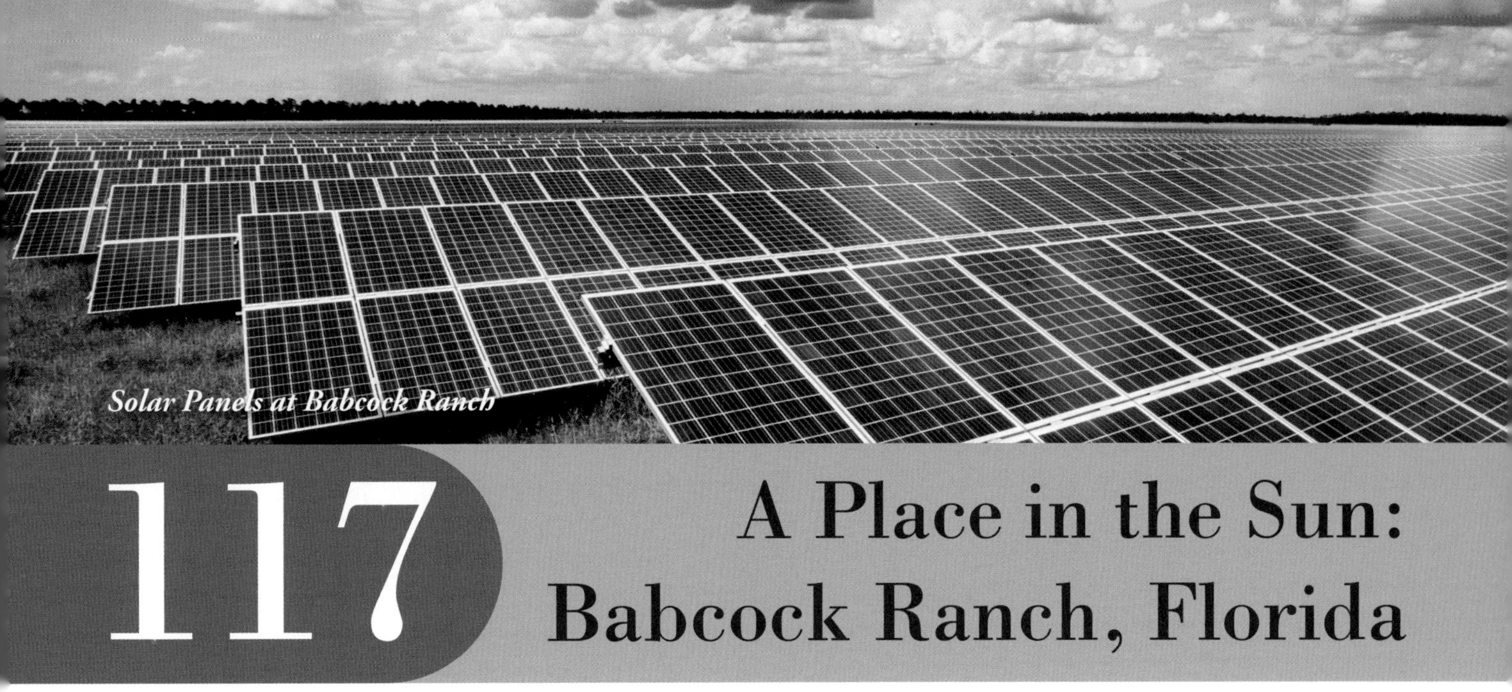
Solar Panels at Babcock Ranch

117 A Place in the Sun: Babcock Ranch, Florida

The day starts out typically.

Mom and Dad come in from their morning walk. Mom prepares breakfast and listens to music on the radio, while Dad checks the news and his email on the computer. Sister turns on the light in her closet to choose her outfit for the day, while Brother goes over his essay on his computer one more time.

In the school building the custodian turns on the lights in the halls and classrooms. The teachers and librarians turn on their computers and audio-visual equipment. Workers in the stores in town turn on their displays and point-of-sale computer terminals. In his office the physical therapist consults his computer to review the list of patients he will see today. A retired couple enjoys a video chat with their daughter and grandchildren who live several hundred miles away.

And so the typical day proceeds in typical fashion in typical homes and offices. However, this community is anything but typical. Despite all of these activities, the town's use of electricity from the utility power grid is almost zero. All of that electricity came from solar power. This is not science fiction or the community of tomorrow. It's happening today in the town of Babcock Ranch, Florida.

As we have said before, one aspect of geography is the weather. Human geography involves how the weather affects human life. The weather where people live affects how they live. Where people live influences the sources of energy they use, whether it is wood, coal, petroleum, nuclear energy, or another source. For Babcock Ranch, the Florida sunshine provides the energy they need for a typical day's activities.

Babcock Ranch is a planned community in southwest Florida near Ft. Myers. It is the country's first all-solar city. The vision for this town began in 2006 in the mind of developer (and former NFL player) Syd Kitson. He and his company spent a decade planning the effort before the first spade of dirt flew.

From Gold Mines to Golden Sunshine

Perry McAdow made a fortune in the gold mines of Montana in the late 1800s. Seeking a warmer climate, McAdow and his family purchased a large tract of property in Florida and named it Crescent B Ranch. McAdow started a bank with $15,000 capital in the town of Punta Gorda. Other businesses followed. McAdow's wife set about beautifying the town.

In 1914 Pittsburgh businessman Edward Babcock purchased the McAdow property and renamed it Babcock Ranch. Babcock's son Fred eventually took over the operations of his father's numerous and diverse business interests, which included selling pine pitch to South African diamond mines to help ward off termites.

Kitson and Partners purchased the 91,000-acre ranch from the Babcock family in 2006. They immediately sold 73,000 acres to the state of Florida for a permanent conservation site. This was the largest preservation land deal in Florida history.

The Babcock Ranch Solar Energy Center began construction in 2015 and opened just over a year later. It sits on 440 acres that Kitson donated and contains 343,000 solar panels. Florida Power and Light (FPL) owns the facility, which can generate 74.5 megawatts of electricity. Ten large battery storage units can each store one megawatt of power and can discharge for four hours. At the time it was built, it was the largest solar-plus-storage system in the country. Babcock Ranch power customers pay the same rate as FPL customers around the state.

The town does have to tap into the power grid at night and in cloudy weather. The technology required for storing excess energy that the solar cells create is still too costly. Generally, however, the facility produces more electricity than the town uses. In addition, the new homes are relatively more energy efficient and require less power. The solar panel facility took a direct hit from Hurricane Irma in 2017 without losing a single panel.

Life in Babcock Ranch

Other construction work on Babcock Ranch got underway in November 2015. Projects included homes, businesses, schools, underground utilities, roads, and parks. Work has progressed on environmental projects as well, including restoring wetlands. Homeowners began taking up residence in early 2018. The town celebrated its grand opening in March of that year. House prices vary from about $200,000 to about $1 million.

Babcock Ranch was built with hiking and biking trails and opportunities for fishing, canoeing, and kayaking.

Founder's Square at Babcock Ranch

Kitson expects the town's 18,000 acres (half of which is set aside for green spaces) eventually to hold about 20,000 homes, mostly single-family dwellings with small yards, and to house a population of about 50,000. Walkways, bikeways, and the availability of driverless shuttle buses encourage people to leave their cars in their garages.

Other environmental goals for the town include efficiency of water use, nurturing native plants, tree preservation and relocation, and meeting Florida Green Building standards. Town planners will encourage residents to plant gardens in community plots. Babcock Ranch provides community WiFi and gigabit Internet in homes and offers car and bike sharing programs.

Can This Happen Elsewhere?

Kitson recognizes the benefit that comes from starting from scratch. "We had the advantage of a green field, a blank sheet of paper. When you have a blank sheet of paper like this, you really can do it right from the beginning." Developers didn't have to remove an older electricity-producing plant or retrofit older construction with new technology.

Babcock Ranch is not the only place that has gone fully solar. Cochin International Airport in the southern Indian city of Kochi is the world's first fully solar-powered airport. The banks of solar panels on adjacent land, the roof of the parking garage, and above a nearby canal produce 29 megawatts of electricity. Officials made plans to increase the output to nearly 40 megawatts as airport traffic increased. Excess energy produced during the day is fed back into the city's grid and then used at night. The airport's managing director said that savings on electricity enabled the airport to recoup its investment in less than six years. In addition, people grow organic vegetables under the panels and sell them to airport staff—sixty tons in one year! Airports in other places have been looking at the Kochi facility to see what might be possible in those locations.

In 2020 General Motors announced plans to make its assembly plant in Spring Hill, Tennessee, completely solar powered by 2023. The facility is GM's largest manufacturing plant in North America. GM plans to buy electricity from a solar farm near Columbus, Mississippi, about 175 miles away. The company's Spring Hill facility sits on 2,100 acres, of which 700 are used for farming and 100 for a wildlife habitat.

Can people use solar power to this extent anywhere? Obviously not. Areas with fewer daylight hours part of the year or with consistently cloudy weather don't provide the same potential for solar-generated electricity that (mostly) sunny Florida does. However, as technology develops, more will be possible in more places. Of course, building, installing, and maintaining solar panels requires energy also; but the goal is to use less net fossil fuels overall.

Psalm 19 speaks of the sun as a mighty work of God. In the heavens . . .

. . . He has placed a tent for the sun,
Which is as a bridegroom coming out of his chamber;
It rejoices as a strong man to run his course.
Its rising is from one end of the heavens,
And its circuit to the other end of them;
And there is nothing hidden from its heat.
Psalm 19:4b-6

Assignments for Lesson 117

Gazetteer Read "The Song of the Chattahoochee" (pages 321-322), and answer the questions in the *Student Review Book.*

Worldview Copy this question in your notebook and write your answer: If you could choose one place to accompany Paul in his ministry, where would it be and why? Would it be at the Areopagus (or on Mars Hill) in Athens, in another major city he visited, on his voyage to Rome, or some other place?

Project Continue working on your research paper (or project).

Literature Continue reading *The Country of the Pointed Firs and Other Stories.*

Student Review Answer the questions for Lesson 117.

Diomede Islands Seen From Alaska

118 Together and Apart: The Diomedes

These relatives hardly ever see each other. Their ancestors once lived close to each other and visited often, but today's family members live in different countries; in fact, they live in different days of the week.

One Country Becomes Two

The Danish explorer Vitus Bering, sailing on behalf of Russia, on August 16, 1728, sighted two islands in the strait that came to bear his name. That day was the feast date of the Orthodox saint Diomede (or Diomedes), who lived in Asia Minor and died around 300 AD. Thus Bering gave the islands their names. Bering sighted the Alaska mainland on another voyage in 1741 and claimed it all as Russian territory.

In 1867 the United States bought Alaska from Russia. Many Americans ridiculed the purchase, calling it Seward's Folly and Seward's Icebox because Secretary of State William Seward encouraged the deal. Today most Americans see the transaction as one of the best land deals in American history. Besides, the czar of Russia needed the money.

As a result of the purchase, Russia ended in Siberia and American territory began in Alaska. At its narrowest point, the Bering Strait is 51 miles wide. There had to be a dividing line between the two countries, but where? Obviously the line would be in the Bering Strait that divided Asia from North America, but where exactly? The decision was made to draw the border between the two islands in the Bering Strait. Big Diomede to the west belonged to Russia. It is the easternmost point in that country. Little Diomede to the east belonged to the United States.

The Russian-American border coincided with the International Date Line that ran between the two islands. Thus these two largely granite islands, less than three miles apart, became parts of different countries, are considered parts of different continents, lie in different time zones, and even exist in different days of the week. It is all another example of the interaction of people and geography.

The People of the Diomede Islands

However, the two governments involved in the land deal didn't bother to talk to the people who lived on the islands. They were members of the Inupiat nation and had lived in harmony with each other for centuries. They continued to visit with each other for almost a century after the sale, but in 1948 the Communist government of the Soviet

Union closed the border. What is more, the Soviets moved all civilians off of Big Diomede and resettled them along the Siberian coast. The visits among the relatives who had lived on the two islands ceased.

The Soviets built a weather station on Big Diomede as well as a military installation to keep an eye on that part of the United States, their enemy in the Cold War. The U.S. kept an eye on Russia from Little Diomede also. The line in Europe between Communist countries and democratic countries came to be called the Iron Curtain. The line between Big Diomede and Little Diomede came to be called the Ice Curtain.

Life on the Diomedes

The only village on Little Diomede is Inalik, also called Diomede. It lies on the western shore of Little Diomede facing Big Diomede and is home to about 80 people. The residents live in homes huddled between the shore and a tall cliff behind them. They have no roads or vehicles. The people get around on walkways and steps. They receive mail and supplies from a helicopter that comes once a week, weather permitting. During the winter, villagers clear a short runway on the ice sheet so a small plane can make deliveries. Since the surface of the ice sheet changes from year to year, the location of the runway is never the same.

On Big Diomede, Russian soldiers and government workers rotate in and out for their tours of duty. Relations between Russia and the U.S. can blow warm and cold, but this border is fairly relaxed. The troops there do not live in a state of high tension. As one soldier recalled, there was mostly a lot of white and snow. Again, helicopters bring in supplies, weather permitting.

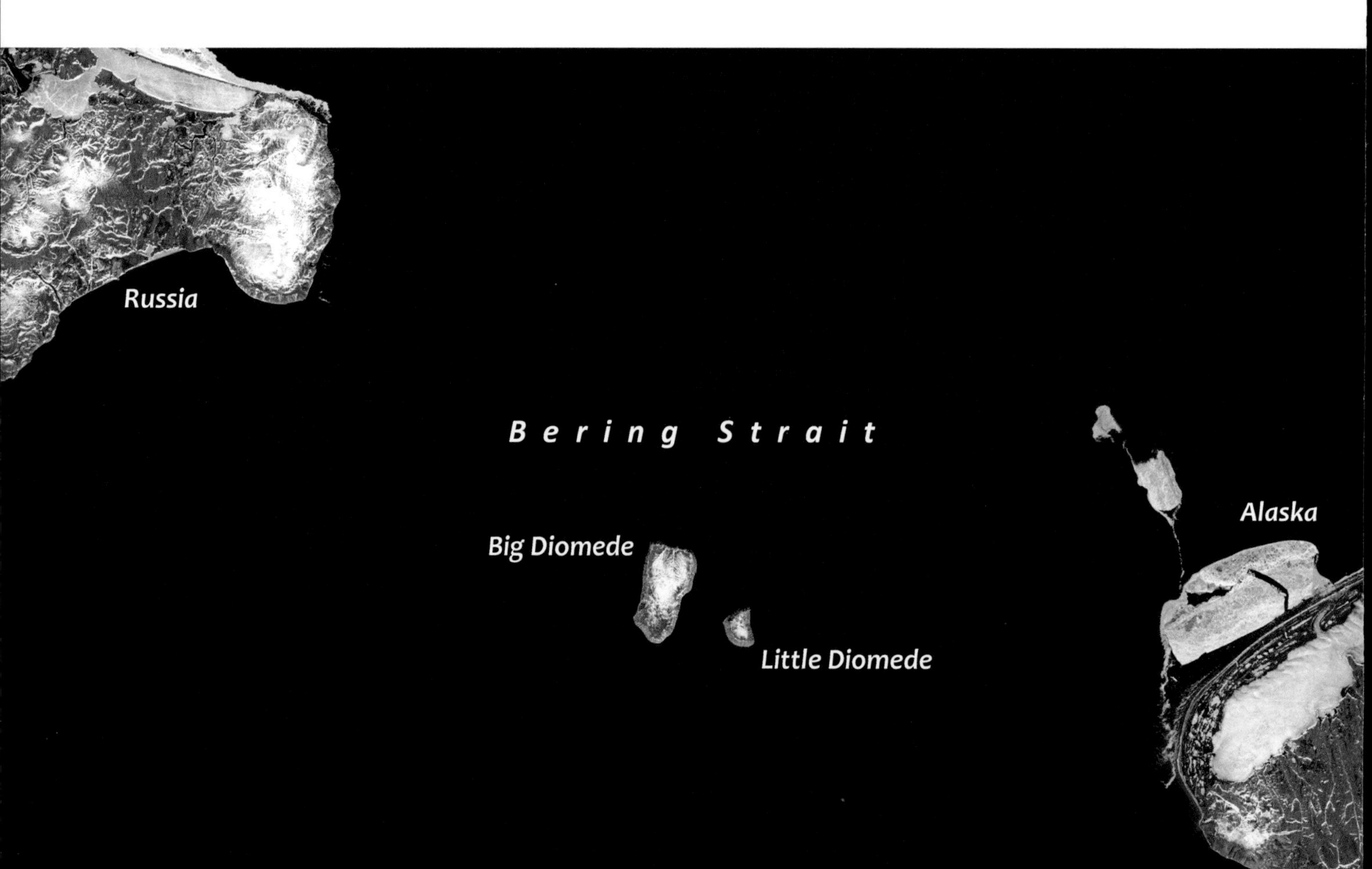

Inalik, Alaska, on Little Diomede

The weather on both islands is usually windy and is often foggy when there isn't snow. In terms of topography, Big Diomede is about the same shape as Little Diomede: steep cliffs that rise from the ocean to a plateau that covers most of the island. The difference is their size. Big Diomede covers about eleven square miles; Little Diomede, about three square miles.

Relations between the two islands are calm, but the residents of Little Diomede long to be able to see their cousins who once lived on Big Diomede. After the Soviet Union ended in 1991, the people on Little Diomede were hopeful that travel restrictions would be eased. One expedition did take place, and the Inupiat who went found relatives in the villages scattered along the Siberian coast. However, easy travel has not developed for the people there.

Potential Issues

Things are calm in the Bering Strait now, but that could change because of two developments in the Arctic. For one, the region is believed to have significant oil and natural gas reserves. If countries agree on how these reserves might be recovered, the strait might see a significant increase in the traffic of oil tankers and well construction and supply vessels.

In addition, a Northwest Passage has opened in the usually frozen waters north of Canada, which vessels can use during a few months of the year (see Lesson 111). Since the Bering Strait is the western sea exit for the Arctic, the warmer months (relatively speaking) have seen more transport ship traffic coming through there. Increased regulation, heightened security, and potential conflicts could occur.

Another development lies many years in the future, if it happens at all. Russia has proposed digging a 64-mile tunnel under the Bering Strait to connect the two continents. This tunnel would be twice as long as the Chunnel beneath the English Channel that connects Great Britain and France. The American government is open to the idea, but so far no real action has taken place. Such a tunnel would be a massive undertaking.

One Way to Connect the Islands

In 1987 American long-distance swimmer Lynne Cox swam from Little Diomede to Big Diomede. She took on this challenge to try to ease Soviet-American relations. Cox jumped into the 40-degree water that was shrouded in fog. She had a medical team and advisers with her and received help from Inupiat guides in kayaks, but they weren't entirely sure of the route. Without accurate guidance, she might have missed Big Diomede.

Finally a Russian skiff emerged from the fog and guided her to shore. Two Russian soldiers pulled her from the water. The Soviets had a small welcoming party waiting for her. A female Russian doctor led her into a tent, covered her with hot water bottles, put her inside a sleeping bag, and hugged her. To Cox, this welcome and embrace symbolized the reason for her making the trip.

When Soviet president Mikhail Gorbachev traveled to Washington to sign a nuclear weapons treaty later that year, he and President Ronald Reagan honored Lynne's accomplishment. Gorbachev said:

> Last summer it took one brave American by the name of Lynne Cox just two hours to swim from one of our countries to the other. We saw on television how sincere and friendly the meeting was between our people and the Americans when she stepped onto the Soviet shore. She proved by her courage how close to each other our peoples live.

The people of the Diomedes have given the islands nicknames. They call Big Diomede Tomorrow Island, and Little Diomede Yesterday Island.

Wherever we are in space and time, Jesus is the same:

Jesus Christ is the same yesterday and today and forever.
Hebrews 13:8

Assignments for Lesson 118

Worldview Copy this question in your notebook and write your answer: God gave a place for Israel, and He plans a place for the church in heaven. Between these, the church does not have a particular geographic setting on the earth that is its place. Why do you think this is true?

Project Continue working on your research paper (or project).

Literature Continue reading *The Country of the Pointed Firs and Other Stories.*

Student Review Answer the questions for Lesson 118.

Prairie, Southwest Minnesota

119 Little House on Geography

Millions of readers around the world know about Laura Ingalls Wilder's growing-up years from her accounts of life on the American prairies. Laura lived on stretches of geography that gave her grasshopper plagues and severe blizzards but also beautiful wildflowers and long summer days.

The backdrop for Laura's stories, the places where those stories occurred, is the geography on which she lived. That geography was at once comforting and challenging, enticing and frightening. The places where she lived shaped her stories and her life.

Laura experienced life in several different biomes. A biome is a naturally occurring environment with distinctive characteristics that serves as a home for a community of plants and animals. A biome includes the habitats of particular plants and animals. Ma (Caroline) Ingalls commented on this reality in *By the Shores of Silver Lake*, when she said, "We're west of Minnesota, and north of Indian Territory, so naturally the flowers and grasses are not the same." Laura probably never used the term biome, but she understood the principle.

Caroline ("Ma"), Grace, Laura, Charles ("Pa"), Carrie, and Mary Ingalls in De Smet, South Dakota (c. 1894)

The Great American Desert

America was born in the heavily wooded landscapes of the Atlantic coast. It grew across the Appalachian Mountains where forests were still the norm. An old saying held that a squirrel could jump from tree to tree from Maine to Mississippi and never touch ground, and that was not far from the truth. As America stretched further west across the Mississippi, however, settlers encountered a different kind of landscape.

America's purchase of the Louisiana Territory presented the country with a vast expanse of geography about which most Americans knew little. The Lewis and Clark Corps of Discovery revealed to the country the nature of the upper reaches of the Louisiana Purchase, but much of what lay west of the Mississippi was still unknown. In 1819-1820 U.S. Army Major Stephen Long surveyed parts of Oklahoma, eastern Colorado, and Nebraska to help the federal government and the American people know more about what they had bought.

Long's report was not favorable. He described it as "wholly unfit for cultivation, and, of course, uninhabitable by a people depending on agriculture." His term for the region was "the great American desert." The specific area that Long surveyed was similar to and not far from the geographic places where Laura Ingalls lived during her childhood.

Plains are any relatively level area of the earth's surface. They usually have a gentle slope with small local areas of hills or valleys. Plains might be covered with trees or grass, bushes or even desert. About one-third of the earth's surface is identified as plains. The Great Plains of North America extend from the Rio Grande on the southwest Texas border to the Arctic. These plains cover about one-third of the land area of the continental United States and parts of ten states. The High Plains, which feature a higher average elevation above sea level, are part of the Great Plains.

Prairie is from the French word for meadow. It is the term for a level or rolling grassland. One characteristic of the American plains and prairies is a relative scarcity of trees. Prairies cover much of the Great Plains but also extend across much of the upper Midwest that lies east of the Mississippi, across Ohio, Indiana, and Illinois. One nickname of Illinois is the Prairie State.

Weather features of the plains and prairies include an almost constant wind (no tall hills to serve as windbreaks) and for places further north, severe winter blizzards. With these geographic features and the temperature differentials provided by Lake Michigan, it is no surprise that Chicago is called the Windy City. These features suggest some of the challenges that the Ingalls family faced as they carved out a life on the Great Plains of the United States. Ironically, Laura's adult life took place primarily on another kind of geography in the Ozark Mountains of Missouri.

When Laura's older sister Mary became blind, Pa (Charles) Ingalls told Laura that she would have to become Mary's eyes and describe to her sister what things looked like around them. Many believe that this practice helped Laura develop into the skilled descriptive writer she became. Her books give vivid accounts of the landscapes she experienced. She described not just wildflowers but dozens of specific kinds of wildflowers. She wrote about the majestic sunsets she saw and the lonely beauty of her little houses set on vast stretches of rolling farmland.

The Big Woods: Pepin, Wisconsin

Laura was born on February 7, 1867, in her parents' cabin seven miles north of Pepin, Wisconsin. Pepin is on the eastern shore of the Mississippi River. On the western shore is Minnesota. Just south of Pepin, the Chippewa River flows into the Mississippi. The delta of the Chippewa has formed Lake Pepin in the Mississippi. The lake is two miles wide at most and about twenty miles long.

Pepin and the big woods around it lie on rolling land amid many tall limestone cliffs along the river. A few miles upstream from Pepin is Maiden Rock, which looms 400 feet above the river. The legend associated with the rock is that the daughter of

Reconstructed Ingalls Cabin, Pepin, Wisconsin

the Dakota Sioux chief leaped to her death after her father had her beloved, a member of the rival Chippewa tribe, put to death. It makes for a dramatic story, one that is similar to stories attached to several other high points around the country.

Pepin is at the northern edge of the American broadleaf forest (oak, hickory, black cherry, walnut, and hazelnut trees). To the north and east of Pepin begins the boreal forest of pine, spruce, tamarack, birch, beech, and maple that extends into Canada. Where Pa built this little house was near the western edge of the line of cleared trees at the time, as settlers moved into the still-growing United States.

A reproduction of the Ingalls house, which sits on land that was once part of the Ingalls' farm, has a few trees immediately around it, but most of the area is now cleared farmland. Farming has been a major economic activity in that area for many years. The wheat grown in the region was a major reason why the Gold Medal Flour mill and other mills were located nearby in Minneapolis at St. Anthony Falls on the Mississippi River. The Mississippi and its tributaries made up the transportation superhighway that served a large part of the nation. Logging (in the Big Woods) was another major industry in the area that utilized the Mississippi for transportation.

Pepin, Pepin County, and Lake Pepin are named for two French brothers, explorers and traders, to whom Louis XIII made a land grant in the region. Their name became associated with this place. Laura mentions the lake in the first chapter of *Little House in the Big Woods*. She mentions it again in the first chapter of *Little House on the Prairie*, when she describes her family moving west.

Pa came to think that the Big Woods were being built up and getting too crowded (crowded is a relative term; the current population of Pepin County is still less than 8,000). He had a buyer for their cabin, so in 1869 he packed up his family and goods and drove their wagon across the frozen lake in winter. Pa said they had to move then so they could get across the river.

Kansas Wildfire

Little House on the Prairie: Kansas

The Ingalls made the slow trek of over six hundred miles in their wagon to land near Independence, Kansas, in 1869. The area is another place of rolling fields but with much fewer trees except for those that grow along the creeks in the region. Near the site of their house are escarpments, tall ridges also known as *cuestas* (Spanish for slopes). Also nearby is the Verdigris River. As Laura described it, "Kansas was an endless flat land covered with tall grass blowing in the wind. Day after day they traveled in Kansas, and saw nothing but the rippling grass and the enormous sky." A hardship they encountered more than once in places they lived was the danger of prairie wildfires, caused either by lightning or by human carelessness.

To young Laura, their land in Kansas was a place where "no one lived." Coming from the woods of Wisconsin to which many people were moving, it must have seemed so. Their nearest neighbor was two miles away. Unfortunately for them, people did live there. The Ingalls had moved onto the Osage Indian reservation. This made the Ingalls squatters, which meant that they had to move again after less than two years in Kansas.

The Ingalls returned to their cabin near Pepin. The family that had purchased it was already planning to move west, so the Ingalls moved back in when they did so. They stayed from 1871 until 1874 when, on Laura's seventh birthday, they once again crossed the frozen Mississippi. This time, they traveled pretty much due west for about two hundred miles and came to the community of Walnut Grove in southwest Minnesota.

On the Banks of Plum Creek

Walnut Grove got its name from a hundred-acre grove of black walnut trees about a mile from the railroad depot. This was about the northern limit of the tree's habitat.

The Ingalls settled north of town in certainly their most dramatic encounter with geography. They lived in a dugout on the bank of Plum Creek that the previous owner of the land had created. A roof of willow branches and sod extended out from the bank; the front wall that contained the door and a window was made of boards. Laura recalled that their "house" scarcely had more room than their wagon. But the dugout was another example of people using the materials they had to build their homes. Where there were trees, they built log cabins. Where they had mud, they built pueblos. Where they had blocks of sod from breaking up the prairie, they built homes of sod blocks. The roots of prairie grass went down several feet below ground level, which made plowing up the fields difficult but made for strips of sod that could be cut into blocks that held together well.

Plum Creek, Walnut Grove, Minnesota

Although the Ingalls enjoyed many things about where they lived on Plum Creek, including eventually building a real house on top of the ground, a deciding factor for them to move was the recurrent grasshopper infestations. Millions of the insects would darken the sky, cover the ground, and eat every growing thing. The newly turned ground was a perfect place for them to lay their eggs. Thanks to insecticides such plagues are largely a thing of the past, but in the late 1800s they were a disheartening reality on the American prairie. After a second infestation in two years, Pa Ingalls decided in 1876 that he had had enough.

Sad Times in Burr Oak, Iowa

This time Pa moved his family east. They spent a little time in southeastern Minnesota with the family of Charles' older brother, who had married Ma's younger sister. This other Ingalls family lived in Wabasha County on the Zumbro River. This began a sad time for the Ingalls family. Laura's little brother Freddie, who had been born on the banks of Plum Creek, died when he was less than a year old in August of 1876.

Charles moved his family again in the fall, this time south, just over the Iowa state line to the town of Burr Oak. Burr Oak had once been a key hub on the busy routes of western migration, but by 1876 the town had declined in importance. The Ingalls family lived and worked in a hotel, and then Charles worked in a mill. In their work at the hotel, the Ingalls came into contact with some rough characters, especially because of the saloon that was nearby. The family lived in three places in town, and life was not easy for them there. Because of the loss of little Freddie and the difficulties in Burr Oak, Laura did not write about this period of their lives, even though little sister Grace was born there in 1877. They moved back to Walnut Grove late that year and lived in town for about two years.

Little Town on the Prairie: De Smet, Dakota Territory

Miners had discovered gold in the Black Hills of western Dakota Territory in 1874, and a large number of settlers began to move to the region. Railroads capitalized on the movement west and built lines to carry passengers and freight. Pa Ingalls took the opportunity to go to work for one of the railroad companies. In 1879 he went ahead by rail to Dakota Territory, then came back by wagon to Tracy, Minnesota, to meet his family. They settled in a railroad camp that became the town of De Smet. The town was named for a beloved French priest who ministered to native nations in the area.

Their first winter in this little town on the shores of Silver Lake (the lake has since dried up) they lived in the railroad surveyor's house. In February of 1880 Pa filed a homestead claim for 160 acres outside of town that featured a gentle rise. The 1862 Homestead Act allowed people to file a claim for a small fee. If they lived on the claim for at least six months every year for five years and produced a crop, the land was theirs. Pa also bought property in De Smet and built storefronts. The family lived in one of them during the harsh winters.

The area was dotted with small lakes and sloughs (swampy areas) filled with six-foot-tall prairie grass. Another feature were buffalo wallows, amazingly round depressions that buffaloes created by rolling around to get dirt on their backs to prevent insect bites. Some wallows were as large as two acres.

It didn't take long for the Ingalls to encounter the harshness of life in the Dakota Territory. A blizzard struck the area in mid-October of 1880. Numerous snowstorms followed over the succeeding months of that long winter. Trains were not able to bring supplies to De Smet and other towns on the rail line. Almanzo Wilder and Cap Garland made a brave run across the frozen prairie to locate a farmer who had wheat they could buy that helped the town survive. The first train made it through to De Smet in mid-May of 1881.

Ingalls Homestead Site, De Smet, South Dakota

One day little Grace asked, "What's a tree?" In response, Pa planted five cottonwood trees on his claim, one each for Ma and their four girls. Four of the now-large cottonwoods still stand. Pa still had itchy feet and wanted to move to Oregon, but Ma refused. She had gone far enough west. Pa eventually built a house (which still stands) in town, and Ma and Pa lived out their later years there. Five of Laura's books are set in De Smet.

Little House in the Ozarks

On August 25, 1885, Laura Ingalls and Almanzo Wilder became husband and wife. They lived on Almanzo's claim to the north of De Smet, which was next to his tree claim. An 1873 law allowed a homesteader to own a second 160-acre claim if he agreed to plant forty acres of trees on it. Settlers on the plains felt a great need for trees to provide building materials and fuel. Homesteaders often planted a line of trees near their homes to serve as a wind barrier and to provide shade.

After the birth of their daughter Rose in 1886, life for the young couple was tremendously difficult. A hailstorm ruined their crops. Their house burned in a fire. Laura and Almanzo both contracted diphtheria, and Almanzo never fully recovered. Saddest of all, their second child, a boy, died soon after he was born.

Laura and Almanzo decided to leave South Dakota in 1890. They lived for a while with Almanzo's parents in Minnesota. The older Wilders had moved from Malone, New York, where Almanzo

Shown above are Laura and Almanzo Wilder with neighbors at Rocky Ridge. At right are photos of Almanzo plowing a field and Laura working in their garden.

had grown up as described in *Farmer Boy*. Then Laura and Almanzo moved to the Florida panhandle and lived near Laura's cousin Peter and his family, but that didn't work out well. At some point they lived for a brief time in central Missouri, a stay we know little about. Then in 1892 it was back to De Smet, but life for them there was too difficult.

In 1894 the Wilders succumbed to Ozark Fever, the fervent desire of many people to move into the beautiful Ozark Mountains area of Missouri. They traveled 650 miles in their wagon alongside another family. The Wilders bought a forty-acre tract near Mansfield, a town that had a railroad stop. The tract offered trees, creeks, springs, valleys—and hills! But the land was rocky and would require intense work to be made productive. The Wilders named their place Rocky Ridge Farm. They bought adjacent tracts and eventually owned two hundred acres. Almanzo and Laura built their home with their own hands, using lumber and stones from their own land. They farmed, kept livestock, and planted a vineyard. After so much trial and hardship, Laura and Almanzo had a good life at Rocky Ridge, although it was by no means easy.

Laura became a contributing writer for the *Missouri Ruralist* farm magazine, writing her articles by hand in the evening after working with Almanzo on the farm. At Almanzo's encouragement she began writing down memories of her childhood. When she was 65, she published *Little House in the Big Woods*, the first of eight novels about her growing-up years. Another book, *The First Four Years*, described their early married life. Almanzo, who was ten years older than Laura, died in 1949 at the age of 92. Laura died in 1957, three days after her ninetieth birthday.

Farming is the most basic way that people interact with geography. A farm family takes the land before them, clears it to produce crops, finds or brings water, and makes and keeps the land productive. Almanzo's father once told him, "Don't forget it was axes and plows that made this country." The lives of Almanzo and Laura Ingalls Wilder are testimony to the truth of this statement. In her writing, Laura recognized the hand of the One she called the Great Architect of the world in which she lived, the places she enjoyed so much, and which she observed so closely and described so well.

O Lord, how many are Your works!
In wisdom You have made them all;
The earth is full of Your possessions.
Psalm 104:24

Assignments for Lesson 119

Geography	Complete the map skills assignment for Unit 24 in the *Student Review Book.*
Project	Continue working on your research paper (or project).
Literature	Continue reading *The Country of the Pointed Firs and Other Stories.*
Student Review	Answer the questions for Lesson 119.

Laura visited her daughter, Rose, and son-in-law, Gillette Lane, in California in 1915.

Sea of Galilee, Israel

120 Geography in the Bible

Geography is the study of places, spaces, and faces.

A place is where something exists and/or where people live. Berlin, Germany, is a place; so is Victoria Falls on the border between Zambia and Zimbabwe in Africa.

Places have spaces, a term that refers to the spatial or physical characteristics of a place, such as rainforest or mountains, arid or humid. Spatial characteristics also include plant and animal populations, environmental factors, and any human population. The humans are where the faces come in.

In other words, we can describe a place by telling what it is like, what makes that place special or unique. The interaction of humans and places is what makes human geography.

Places have locations in two senses. They have absolute locations, which we can identify on a map or globe. For example, Chicago is located at 41.8781° north latitude, 87.6298° west longitude. Places also have relative locations in relation to other places. Chicago is located next to Lake Michigan and about 712 miles west of New York City.

A region is an area that is made up of places that share characteristics. Chicago is in the Midwest region of the United States, which features relatively flat land, agriculturally productive soil, and abundant water. The Sahara region of Africa, by contrast, has vast sandy deserts and only occasional sources of water.

This technical information and these formal definitions come alive when we see the importance of places in human life. We can clearly see the relationship between people and places in the story of the Bible.

Skeptics say that the Bible is a fairy tale; but fairy tales take place in fictional places, such as a faraway kingdom, an enchanted forest, or a rabbit's den. The Bible is not a story of some never-never land. Its narrative occurs in specific, known, identifiable places on the earth. Those places play an important role in the spiritual journey of God's people. This lesson makes repeated use of the term *place* to emphasize the role of geography in the narrative of the Bible.

Places in the Old Testament

When God created Adam and Eve, He gave them a place to live: the Garden of Eden. One of the consequences of their sin was that they had to leave that specially prepared place and go to some other place to live.

The significance of the flood was that it did not occur in just one place. It covered all places on the earth, in order to destroy all of mankind except the family of Noah. The ark came to rest in a certain place, and Noah and his descendants repopulated the earth. The particular place that the Bible mentions where people settled was the plain of Shinar. People built a city there and began to build a tower to heaven in that place, but the Lord confused their language and scattered them across the face of the earth. Thus people began to live in various places.

When God called Abram (Abraham) to begin the process of creating His chosen people, the Lord told Abram to leave the place where he was and go to the place He would show him. One aspect of the covenant God made with Abraham was that He promised to give Abraham's descendants a place, a Promised Land, where they would dwell (Genesis 15:18). That place, sometimes called Canaan, became crucial to the life and identity of God's chosen people.

The places that Genesis describes had a significant role in the events that the book relates. Abraham and Lot could not live in the same place because it could not sustain both of their households, so they separated and lived in different places. God later destroyed the place where Lot lived because of the wickedness there. Abraham wanted a place where he could bury Sarah after she died, so he purchased the cave of Machpelah (Genesis 23:8-9). This place was so important to the family that Abraham's grandson Israel (Jacob), while living in Egypt many years later, gave instructions to his family to bury him in the same place (Genesis 50:13). Abraham's sons, Isaac and Esau, developed great bitterness in their relationship; and they dealt with this division by residing in different and distant places as adults.

Through the working of God, Joseph lived for many years in Egypt and became a high official there. His brothers and their families went to live in that place to escape the famine in Canaan. Egypt thus became a place of restored hope, but it then became a place of oppression years later when the Egyptians forced the Israelites to become slaves. Ever after, Egypt was a place that the people of Israel associated with slavery and oppression. People often associate meanings and emotions with particular places. Just as Canaan became a place that meant home to Israel, Egypt became a place that represented bondage to the people of Israel.

Sinai Desert, Egypt

When God led the Israelites out of bondage in Egypt, He led them to and through many places: the Red Sea, Mount Sinai, the wilderness of Sinai for forty years, the Jordan River, and into Canaan, the place He had promised Abraham that He would give to Abraham's descendants. God described Canaan as a land flowing with milk and honey (Exodus 3:8).

God made a covenant with Israel at Mount Sinai. He said that He would be their God and they would be His people. Their relationship was paramount in the covenant, but an essential part of the covenant was the geographic place of Canaan that God gave them in which to live. Canaan provided Israel with the opportunity to live every day in the three-dimensional proof of His promise of identity, protection, and provision. God's offer to Israel was simple: stay faithful to Me and you can live in this place, but if you become unfaithful you will lose it.

Because of Moses' disobedience, God did not allow him to enter the Promised Land; but Moses was able to see it from another place, Mount Nebo, which is east of the Jordan River.

The Israelites conquered the Promised Land (albeit imperfectly) and each tribe received its own place where the families of those tribes could live generation after generation. Joshua used detailed geographic descriptions to identify the allotments of Canaan that God gave to the tribes of Israel when they took possession of the land. Joshua 13-19 gives a detailed description of the place set aside for each tribe. Joshua used such geographic features as rivers, valleys, mountains, springs, brooks, the Sea of Galilee, the Dead Sea, and established cities to describe the boundaries of the tribes' allotments.

The Israelites lived in these geographically defined areas for centuries. The place that God gave to Israel had streams, mountains, valleys, cropland, pastureland, and a small sea (the Sea of Galilee) where fishing took place. The place that became the heart of Israel was the geographically elevated place of Jerusalem in the hill country of Judea. There, on a still more elevated place, God chose for His name to dwell in the temple that Solomon built.

God offered and Israel accepted the simple arrangement regarding faithfulness and the place He gave them, but Israel violated that agreement. Israel's unfaithfulness to God also involved places. Following the example of the pagan Canaanites, the unfaithful Israelites conducted pagan worship on the high places of the land (see Numbers 33:52 and 1 Kings 13:32).

As a result of Israel's unfaithfulness, the invading Assyrians carried most of the people of the Northern Kingdom into a place of captivity. Later, invading Babylonians carried people of the Southern Kingdom into captivity in another place, Babylon. Babylon became the place that epitomized trial and punishment for the Jews. There the displaced Jews lamented their loss of the place that God had given them (see Psalm 137).

The books of Jeremiah and Daniel place special emphasis on Babylon. God kept His promise to the captives in Babylon. After 70 years, they were able to return to Jerusalem and the area around it, which came to be known as Judea.

Saddam Hussein, leader of Iraq from 1979 to 2003, oversaw an attempted reconstruction of the palace of Babylonian King Nebuchadnezzar II in the 1980s. He had inscriptions such as "This was built by Saddam, son of Nebuchadnezzar, to glorify Iraq" put on the bricks.

Jordan River Valley, Israel

Places in the New Testament

The story of Jesus and the early church in the New Testament takes place in the context of the Roman Empire. This largest empire in the ancient world included many places, from Spain in the west to the Caspian Sea in the east and from the island of Britain in the north to northern Africa in the south.

The gospel accounts begin in specific places: Nazareth and Bethlehem (Matthew), John baptizing in the Jordan River (Mark and John), and the temple in Jerusalem and the hill country of Judea (Luke). Shepherds came to the manger from their place in the fields near Bethlehem, and wise men came from a distant place in the east to visit the Child in Bethlehem (Luke 2:8-20, Matthew 2:1-12).

The ministry of Jesus unfolded in specific places, which are significant in His ministry. It began with John baptizing Jesus in the Jordan, a river that always had deep meaning for the Jews. Jesus then went into the wilderness for forty days, where He was tempted by Satan (the period of forty in a place of wilderness connects to the place of Sinai where Israel spent forty years). The best-known teaching of Jesus is a message He gave at a specific place, traditionally along the northwestern shore of the Sea of Galilee. We call the message the Sermon on the Mount.

The gospels describe Jesus spending a great deal of time in Galilee, where He grew up. To many Jews, Galilee was the backwater, unsophisticated part of Israel; notice Nathanael's snide comment, "Can any good thing come out of Nazareth?" (John 1:46). Jesus had contact with people in Samaria, the region where once stood the capital of the unfaithful Northern Kingdom of Israel. The Assyrians had repopulated this place with pagan Gentiles at the time of the Assyrian captivity. Jesus' ministry there was remarkable since Jews for the most part had no dealings with Samaritans because the Jews saw them as unfaithful half-breeds (John 4:9). Jesus also carried out ministry in the largely Gentile region east of the Sea of Galilee.

Jesus' contact with people in Galilee, Samaria, and the region east of the Sea of Galilee showed that He was willing to reach out to anyone and everyone, not just to the Jews and the Jewish leadership in Jerusalem. In fact, Jesus had His sharpest words about and His most difficult interactions with the leaders in that religious center of Judaism.

The ministry of Jesus culminated in Jerusalem, where Jesus had gone (as was His custom) to observe the Jewish festival of Passover. On the night He was arrested He went to a place that was special to Him, the Garden of Gethsemane on the Mount of Olives, across the Kidron Valley from Jerusalem. He was crucified on the hill of Calvary (the Latin name for "place of a skull"; the Hebrew name was Golgotha) outside of the city. He was buried in a tomb cut out of rock. Three days later, Jesus rose from the dead.

The gospel of Luke gives special attention to the place Jerusalem. When Mary and Joseph took the baby Jesus to the temple for purification, the prophetess Anna spoke about Him "to all those who were looking for the redemption of Jerusalem" (Luke 2:38). After His resurrection, Jesus told His disciples that they would proclaim His name "to all the nations, beginning from Jerusalem" (Luke 24:47). After Jesus ascended into heaven, the disciples "returned to Jerusalem with great joy, and were continually in the temple praising God" (Luke 24:52-53). Jesus had indeed redeemed Jerusalem, not from the Romans as most Jews expected the Messiah to do, but from the hopeless burden and slavery of sin. Luke's emphasis on the place Jerusalem was not as a political or religious capital but as a symbol of the redeemed people of God.

Acts begins with Jesus telling the apostles that, empowered by the Holy Spirit, they would be His witnesses throughout an ever-expanding geographic area: Jerusalem, Judea, Samaria, and even to the remotest part of the earth (Acts 1:8). Peter preached the first gospel sermon to people whom the text describes as coming from fourteen different places (Acts 2:9-11). In Acts we see the gospel spread to cities in Israel such as Caesarea and Joppa, as well as to Samaria, Ethiopia, Antioch, and Damascus.

Saul of Tarsus, a city in Asia Minor, was God's chosen instrument to take the gospel to the Gentile world. He began by going with Barnabas of Cyprus to Barnabas' home region and then to cities in Paul's home region of what we often call Asia Minor. He spent considerable time in the major city of Ephesus

This view of Jerusalem is from the Mount of Olives, with the Jewish Cemetery in the foreground.

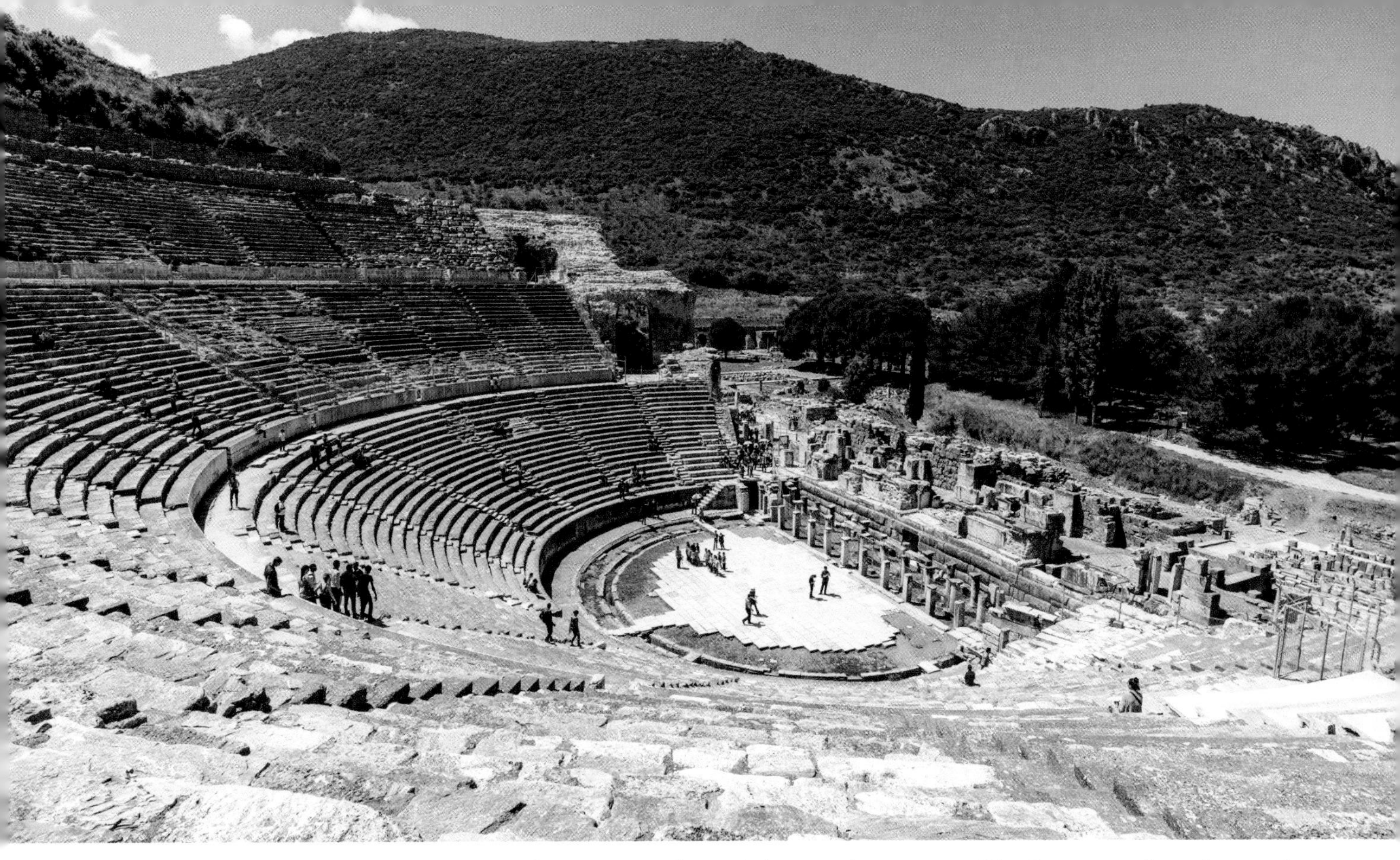

The Greeks built an amphitheater in Ephesus, in modern Turkey, in the 200s BC*. The Romans enlarged it in the 1st century* AD *and later. Eventually it could seat 25,000 people. This was the scene of the riot described in Acts 19.*

in Asia Minor. Paul later carried the gospel to several cities on the Greek peninsula, including Philippi, Thessalonica, Berea, Corinth, and Athens. He returned to Jerusalem and the region of Israel for a time, and then made a dramatic voyage to Rome. Paul's letter to Titus mentions him leaving Crete and spending the winter in Nicopolis, a journey that Acts does not include (Titus 1:5, 3:12).

Paul expressed his zeal to spread the gospel in geographic terms. He said that he had preached the gospel from Jerusalem to Illyricum, a Roman province in modern-day Dalmatia on the Adriatic Sea (Romans 15:19). In the same letter to the Romans, Paul stated a desire to go to Spain, the western extremity of both the Roman Empire and the Mediterranean world (Romans 15:24, 28). Almost all of the letters in the New Testament address churches in specific places or individuals in specific places. The last book of the Bible, Revelation, records letters to churches in seven places in Asia Minor.

Worldview Lessons from the Geography of the Bible

1) God created the diverse geography of the world. On it, in real times and places and with real people, He worked out and continues to work out His story of redemption.

2) The story of Israel and the story of the Jews focused primarily on one place, the Promised Land of Canaan. By contrast, Christians carried the story of the gospel to people in many different places, each of which had its own geography, history, and culture. Ephesus was different from Antioch, and Athens was different from Crete, in significant ways. We read about Paul preaching the gospel to Jewish synagogues in many cities, in the pagan Areopagus in Athens, and to the Roman governors Felix and Festus in Caesarea.

A similar variety of places exists in our world today. The gospel is not tied to any one place, culture, people group, or social or economic class. Lagos, Nigeria, is different from Beijing, China; and

Glasgow, Scotland, is different from Ball Ground, Georgia. But the same gospel speaks a message of hope, reconciliation, and redemption to people of every place, culture, background, and ethnicity.

3) The first books of the Bible tell of God's promise to His covenant people Israel regarding a place to which He would guide them and where He would sustain them. God intended for that promise to affect their view of themselves and their worldview. The last book of the Bible tells of a different kind of place God has prepared for His new covenant people, the church. That place is the heavenly Jerusalem, and that promise and expectation should influence our view of ourselves and our worldview until that promise becomes a reality.

Jesus spoke about this same promise:

In My Father's house are many dwelling places; if it were not so,
I would have told you; for I go to prepare a place for you.
If I go and prepare a place for you, I will come again and
receive you to Myself, that where I am, there you may be also.
John 14:2-3

Assignments for Lesson 120

Worldview	Recite or write the memory verse for this unit.
Project	Continue working on your research paper. ***or*** Finish your project for this unit.
Literature	Continue reading *The Country of the Pointed Firs and Other Stories.*
Student Review	Answer the questions for Lesson 120. Take the quiz for Unit 24.

Lake Coatepeque, El Salvador

25 Central America

Mexico is not easily defined and categorized; we take a look at its complexities. Guatemala faces serious issues, only one of which is the illegal harvesting of rosewood. We define what ecotourism is and how it has grown in Costa Rica. Belize is a small country but it has great diversity. The worldview lesson considers how the physical world (i.e., geography) affects worldview.

Lesson 121 - The Mixture That Is Mexico
Lesson 122 - Rosewood in Guatemala
Lesson 123 - Ecotourism in Costa Rica
Lesson 124 - Belize: A Small Country with Big Diversity
Lesson 125 - How the World Affects Your Worldview

Memory Verse Memorize Romans 12:1-2 by the end of the unit.

Books Used

The Bible
Exploring World Geography Gazetteer
The Country of the Pointed Firs and Other Stories

Project (Choose One)

If you are writing a research paper, continue to work on it following the schedule outlined on pages xviii-xix of *Part 1* of this curriculum.

If you are not writing a research paper, choose one of the following projects and plan to complete it by the end of this unit.

1) Write a 250-300 word essay on one of the following topics:
    - Choose one aspect of life in Mexico (such as food, music, clothing, immigration, Spanish heritage, or religion) and write a report on it. (See Lesson 121.)
    - How has the world (geographic and cultural) in which you live influenced your worldview and the worldviews of others who live there?
2) Imagine you are planning an ecotour of Costa Rica. What would you like to see? Develop an itinerary. (See Lesson 123.)
3) Design a brochure recruiting people to come to Belize. What would you encourage people to see? What business opportunities are there? Why would someone want to go? Include some pictures. (See Lesson 124.)

Agua Azul Waterfalls, Chiapas, Mexico

121 The Mixture That Is Mexico

To describe something simply as "Mexican" is to oversimplify the country. Mexico is a mixture in many ways.

To begin with, Mexico can be counted as part of several regions of the world. It is geographically part of the continent of North America. It is also historically and culturally part of Central America. It has strong economic ties to both regions.

Mexico is also part of Latin America. Latin America consists of the countries and dependencies in the Western Hemisphere south of the United States where Romance languages predominate. Romance languages are based on Latin, the language of Rome. In Latin America these languages are specifically Spanish, Portuguese, and French. Mexico is the third largest country in Latin America after Brazil and Argentina.

Geography

Mexico is a mixture of geography. About three times the size of Texas, Mexico has deserts in the north, a high plateau in the central region, tropical rain forests in the south, mountain ranges, and coastal lowlands along the east and west coasts. It has few major rivers or natural lakes.

The country is one of the world's most biologically diverse countries. It lies on the migration routes of many birds, geese, and ducks. Millions of monarch butterflies make their winter home there.

Mexico is part of the Ring of Fire surrounding the Pacific Ocean. This means that it has significant earthquake and volcano activity. The 1985 earthquake was especially devastating to Mexico City, when 5,000 people lost their lives and 100,000 homes were destroyed. As difficult as it was, the city has largely recovered from the destruction.

People

Mexico has a mixture of ethnic groups and cultures. It was home to several ancient civilizations, including the Olmec, Maya, and Aztec. In several aspects of human activity, such as architectural designs, irrigation technology, urbanization, and social organization, many scholars believe that these civilizations were more advanced than the Spanish who conquered them. Despite these advanced elements, however, the Maya and Aztecs practiced human sacrifice in their religious ceremonies.

Spain ruled Mexico for about three hundred years. The revolutionary period of the 1800s, beginning in 1810 when Mexico broke away from

Spain and continuing through the 1800s, saw many years of turmoil as several different leaders and groups tried to gain the upper hand in Mexico's government. The modern era that began in the 1900s and continues today has seen Mexico make significant advances in its economy, technology, and democratic institutions.

About 60% of Mexico's population are *mestizos*, people who have both indigenous and Spanish heritage. Other Mexicans are purely of indigenous descent. These facts mean that many of Mexico's people are descendants of the ancient civilizations that arose there. Some of the indigenous peoples today live largely as their ancestors did centuries ago. Still other Mexicans are of European descent. The population of Mexico is about 80% urban. About half of the population lives in the central region. The northern desert and southern tropical regions are more sparsely populated.

The Dance of the Voladores is a centuries-old custom among indigenous peoples in Mexico. Dancers start at the top of a pole and together fall off. They spin down to the ground as the ropes tied to their feet unwind.

Because of the Spanish influence, some 80% of Mexico is at least nominally Roman Catholic. The country has a much smaller but growing presence of evangelical and Pentecostal believers. Many people in Mexico hold to the Christian faith and to traditional folk beliefs at the same time. This results in a syncretistic belief system in which people honor Catholic saints, their own familial ancestors, and spirits all at the same time.

Mexico has the largest population of Spanish-speaking people of any country in the world, about 2 1/2 times as many as are in Spain. Almost everyone in Mexico speaks Spanish, but people also speak some 50 indigenous languages.

About half of Mexico's population of about 130 million people live in the central plateau. The metropolitan region of Mexico City, which is in this region and lies about 7,240 feet above sea level, has about 22 million residents, making it one of the largest population centers in the world. Other large cities in Mexico include Guadalajara and Monterey, which have about five million each. Cities that lie just across the border from the United States have also grown significantly, such as Juarez near El Paso, Texas, and Tijuana, across the border from San Diego, California.

Economy

Mexico has the eleventh largest economy in the world. Its manufacturing base has increased significantly since the North American Free Trade Agreement went into effect in 1994. The United States is Mexico's biggest trading partner, and the goods and services produced in Mexico are important to the U.S. economy as well.

Cars Ready for Export from Veracruz, Mexico

However, as with much else, Mexico is a mixture economically. The country has a few very wealthy people, a small but growing middle class, and many millions who live in extreme poverty. Per capita income in Mexico is one-third that of the United States. The fact that people in Mexico earn so much less than people in the U.S. is a big reason why manufacturing has grown there in recent years. Another negative is the illegal drug traffic, which generates significant income for the drug cartels but causes great misery for others, including those affected by drug wars among competing drug lords.

Another unresolved issue is illegal immigration from Mexico to the United States. Many of these are not Mexicans but are people who have come from Central American countries into Mexico illegally and then enter the U.S. illegally from Mexico. This issue has many aspects to it: respect (or lack of respect) for the law, the health and safety of those coming into this country and American citizens, how to pay for public services in the U.S. for those entering illegally and not paying their fair share of taxes, the cost of border security and the respective responsibilities of the U.S. and Mexico to provide that security, and other matters that need to be resolved.

Taking In Tabasco

Mexico consists of 32 states. The state of Tabasco in Mexico has nothing to do with the hot sauce. Tabasco hot sauce originated in Louisiana just after the Civil War. The variety of peppers used in the sauce doesn't even grow there.

Tabasco is in southeastern Mexico on the Gulf of Mexico and enjoys 120 miles of coastline. The state is about the size of Maryland with a population of 2.5 million. Unlike Mexico as a whole, nearly half of Tabascans live in rural areas. Many of them are indigenous descendants of the pre-Columbian Chontales civilization.

Tabasco is mostly low and flat with some higher elevations to the south. The countryside has many lakes, wide rivers, wetlands, and lush rainforests. Villahermosa ("Beautiful Village") is the capital, the largest city in the state, and a manufacturing and petroleum refining center. It has a metro population of some 750,000, over one-fourth of the population of the state.

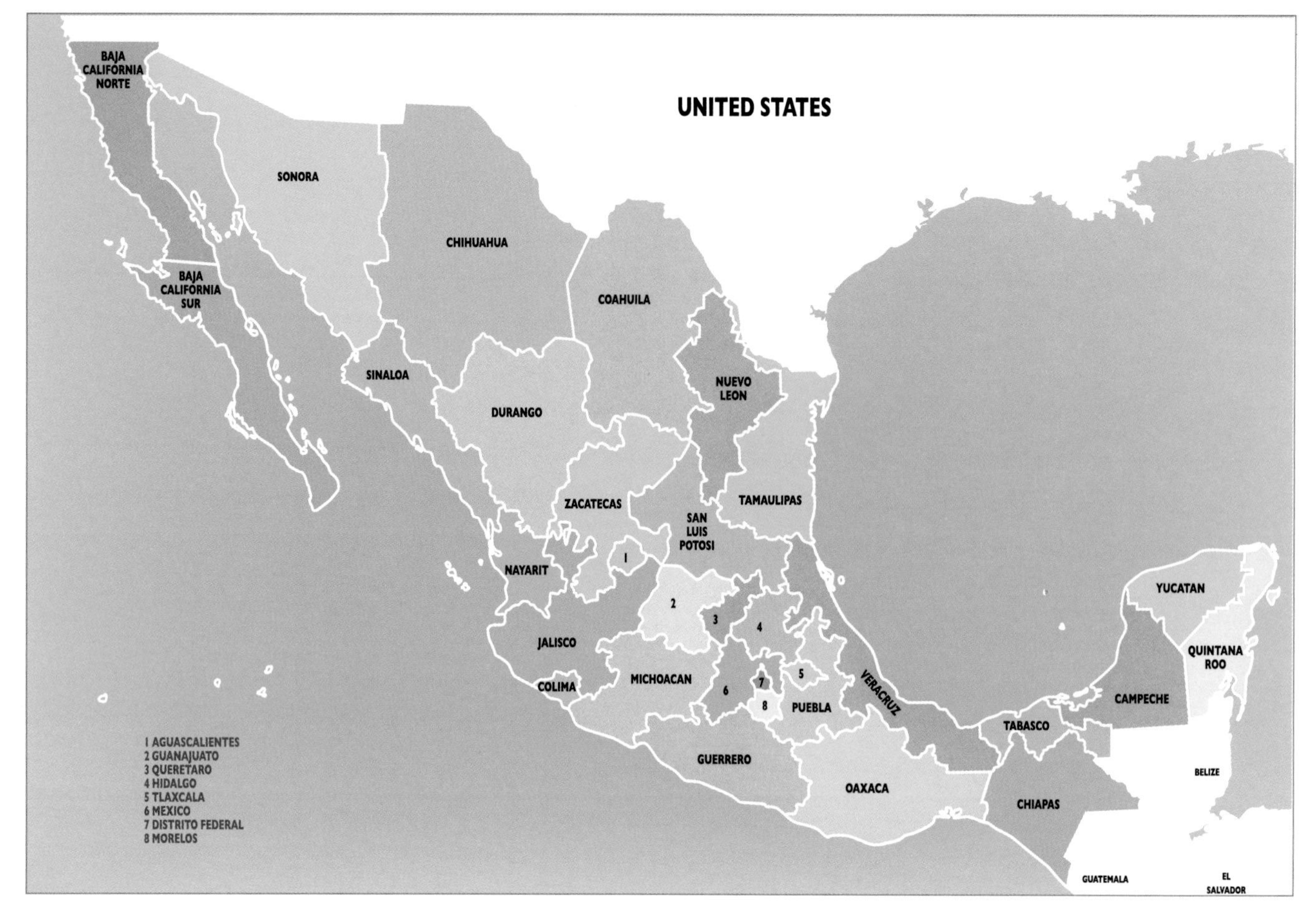

Mexican States

History

The Olmec civilization developed here from about 1500 BC to its high point around 500 BC. LaVenta, with an estimated population of about 18,000, was the largest Olmec city. They created stone and jade carvings that survive today. The Maya people arose in the centuries after Christ. Significant Maya ruins remain in Tabasco. Then the Toltecs developed in the 1200s AD. They were known as skilled architects, metalworkers, and stonecrafters. They also had extensive astronomical knowledge.

The Chontales, a Maya people from Nicaragua, immigrated to the area in the 1300s. The Spanish explorer Juan de Grijalva made the first contact by Europeans in what became Tabasco in 1518. Hernan Cortes invaded and defeated the Chontales the next year. The Spanish population remained small and the Chontales population dwindled due to disease, so in the early 1600s the Spanish brought in African slaves to increase agriculture in the area. The Chontales and enslaved people often rebelled against their harsh treatment and terrible living conditions into the 1700s, so the Spanish were not eager to develop the area. Developments in Tabasco paralleled those in the rest of Mexico during the period of revolution from the Spanish in the 1800s.

Agriculture has long been the major economic activity in Tabasco. The state produces cocoa, copra (from coconuts), corn, sugarcane, and tropical fruits such as bananas and papayas. Tabasco is also a source for red cedar and mahogany; and its residents practice beekeeping, fishing, and raising cattle. The discovery of petroleum in Tabasco has transformed the economy. Extracting and refining petroleum is now a major portion of the economic activity.

But still no Tabasco hot sauce. However, many people credit Tabasco with being the place where chocolate started. Apparently native nations knew the confection before the arrival of the Spanish. Quite a different taste from hot sauce.

3D Printed Houses for the Poor

Despite the growth of the petroleum industry and the long-term farming activity in the state, many Tabascans remain poor. These are often indigenous people who are some of the poorest in Mexico. Their income averages less than eighty dollars per month. Options for housing have been few and primitive. An initiative by a non-profit organization has provided hope for quality, affordable housing in Tabasco.

The organization, New Story, has partnered with other groups to build 500-square-foot homes in rural Tabasco using 3D printing technology. The 3D printer does not print with ink on paper, but it lays down strips of concrete to build walls according to a pre-programmed pattern. The houses have two bedrooms, a kitchen, a living room area, and a bathroom with indoor plumbing—a great advance for the poor. The homes are also built to withstand the earthquakes that sometimes strike the area. The "printing" takes a total of about 24 hours spread over several days. Construction crews finish each home by more traditional methods. Residents have input in the design of the homes, and they pay about $20 per month for seven years. The money goes into a community investment fund that the residents oversee.

New Story has already built traditional houses for the poor in El Salvador, Haiti, Bolivia, and Mexico. The group has an ambitious goal of using new technologies to eliminate inadequate housing for the poor all over the world.

Remembering the poor was important for the first Christians. When Paul first met with the leaders of the church in Jerusalem:

They only asked us to remember the poor—
the very thing I also was eager to do.
Galatians 2:10

Assignments for Lesson 121

Gazetteer	Study the map of Mexico and Central America and read the entry for Mexico (pages 202 and 208)
Worldview	Copy this question in your notebook and write your answer: What would be some factors in a house's location that would make you want to live there or make you not want to live there?
Project	Continue working on your research paper. Plan to finish it by the end of this unit. ***or*** Choose your project for this unit and start working on it. Plan to finish it by the end of this unit.
Literature	Continue reading *The Country of the Pointed Firs and Other Stories*. Plan to finish by the end of this unit.
Student Review	Answer the questions for Lesson 121.

Rainforest and Maya Ruins at Tikal, Guatemala

122 Rosewood in Guatemala

Thirty-seven-year-old Nevy Banegas supports his family by going regularly into the dense tropical forests of Guatemala. With his machete he harvests the glossy green xate leaves (pronounced shat-ay) from short palm trees. Nevy takes his day's work into his small village, where a worker sorts the leaves and wraps them in brown paper. The leaves are then shipped to the United States, where they become the greenery for flower arrangements.

Nevy and the dozens of other xateros who do this work get paid by their local co-operatives. For 2,100 quality leaves Nevy earns $16.00. In a good month, he can earn as much as $400, which is well above the minimum wage in his country.

Guatemala

The Northern Triangle of Central America consists of the countries of Honduras, Guatemala, and El Salvador. These three countries come together at a single point. They share many traits. Geographically, they have mountainous regions and coastal areas. Ethnically, they have a large percentage of people who are mestizos or indigenous. Politically, they have endured many years of warring factions and instability. Economically, they have struggled because of the political situation. Socially, they have seen much illegal drug activity, and many people have left the country out of a concern for their safety or to look for work elsewhere. In this lesson we will focus on Guatemala, although much of what we say applies to all three countries.

Geography

The two mountain ranges that cross Guatemala northwest to southeast divide the country into three geographic regions: the Peten lowlands in the north, the mountainous highlands in the center, and the lower Pacific coast region south of the mountains. Guatemala has one coast on the Caribbean and another on the Pacific. The country has no natural harbors on its Pacific coast.

Guatemala is about the size of Tennessee and lies on the Ring of Fire that surrounds the Pacific Ocean. As a two-coast country, Guatemala also has to deal with the occasional hurricane. Despite its beautiful landscapes, Guatemala has little tourism. Because of the political instability of the country, few people want to go there on vacation.

Scientists have identified fourteen distinct ecosystems in Guatemala. These include warm tropical coasts, hot lowland jungles, cooler volcanic mountain regions, limestone mountains and forested hills, semiarid flatlands, rich farming areas, and wetlands.

The People

Just over half of Guatemalans are mestizos (Ladinos is the favored term here). About forty percent are of Maya background. One in nine Guatemalans are of the Quiche or K'iche ethnic group. A small percentage are African or Garifuna (a mixture of African and Caribbean backgrounds).

Guatemala has the highest population in Central America south of Mexico, about 17 million. The majority live in the southern half of the country, and many of those live in the mountains. Over half of the population lives in rural areas.

Spanish is the official language and is the main language for 70% of the population. The government recognizes 23 indigenous languages.

Dancers in traditional costumes outside the cathedral in Antigua, Guatemala.

About 70% of Guatemalans are Roman Catholic. Most of the rest are Protestant, Evangelical, or Pentecostal; however, many hold to some level of folk beliefs in their everyday lives even as they claim to be Christians.

The Web of Problems

Guatemala and the other Northern Triangle countries deal with a complex web of problems. It is hard to know which problem came first, how they got to this point, and how they can improve their situation.

On the most fundamental level, the problems in Central America are a worldview problem. These countries do not have a strong democratic tradition or a strong sense of the value of the individual. Their legal systems do not provide a consistent defense of property rights. To an extent, the web of problems and the typical worldview are vestiges of colonialism when only a few held power. At some point, however, people have to move beyond their past in the way they interact with each other and toward a pattern of justice and equality.

Political corruption is a problem. The government too often does not provide the leadership required to improve matters. Instead, those in power tend to focus on lining their own pockets with tax revenue or bribes. The party in power works to eliminate the opposition, sometimes violently. The way the government operates—if it operates at all—leads to instability in the country, which results in a difficult situation in which to live and work.

The lack of education is a problem. Education is not generally valued—another aspect of the common worldview. In Honduras, for example, the government only provides public education through the sixth grade and only with limited resources. Most families cannot afford the books and tuition for their children to continue schooling beyond that level. Most families have few books and other learning resources at home. The situation in Guatemala is similar. The result is that people have little vision for change and lack the tools to change their life path. To be a subsistence farmer like their parents and grandparents is all that many children can envision.

Economic instability is another problem. Unemployment is high. Sixty percent of the people live below the official poverty line. People can find little work, and when they do, it is usually for low wages. The annual average per capita income in Guatemala is about $4,000. This is half the average for Latin America and the Caribbean. The indigenous population is especially poor. Many don't speak Spanish, and for this reason and because of prejudice they are excluded from power and society. In this situation, in order to have a little extra income many are willing to do the work described at the beginning and the end of this lesson.

Crime is a problem. The major issue here is the lucrative illegal drug traffic. Warring factions of drug dealers kill their opponents and often innocent civilians. Drug lords pay off government officials to look the other way, and dealers can collect extortion money (sometimes called a war tax) from legitimate businesses. Honduras is considered the most violent country in the world that is not at war. It has the highest murder rate in the world. The other Northern Triangle countries are not far behind. Legitimate domestic business and potential foreign investors are discouraged from investing in these countries. How can they know that the government will protect their property and that drug lords will not seize it? These countries have experienced a breakdown of law and order. The situation has been called a criminal dictatorship: the criminal element are those who really rule the country.

All of these problems contribute to the problem of illegal immigration into the United States. These issues cause many people to want to leave their country to escape the poverty, corruption, and crime. Entering Mexico and eventually the United States illegally is a dangerous journey; but if they can make it to the U.S., even a steady minimum wage job seems wonderful compared to what they

Panajache, Guatemala

had. Many Central American emigrants send part of their earnings back to their families. Estimates are that these remittances make up about one-fifth of the economic output of the Northern Triangle countries. These funds help those families have more income, but emigration takes willing workers out of the country and remittances don't produce real economic growth within the country. This emigration also separates families, which is not good for those families.

Can the United States help? Should the United States help? Our primary concern has been and should be the security of our borders, but that doesn't address the problems that lead to illegal immigration. Stemming that flow is properly the responsibility of those nations' governments, but those governments have not been willing or able to do that. Government officials there have little incentive to do anything about it.

Perhaps an approach similar to the Marshall Plan could help. As we saw in Lesson 54, the United States government and private American businesses invested in Europe after World War II to strengthen markets for our products and to discourage the influence of Communism in those countries. Perhaps in the same way, American investments in Central America—closely administered to prevent corruption and mismanagement and including legal incentives for participation—could help produce jobs, create economic stability and opportunity, reform government, and work against crime. This idea is certainly not a magic bullet that would solve all the deep-seated problems in Central America, but perhaps it would lessen the desire that so many have to enter the U.S. illegally. This would help our country while it would help people in other countries as well.

The problems in Guatemala and much of Central America are difficult and complex. The illegal harvesting of rosewood is one example of how these issues affect the land and people of Guatemala.

Trafficking in Rosewood

Rosewood is a dense tropical hardwood. It is durable, fragrant, and has a rich color. Rosewood is most commonly used in musical instruments and in upscale furniture that is produced in China and is especially popular among newly rich Chinese. The rosewood that grows in China has been largely depleted. The next most common source at one time was Southeast Asia, but that supply is dwindling also. Dealers in rosewood have turned to the forests of Guatemala.

International agreements are supposed to regulate the legal trade in rosewood. However, as has happened with many plants and animals, the human desire for ownership and for profit at any cost has resulted in rosewood trees being overharvested. Because of the demand for rosewood, illegal trafficking in it has become a major problem. It is, in fact, the most illegally trafficked wild product in the world, more than ivory or rhinoceros horn. One estimate is that worldwide illegal trafficking in rosewood is a one billion dollar per year industry.

The trees are a protected and regulated commodity in Guatemala. The illegal harvesting of rosewood trees there results from many of the issues the country faces. Rosewood trees grow there because of its geography. The harvesting takes place because people are poor and are willing to do the work. It continues because of government corruption. The harvesting is also part of the deforestation problem that Guatemala faces.

A rosewood tree grows less than a half inch per year and takes over one hundred years to reach maturity. The demand leads poachers to cut down younger trees, which further diminishes the long-term supply.

Stands of rosewood in Guatemala grow in the poorer sections of the country. Smugglers can convince local people to help them by offering them guaranteed income. Poor people who work for traffickers go into a forest (sometimes on private land), cut the trees and saw them into planks, then drag the wood to a roadway. Trucks come by to

Honduran Rosewood is one of the varieties found in Guatemala and nearby countries.

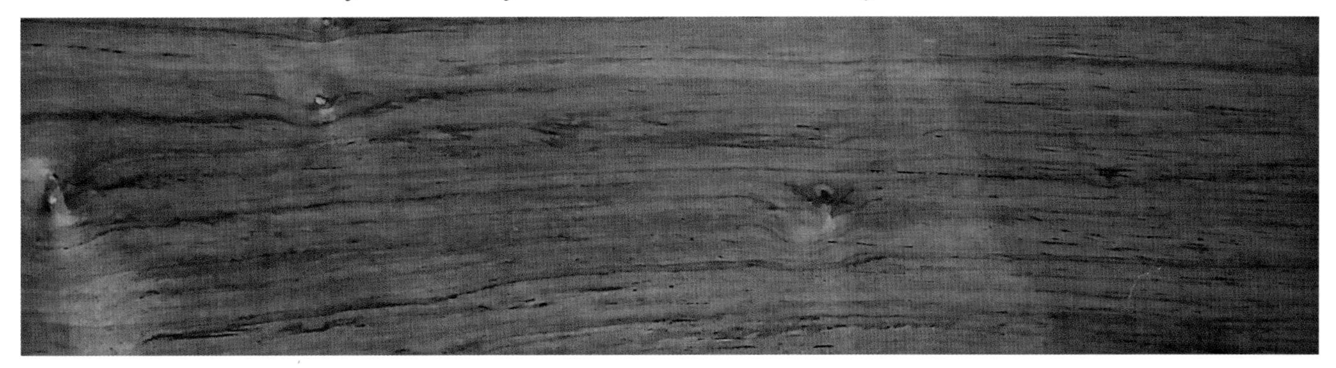

pick up the wood and take it to a seaport. There it is loaded into sea containers and shipped to China.

As of 2019, China and most other countries had no law against importing illegally sourced timber. To make matters more confusing, smugglers put erroneous labels on containers, labeling them as holding such materials as rubber, materials to be recycled, or packing materials. One official got suspicious when a container was heavier than it should have been given the stated contents. She was right: it held rosewood.

The enforcement system in Guatemala is extremely complicated. Several levels of officials, from the forest service to local officials to national enforcement to international inspectors, are involved. Any corruption, usually in the form of smugglers paying off officials, creates hurdles that often block enforcement of the law. A related issue is the problem of limited government funding that is available for such enforcement, which leads to a relatively few overworked officials trying to cover too much territory. When a case against a smuggler reaches a court, the trial may be delayed, or the judge might be corrupt, or the fine assessed is too small to deter a smuggler.

The problem of illegal trafficking in rosewood, along with the other problems that Guatemala faces, do not admit to an easy answer. If they did, people would likely have solved them by now. These problems remind us of the need to overcome sin and injustice, to provide people hope for a better way, and to give practical assistance to help people leave one way of life for a better one. The change in outlook or worldview that Christ offers points the way:

And do not be conformed to this world, but be transformed by the renewing of your mind, so that you may prove what the will of God is, that which is good and acceptable and perfect.
Romans 12:2

Assignments for Lesson 122

Gazetteer Read the entries for El Salvador, Guatemala, and Honduras (pages 205, 206, and 207).

Worldview Copy this question in your notebook and write your answer: How has traveling to other places broadened your outlook on people and the world?

Project Continue working on your research paper (or project).

Literature Continue reading *The Country of the Pointed Firs and Other Stories.*

Student Review Answer the questions for Lesson 122.

Tenorio Volcano National Park

123 Ecotourism in Costa Rica

In northern Costa Rica, tucked away in a rainforest, is the beautiful, majestic, 230-foot-high La Fortuna waterfall. A private group is reclaiming the rainforest to return it to its original state. The group has built a wooden viewing stand where people can admire the falls. The admission fee that visitors pay has financed the construction of a road, a parking lot, a cafe, and a souvenir shop.

Two hours away by a treacherous road is Monteverde Nature Reserve, established by American Quakers who were avoiding the draft during the Korean War. The reserve includes a cloud-forest habitat considered to be one of only twelve true primary rainforests in the world. Parts of the forest have never been cut.

Near Monteverde is the Children's Eternal Forest. In the 1980s, thousands of children around the world contributed money to buy the property in order to save it from logging. No tourists may enter the forest except along a single trail.

The country of Costa Rica, which is slightly smaller than West Virginia and lies just to the north of Panama, is considered the world's leading destination for ecotourism. The population of Costa Rica is 5.1 million persons. The Costa Rican government reported that 3.1 million tourists visited the country in 2019. Ecotourism is important to Costa Rica, but it is only the most recent part of the country's story.

The Rich Coast

In 1502, on his last voyage to the New World, Christopher Columbus became the first European known to visit the area that became Costa Rica. Seeing people there wearing gold and silver jewelry, Columbus thought that he or others could find treasure galore here, so he named it Costa Rica, Rich Coast.

Columbus was wrong about the precious metals. The Spanish showed little interest in the land before they established their first settlement there in 1565. Growth continued to be slow and interest was limited. The indigenous people who did not die off from contact with the Spanish by combat or disease resisted Spanish rule. In 1719 the governor at the time described Costa Rica as the poorest and most miserable Spanish colony in all the Americas.

As a result, when Costa Rica, along with several other Latin American countries, declared its independence from Spain in 1821, it did not have as much of the legacy of colonialism as did other countries that had experienced stronger colonial control. The result is that Costa Rica today has a more stable government, a stronger democracy, and a more successful economy than other countries in Latin America. The population enjoys a high standard of living, a high literacy rate, and wide private ownership of land. San Jose, the nation's capital, was one of the first three cities in the world to have electricity. In recent years Costa Rica has placed great emphasis on renewable energy. Hydroelectric facilities supply about one fourth of the country's energy use. Columbus' name for the country has become true, but not in the way he expected.

Half of all Costa Ricans live in urban areas. San Jose contains about one-fifth of the country's population. Over 83% of the people are either white or mestizo, 2.4% indigenous, and about 8% black or mulatto. This last group is largely the legacy of Jamaican workers brought in during the late 1800s. Almost three-fourths of the population are Roman Catholic; another 15% are either evangelical, Protestant, or Pentecostal. Spanish is the official language but English is widely spoken. Many black Costa Ricans speak Creole, which is a dialect that developed from the contact the Spanish, English, and indigenous peoples had with each other in the Caribbean region. Creole developed so that they could communicate with each other.

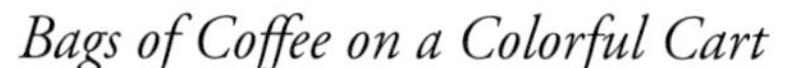

Bags of Coffee on a Colorful Cart

Tourists in Tortuguero National Park

Coffee

The primary indigenous wealth of Costa Rica is not gold in color but brown. The country has a highly successful coffee growing industry. In the late 1700s the Costa Rican colonial government offered free land to anyone who was willing to grow coffee for export. This encouraged the development of a class of coffee barons who became wealthy but who also eventually financed a railroad system, a postal system, a national theater in the capital of San Jose, and other projects.

Costa Rica has the geography for a successful coffee growing industry: mineral-rich volcanic soil, shaded fields, and a cool mountain climate. The country is the thirteenth largest coffee producer in the world, even though now 90% of the farmers own less than twelve acres each. Cooperatives operate most of these farms. Costa Rican coffee growers use only Arabica beans, although different regions of the country grow different varieties of them. As big as coffee growing is in Costa Rica, it is only the second largest segment of the country's economy, coming in behind ecotourism.

Ecotourism

When you think of tourism, what images come to mind? Perhaps loud, artificial, man-made attractions; crowds of people filling sidewalks and shops and generating tons of garbage; lines of cars or RVs in traffic jams coming and going. Or perhaps your idea of a vacation involves visiting an historic site, maybe with costumed characters.

Now think about a different kind of tourism, one that involves quiet trips to exotic natural areas to appreciate the landscapes and observe animals in their natural habitat. This is the focus of a different kind of vacationing called ecotourism.

Ecotourism involves traveling to natural areas in a way that conserves the environment; sustains the wellness of the local human, animal, and plant populations; allows people to learn about and eventually give assistance to the environment; and involves both staff and guests. Examples of ecotourism include hiking through a rainforest on a single trail, quietly observing the movements of bird species, and watching giant tortoises and learning about their habitat.

Ecotourism companies help people appreciate our natural world. They also hope to generate financial resources through fees and sponsorships that can preserve and restore natural areas and make possible the purchase of additional territory. The ecotourism industry helps support local economies without creating many of the problems that often accompany economic growth.

An important guideline for ecotourism is to leave no footprint. Ecotourists try to be aware of what they bring into an area and to take out what they bring in. They seek to cause minimal impact of light, sound, and odor. They commit themselves to leaving behind all of the rocks, shells, leaves, and other artifacts of the places they visit.

The Setting for Ecotourism in Costa Rica

Costa Rica has an amazing geographic diversity in a relatively small area. Experts estimate that the country contains 5% of the world's biodiversity on only .1% of the world's landmass. The country lies in the earth's neotropic ecozone, one of eight such zones on the earth's surface. Costa Rica can be divided into four main geographic areas: Coastal Tropical Lowlands (which include its 800 miles of Caribbean and Pacific coastlines), the Northern Central Plains, the Central Valley, and the Northwestern Peninsula. The country has multiple microclimates, which are small areas with distinct climates. Examples include places that generally have sunny weather at the base of a mountain but cloudy and misty skies a few miles up the mountain.

Fourteen major river systems originate in the Costa Rican mountains. Costa Rica has about 60 volcanoes, although only five are active. The tectonic plates that lie underneath the country have also caused occasional earthquakes.

Ziplines are a popular way to experience the forests of Costa Rica.

Costa Ricans are committed to preserving their environment. About one-fourth of its land area is protected from future development. These protected areas include 28 national parks, over 50 wildlife refuges, over 30 protected zones, fifteen wetlands, eleven forest reserves, and eight biological reserves.

Try to envision an area the size of West Virginia including all the geographic features we have mentioned as well as hot springs; butterfly gardens; excellent sites for whale watching (both Northern and Southern Humpbacks migrate through the area); places for snorkeling and scuba diving through coral reefs; and multiple locations for birdwatching, windsurfing, and rappelling. That is Costa Rica.

Animals of Costa Rica that are popular among ecotourists include the resplendent quetzal, described by noted ornithologist Roger Tory Peterson as the most spectacular bird in the New World. The Maya once used its feathers as currency. The country also has a remarkable diversity of sea turtles, especially in Tortuguero National Park. Green sea turtles, which can weigh up to 300 pounds, roam the Caribbean but return to Tortuguero to reproduce.

Ecotourism has its downside also. Most ecotourists take passenger jets to their destination, which requires the burning of much jet fuel. An ecotour is not usually cheap; a ten-day trip can cost $11,000 or more depending on the destination and what the trip includes. The increased wealth that ecotourism generates in less developed countries can increase the demand for new cars, which can increase oil consumption. The long-term impact of

Resplendent Quetzal

hundreds of tourists on plants and animals can be hard to determine in the short run. Sometimes the line between tourism with an environmental flavor and environmentally based ecotourism can be pretty narrow.

Costa Rica is an example of how even a little country can have a big story. The country is home to many amazing creations of God, and people there are caring for them as His gifts and our heritage on this earth.

Let them praise the name of the Lord,
For His name alone is exalted;
His glory is above earth and heaven.
Psalm 148:13

Assignments for Lesson 123

Gazetteer	Read the entries for Costa Rica, Nicaragua, and Panama (pages 204, 209, and 210). Read the article "Going West by Going South" (pages 323-324) and answer the questions in the *Student Review Book.*
Worldview	Copy this question in your notebook and write your answer: What is your outlook on the world as an American?
Project	Continue working on your research paper (or project).
Literature	Continue reading *The Country of the Pointed Firs and Other Stories.*
Student Review	Answer the questions for Lesson 123.

Howler Monkey in Belize

124 Belize: A Small Country with Big Diversity

Belize is a small country with big diversity. It is located on the northeast coast of Central America, with Mexico to the north, Guatemala to the west and south, and the Caribbean Sea to the east. Belize is the only country in Central America that does not have a Pacific coast.

Geographic Diversity

Belize has lowlands along its 174 miles of Caribbean coast, the Maya Mountains in the south, and swamps to the north. Its many rivers run through mountain valleys and swamp forests to the sea. February through May is the dry season, while June through November (with the exception of August and September) is the rainy season. Even the rainfall amounts are diverse, from 50 inches per year in the north to 175 inches in the south. Hurricanes are a real threat, and the country has been devastated by several tropical cyclones through the years.

The country is about three-fifths forested, with some 50 species of trees. However, savanna grasslands grow on the southern coastal plains and inland. About one-fifth of Belize is set aside as nature reserves. The Belize Barrier Reef is part of the Mesoamerican Barrier Reef, the second largest coral reef in the world.

Wildlife in Belize includes deer, pumas, crocodiles, manatees, turtles, birds, reptiles, and fish. Baird's tapir, the national animal, can weigh 600 pounds. The Cockscomb Basin Wildlife Sanctuary, which covers 150 square miles, has the largest concentrated jaguar population in the world.

A Little Background

The Maya civilization flourished in this region around 1000 AD, as evidenced by the concentration of Maya ruins in Belize. Christopher Columbus probably sighted the land that became Belize but didn't land there.

English buccaneers and loggers who came in the mid-1600s were the first to establish settlements. The British brought slaves into the region from the early 1700s to cut logs. Slaves revolted against their masters four times, and hundreds escaped into the forests. The abolition of slavery in the early 1800s left many former slaves free but impoverished and with few economic options.

Spain and Britain had conflict for decades because of their rival claims for Belize. This continued until Spain's power waned and British power increased to the point that, by 1798, the land was a British colony in all but name. That became official in 1862 with British Honduras (as it was called at the time) becoming part of the British Empire.

Following its independence from Spain, neighboring Guatemala challenged the British land claims by asserting that it had inherited rights to the area from Spain. Mexico has also had border disputes with Belize. These disputes remain unsettled to the satisfaction of Guatemala and Mexico even today. Belize gained independence from Great Britain in 1981, but it remains part of the Commonwealth of Nations. Belize was the last British colony on the American mainland.

Population Diversity

Belize is slightly smaller than Massachusetts and has a population of about 400,000 (compared to Massachusetts' population of about seven million). The country has the lowest population density of any Central American country. However, this relatively small population contains a diversity of national and ethnic backgrounds.

Creoles of mixed African and British descent were once the majority population; but many Spanish-speakers came in the mid-1800s fleeing war in the Yucatan Peninsula. Mestizos (of Maya-Spanish background) came as immigrants to flee civil wars in Guatemala, Honduras, and El Salvador during the 1980s. Spanish-speakers now comprise about half the population of Belize while Creoles now account for about one-fourth of the population. Maya descendants are about ten percent of Belizeans. In addition, smaller numbers of Garifuna (of mixed Caribbean and African parentage) have come, as have workers from Europe, South Asia, China, and the Middle East. Only about 1% of the population is white. An estimated 16% of Belizeans live and work outside the country and send remittances back to family members who are still in the country.

Over half of the population lives in rural areas. The largest city and former capital, Belize City on the coast, is home to about one-fifth of the people. After a hurricane devastated Belize City in 1961 (which was thirty years after it was severely damaged by another hurricane), Belize built a new capital city, Belmopan, about fifty miles inland.

One group you might not expect to see in Belize are the Mennonites, who began coming in the 1950s from Canada and Mexico to escape religious persecution. They organized themselves into several communities in Belize and have won the respect of others for their hard work and for the way that their farm products have contributed to the nation's economy.

Mennonite Community, Belize

Diversity in Language and Religion

Because of its British background, English is the official language and the language of government; but because of the number of recent immigrants from Central America, Belize has more native Spanish speakers than native English speakers. Many residents are multilingual, speaking Spanish or English as well as a language or dialect that reflects their ethnic background. Culturally, Belize is more Central American than British. The Mennonites speak Plautdietsch, an archaic Germanic language influenced by the Dutch.

Roman Catholics make up about 40% of the population, while Protestants and Pentecostals account for about one-third. Beyond these groups, the people have a wide range of religious affiliations.

Caye Caulker Island, Belize

A Diverse Economy

Like many countries, Belize has come to depend heavily on tourism, which accounts for about half of its gross national product. Most of these tourists arrive on cruise ships. However, about half of the population is involved in one way or another with agriculture. Sugar accounts for 60% of its agricultural exports, which also include citrus fruits, bananas, corn, kidney beans, and rice. A large part of the fishing exports goes to the United States.

Timber once dominated the country's exports, but lumber output has decreased. Unfortunately, logging has done significant damage to the country's forests. Recent oil discoveries have brought revenue into the country; the crude oil is shipped to the United States for refining. The Belizean government is working hard to preserve the offshore reef and to regulate the fishing industry.

What's It Like to Visit There?

Facts and statistics in a book can give you some idea of a country, but there is nothing like actually visiting a place to get a feel for what life is like there. This is especially true for a country that has a culture which is different from our own.

My daughter, Mary Evelyn McCurdy, has made three trips to Belize. Because of her experiences, I have asked her to answer the question, "What is Belize?"

Belize is goats grazing under palm trees and skinny dogs running through the streets. It is houses built on stilts and it is babies napping in hammocks. It is a woman sewing on her back porch using a treadle sewing machine. Belize is mosquito bites. It is buying eggs from a crate on the floor of the grocery store that is owned by a Chinese family. It is visiting a shop with the words "Stretch Yu Dolla" painted on the outside.

Belizeans can ride three people to a bike with no problem. Cyclists sell a variety of items, from peanuts and plantain chips to shoes and plastic buckets. Belizeans often find a reason to have a parade through the street.

Belize is children walking to school in matching uniforms. Some are green or blue, others are yellow or gray, depending on which school the children attend. These children and their families live in homes painted with bright and beautiful colors, some with thatched roofs.

Belize is trucks loaded with sugarcane going down roads that have enormous potholes. It is coconut trees and avocado trees in the neighbor's yard. It is exploring the ruins of Maya temples. Belize is watching a man fish with a net in the turquoise water of a bay. It is a funeral procession moving respectfully through the street with the loved one's coffin in the back of a pickup truck.

Belize is beans and rice. It is cantaloupe juice and watermelon juice on the breakfast menu. It is passion fruit ice cream for dessert. It is an orange for a snack sprinkled with crushed red pepper. It is bananas that taste like bananas should—fresh from the tree. It is salsa made with chicken feet and homemade papaya jam. Belize is more beans and rice. It is a cup of piping hot maize atole from a roadside stand that tastes like drinking corn on the cob. It is apple bananas—a delicious fruit with the look and feel of a banana and the taste of an apple. And Belize is more beans and rice.

In Belize black howler monkeys swing through the trees and frogs feel welcome to come inside. The humidity is so thick you can almost cut it with a knife. In Belize, it gets so hot that people say, "The sun is not joking!" and mean it. Sometimes in the rainy season, yards flood and schools close. At such times the roads are too muddy for the children's feet and bicycles to head out to school.

In Belize the church building might still be locked when it's time for church to start, but it will get unlocked eventually. Things don't always start on time. People might trickle in gradually, but their hearts are there, full force. Lizards might come in, too, and climb up the wall during the sermon. The Christians sing with great gusto. Their voices easily drift through the open windows since the windows have no glass. They sing and they pray like they mean it because they do. They serve a big God—the God of all people, everywhere. —*Mary Evelyn McCurdy*

The diversity in Belize is a reminder of how people from many backgrounds came to Christ in the early days of the church, and how people all over the world are still coming to Christ:

Now there were at Antioch, in the church that was there, prophets and teachers: Barnabas, and Simeon who was called Niger, and Lucius of Cyrene, and Manaen who had been brought up with Herod the tetrarch, and Saul.
Acts 13:1

Assignments for Lesson 124

Gazetteer	Read the entry for Belize (page 203).
Geography	Complete the map skills assignment for Unit 25 in the *Student Review Book.*
Project	Continue working on your research paper (or project).
Literature	Continue reading *The Country of the Pointed Firs and Other Stories.*
Student Review	Answer the questions for Lesson 124.

Santa Lucia, Honduras

125 How the World Affects Your Worldview

Any real estate agent will tell you that the three most important factors in the selling price of a house and how quickly it will sell are: location, location, and location. In other words, the physical condition of a house and its "curb appeal" (how it looks from the street upon initial viewing) make some difference, but by far the main factor in selling a house is the location of the house: inner city or posh suburb, the quality of the schools in the neighborhood where the children of the buyer will likely attend, and so forth.

Geographic location has a major impact on a person's worldview. Geography is not deterministic; someone who lives in a given place will not automatically have a certain outlook, but where a person lives does make a difference in his or her perspective on the world. Your location helps to determine who you are.

The Impact of Physical Geography

Someone who lives in a geographically isolated location will be more likely to have a narrow, limited view of the world. The people of Japan who lived there before American Commodore Matthew Perry opened that country to contact with the world in 1853 knew little or nothing about the rest of the world. The aborigines of Australia, residents of the hollows of the Appalachian Mountains, those who live in small remote villages in Central America, and people of some Pacific islands have a relatively limited view of the world. Most of the people they know have a similar appearance, probably think and talk in similar ways, and quite possibly look with suspicion on people from elsewhere.

On the other hand, people who live in a busy port city on a nation's coast, where many people from remote places come and go, or who live in a university town where people value learning and come from many different places, come to realize that people have different customs and ideas. People who live in such places have to learn how to handle this diversity of worldviews.

People who live in a place that has one main kind of economic activity tied to that place, such as coal mining, automobile manufacturing, or cotton production, are likely to see economic and political developments in terms of what those developments mean for that particular industry and the people who work in it. On the other hand, people who live in a diverse economy might find it easier to understand how various economic activities relate to each other; and they might not be as concerned about a downturn in one part of the economy.

Someone who lives in Iowa might view the use of ethanol (a replacement for gasoline that is made from grain) differently from how someone who lives in Houston, Texas, sees it. Someone who lives on a large farm in South Dakota will probably not have as great a concern about population issues as someone who lives in the inner city of Kolkata (Calcutta), India. Geographic location makes a difference in our worldview.

The Impact of Cultural Geography

Consider these examples of how cultural geography influences the activities and worldviews of people.

For the most part, the political leaders of contemporary Russia believe that the United States and NATO are threats to Russia's security. Most Americans, on the other hand, think that such a view is ridiculous.

Why would the leaders of Russia have this outlook? Foreign powers have invaded Russia many times throughout history. The Mongols invaded in the 1200s. The Poles came in 1605. The Swedes invaded in 1708. Napoleon led an invasion in 1812. Germany invaded in 1914 and 1941. The United States positioned itself as the arch enemy of the Soviet Union in 1945, and countries formed NATO in 1949 specifically to oppose and limit Soviet moves toward the west. Most Americans have little or no interest in controlling any part of Russia (as long as Russia is not aggressive toward other countries), but history and geography influence the worldview of Russia's political leaders.

A young Christian whose father works as a maintenance man at a casino in Las Vegas, Nevada, will likely have a view of gambling that is different from the perspective of a young Christian who lives in the southern United States and is a member of a megachurch that was active in a campaign that urged people to vote against a state lottery in a referendum.

The worldview of an African American slave in Alabama before the Civil War would have been vastly different from the worldview of a contemporary African American school administrator in Alabama who holds an earned doctorate in education.

Market in Kolkata, India

Ruins at Tripitos, Crete

When Paul wrote to Titus, an evangelist working on the Mediterranean island of Crete, he quoted a line from the Cretan poet Epimenides: "Cretans are always liars, evil beasts, lazy gluttons." Paul went on to say, "This testimony is true. For this reason reprove them severely so that they may be sound in the faith" (Titus 1:12-13). Paul was helping Titus to deal with the reality of cultural geography. Not everyone who lived on Crete at the time was a liar, evil beast, or lazy glutton; but the cultural pattern there was strong. New Christians on Crete would have to work hard to overcome this cultural pattern.

Cultures develop in specific times and places that encourage people to produce great works of art, literature, and music. Italy during the Renaissance gave rise to several great artists. Authors in Great Britain in the nineteenth century produced an abundance of literature that people still read and honor today. Musicians in England and the United States during the 1960s produced popular music that still sells well and that we hear as background music in restaurants and elevators. People with computer skills migrated to Silicon Valley in California in the 1980s, and the culture there encouraged many advances in computer science. Geographic places at specific times can be home to significant cultural activity.

The Impact of Religious Culture

The typical resident of Ireland is likely to be Catholic; the typical citizen of India is likely to be Hindu; the typical resident of Greece is likely to be Orthodox; the typical resident of Saudi Arabia is likely to be Muslim; the typical Norwegian is likely to be Lutheran; and the typical resident of Arkansas is likely to be Baptist. This says nothing about the depth or sincerity of their faith, and persons can certainly change their beliefs; but where a person lives often does have an impact on their faith background.

Different religious cultures lie behind the continuing conflicts between Israelis (Jews) and

Palestinians (Muslims) and between citizens of India (Hindus) and Pakistan (Muslims). Even if citizens of these countries have not had contact with many citizens of the other country, they still are likely to have definite views of people from those other countries because of their religious differences.

The Impact of Political Culture

Different political cultures and expectations influence worldview. For example, how might a Christian in China and a Christian in the United States see government differently?

The typical Christian in China does not expect support from the government for his faith; to the contrary, he might expect some persecution for himself or for other Christians. Nevertheless, he might have overall respect for the Chinese government in keeping with Paul's instructions in Romans 13.

On the other hand, a typical American Christian might expect some degree of support from the government for his faith. After all, his government's official motto is "In God We Trust." He may expect to hear prayers at public gatherings, even civic events. He understands that the United States does not have an established religion, but he is still disappointed at court rulings that limit public expression and practice of religion.

Gaza, Palestine

The Impact of Religious Conversion

Conversion often brings some struggles regarding worldview. Here is an example. I once knew a young man from Taiwan who became a Christian while living in the United States. He planned a trip home to see his family, and he struggled with what he should do during a family gathering that was intended to honor his ancestors. He had respect for his ancestors and wanted to show that, but he knew that he was not supposed to worship his ancestors and did not want to. For some of his family members, the gathering was a spiritual event. He wanted to influence his family members for Christ, and he did not want to burn bridges with his family members by refusing to participate in the gathering. His struggle about what to do occurred because his worldview had changed from what it had been, which was the worldview that his family members continued to hold.

Sometimes cultural or religious influences outweigh the impact of geography. For instance, I can think of two places that have a relatively flat, mountainless topography, each with a fairly large population. However, there is a great difference between the typical outlook of someone who lives in Gaza in the Middle East and someone who lives in Indianapolis, Indiana. Location matters, and it can matter in several different ways.

Travel

Traveling to other places, even for brief visits, can change a person's worldview. Not everyone in the world lives in the circumstances that you do; probably few people do. A short-term mission trip from Nashville or Dallas to Haiti or Mumbai can revolutionize a person's understanding of the world and increase a person's love for others in the name of Christ. Travel to another country can help someone appreciate what he has in his own country, the price that was paid for creating the culture where he lives, and the effort required to maintain it.

Eric Weiner has said, "Crossing geographic borders enables us to transcend our own internal ones." Henry Miller observed, "One's destination is never a place but a new way of seeing things." Changing or broadening one's experiences in the world can change a person's worldview.

Physical geography, cultural geography, religious geography, and changing geographic settings can influence a person's worldview. The Lord's guidance of Peter and that apostle's contact with the sincere faith of the Gentile centurion Cornelius affected Peter's understanding of God's will for and His work in the world.

Opening his mouth, Peter said:

I most certainly understand now that God is not one to show partiality, but in every nation the man who fears Him and does what is right is welcome to Him.
Acts 10:34-35

Assignments for Lesson 125

Worldview	Recite or write the memory verse for this unit.
Project	Finish your research paper. ***or*** Finish your project for this unit.
Literature	Finish reading *The Country of the Pointed Firs and Other Stories.* Read the literary analysis and answer the questions in the *Student Review Book.*
Student Review	Answer the questions for Lesson 125. Take the quiz for Unit 25. Take the fifth Geography, English, and Worldview exams.

Willemstad, Curacao

26 The Caribbean

The Caribbean region is not only beautiful; it provides opportunities for entrepreneurs to utilize the resources there. Hurricane Hunters fly dangerous missions above the Caribbean to learn more about hurricanes in order to save lives and preserve property. Sugarcane growing was once a major part of Caribbean island life, but slaves did the work and paid the tragic price. Cuba is close to the United States geographically but far away from it in terms of political freedom and worldview. The worldview lesson tells how the experiences of life can affect a person's worldview.

Memory Verse Memorize 2 Corinthians 5:16-17 by the end of the unit.

Books Used The Bible
Exploring World Geography Gazetteer

Project (Choose One)

1) Write a 250-300 word essay on one of the following topics:
   - Tell about an event, a person, or an experience that profoundly changed your worldview. See Lesson 130.
   - What is an adult role in which you are interested? It might be a profession, a craft, the role of a mother, or some other work. Tell what it would mean for you to serve in this role (education or training needed, the sense of accomplishment you would have, etc.) and how a person in this role can serve others. See Lesson 127.
2) What resource do you know about near you, such as idle land or a river suitable for kayaking, that you could turn into a profitable business? Develop a proposal for what you would like to do with it.
3) Write and deliver a five-minute speech against slavery.

Sargassum on the Beach, St. Lucia

126 The Blue Economy of the Caribbean

Johanan Dujon was a physical education teacher in his mid-20s at a Roman Catholic boys school on the island of St. Lucia, a nation in the eastern Caribbean Sea. For years he had longed to start his own business. He had shown an entrepreneurial drive when he was sixteen. At that time he bought reconditioned Blackberry devices and resold them on Facebook. But Dujon had a bigger vision. He wanted to create wealth for himself and for later generations and get beyond, as he put it, struggling in the rat race.

Dujon saw a problem and turned it into an opportunity. Brown algae seaweed called sargassum is a significant part of the Atlantic Ocean ecosystem. The Sargasso Sea is a concentrated patch of the seaweed in the Atlantic. It is the only identified sea without land borders. Only ocean currents define it. Columbus noted the presence of sargassum on his first voyage to the New World in 1492. Sargassum reproduces on the surface of the water, whereas other seaweeds reproduce on the ocean floor. The Sargasso Sea provides a habitat for turtles, shrimp, crab, fish, and other ocean species. Migrating whales, birds, and fish find food there.

Beginning in 2011, a belt of sargassum began developing south of the Sargasso Sea. The Great Atlantic Sargassum Belt eventually stretched across the Atlantic Ocean from the Gulf of Mexico to the western coast of Africa. Sargassum began washing up on the coasts of St. Lucia and other Caribbean islands in unusually large amounts. The seaweed smelled bad, complicated the work of fishermen, and discouraged tourists from enjoying the beaches. Scientists are as yet unsure why this increased formation has been taking place, but they don't believe the Belt is connected to the Sea.

Dujon had the idea of gathering the seaweed and turning it into an organic fertilizer. He obtained funding through loans from his family and another business at first, then found a credit union who believed in him enough to grant him a loan, and finally Dujon partnered with the St. Lucia Fisherfolk Cooperative Society Ltd.

He hired local people to collect the seaweed. He then developed a zero-waste proprietary technology, meaning a process to which he owns the rights, which primarily involves grinding the seaweed into liquid form. Thus Dujon created Algas Organics Total Plant Tonic. In 2015 he began marketing the fertilizer on St. Lucia and on other Caribbean islands. The fertilizer is now available in the United States and Canada.

Collecting Sargassum for Algas Organics

Total Plant Tonic produces better results than chemical fertilizers in applications involving commercial agriculture, plant nurseries, and private lawns. It also avoids the problems associated with putting artificial chemicals on crops and lawns. Workers have removed hundreds of tons of seaweed from the St. Lucian coast. This has helped the fishing and tourism industries on St. Lucia and has built Algas Organics into a successful international agricultural biotech company in the $6 billion-plus per year organic fertilizer market.

The Geography Around You

What is the main geographic feature near you? It might be mountains, plains, a river, a large lake, a valley, a forest, or a large urban area. This geographic reality defines in large measure what people do there. It impacts their jobs, how they play, how they see themselves, the ease or difficulty with which they interact with the people of other places, and other aspects of their lives.

For the 43.5 million people who live in the Caribbean region, the main geographic reality is the sea. Perhaps when you think of the Caribbean, you think of blue skies and blue waters, beautiful beaches and a laid-back lifestyle—and maybe pirates. Caribbean reality is much more complex, just like the social, political, and cultural reality in which you live is complex. And just like where you live, geography plays a major role.

The Caribbean Sea covers just over one million square miles. In and bordering the Caribbean are about 7,000 islands, but the water surface area is about eighty times greater than the land area in the region. The Caribbean is bounded by South America to the south, Central America to the west, the Greater Antilles islands (Cuba, Jamaica, Hispaniola, and Puerto Rico) to the north, and the Lesser Antilles (including St. Lucia) to the east. The Caribbean islands make up 13 sovereign countries and 12 territories that belong to other countries.

What You Find in the Waters of the Caribbean

Biodiversity. The Caribbean Sea region is home to about 1,000 species of fish, 600 species of birds, 500 species of reptiles, 170 species of amphibians, 90 species of mammals, and 13,000 species of plants.

The Belize Barrier Reef, also called the Mesoamerican Coral Reef, is the second largest coral reef in the world. It runs along the coast of Mexico, Belize, Guatemala, and Honduras.

The Cayman Trench or Trough. This is the deepest point of the Caribbean at 25,220 feet below sea level. It runs from near the southeastern tip of Cuba toward Guatemala. Scientists have discovered volcanic vents or chimneys that release subterranean water hot enough to melt lead.

The Caribbean Current. This surface current flows west from the Atlantic through the southern Caribbean, turns north near Central America, then exits the region through the Straits of Florida between Cuba and the Florida Keys. The water flows at a rate of 15-17 inches per second and moves one billion cubic feet of water per second.

Weather—Good and Bad. When the weather is good, it's beautiful. Both the air temperature and the ocean temperature in much of the Caribbean region average around 80 degrees. When the weather is bad, it's hurricanes, an average of eight per year.

And museums. In places around the world where land features are important, people have founded museums and national parks to preserve those places. The same is true for people surrounded by water. Researchers at Indiana University have joined forces with the government of the Dominican Republic to found the underwater Living Museum of the Sea, in La Caleta Underwater National Park. This is the fifth such underwater museum that the Dominican Republic has established. The museums preserve archaeological and biological treasures where they are found.

The Living Museum of the Sea preserves a Spanish merchant ship that sank in a storm in 1725. Bringing artifacts to the surface hastens their disintegration. Leaving them in place preserves them and lets tourists and researchers see what in this case has been there for about 300 years. Some pieces that had been removed have been returned, and some replicas of missing pieces have been put in place.

The Caribbean Sea might appear to be just a calm body of water; but it is actually full of life, provision, and history.

The "Blue Economy"

When people need to work and are surrounded by the sea, they develop what economists call the blue economy: ways to generate income from the resources they have available to them, as Johanan Dujon has done. Perhaps the longest-standing part of the blue economy is fishing, which provides food as well as products for export. Another economic aspect of the Caribbean is the tourism industry, which welcomes over 25 million visitors to the region every year and generates over $49 billion for the region's annual blue economy. A third economic impact for three Caribbean countries consists of oil and natural gas reserves.

Cruise Ships in Grenada

Fish Caught in the Dominican Republic

The Caribbean setting is also a major inspiration for music, art, literature, hobbies, and other activities. People are even looking into using the ocean to generate energy. The motion of the waves might be harnessed to power turbines that could produce electricity.

When Geography Influenced Piracy

What about those pirates of the Caribbean we've heard about? Piracy really existed there, and geography played a part.

The Spanish were the first Europeans to explore and settle the Caribbean. Spanish galleons carried gold, silver, tobacco, sugar, and other products from the New World back to Spain. The governors of Jamaica and other islands (all of which were colonies of other countries) paid settlers on Hispaniola and Tortuga to raid the Spanish ships. These settlers, who smoked the meat of wild pigs in huts called boucans, came to be called buccaneers. The buccaneers later pirated Spanish ships for their own profit instead of just being government-backed raiders.

The many islands in the Caribbean with their secluded bays provided the pirates with excellent hiding places near the lucrative Spanish trade routes. The pirates could sell their stolen goods at black markets in Nassau and other ports. Both French and English raiders preyed on Spanish shipping.

However, the danger of piracy at sea discouraged the founding of sugar and tobacco plantations. To bring stability to the region, the British Royal Navy increased its presence there as Spanish power declined. By the early 1700s, piracy had decreased sharply. By the early 1800s, the French had lost their primary presence in the region because of a revolution in Haiti, and British naval presence was even stronger. The days of widespread piracy in the Caribbean were over.

Anne Bonny and Mary Reed were captured with other pirates by the British Navy in 1720. They were convicted of piracy in Jamaica and sentenced to death. Both claimed to be pregnant and were held in prison. Reed died in prison, while Bonny mysteriously escaped or was released. Their stories and this illustration are included in A General History of the Pyrates, *published in London in 1724.*

Good Advice

Johanan Dujon accomplished business success in the beautiful Caribbean. Pictured at right, along with his product, he has some advice for anyone who wants to start his or her own business. Be passionate about the business you want to undertake, he says, so that when (not if) the challenges come, you will press on to achieve your dreams despite those challenges. He also says, "Be meticulous, take pride in your brand, don't cut corners, ignore the naysayers, . . . and keep your customers happy." These are wise words from a young entrepreneur who built success out of a problem on a little island in the Caribbean.

Proverbs has much inspired advice for those who operate a business, such as:

Poor is he who works
with a negligent hand,
But the hand of the diligent
makes rich.
Proverbs 10:4

Assignments for Lesson 126

Gazetteer Study the map of the Caribbean Sea and read the entries for Anguilla, Antigua and Barbuda, Aruba, the Bahamas, and Barbados (pages 211-214).

Worldview Copy this question in your notebook and write your answer: How did the COVID-19 pandemic affect your worldview?

Project Choose your project for this unit and start working on it. Plan to finish it by the end of this unit.

Student Review Answer the questions for Lesson 126.

Cienfuegos, Cuba, During Hurricane Irma (2017)

127 He Does It to Save Lives

Scott Price flies airplanes into hurricanes. He is not crazy. It is his job. Flight training for commercial pilots includes lessons on how to avoid bad weather. Pilots like Scott Price, on the other hand, fly directly into the worst storms on earth. Price risks his life to save the lives of others.

Commander Price is part of a Hurricane Hunter team with the National Oceanic and Atmospheric Administration (NOAA). His father was a pilot, and Price served as a Navy pilot who flew missions in many parts of the world, from South America to Afghanistan. During his enlisted career, he learned about NOAA's hurricane research and reconnaissance missions and signed up.

The best way to train for this kind of flying is simply to take part in this kind of flying. As Price puts it, "It's impossible to accurately simulate a hurricane eyewall penetration [entering the eye of a hurricane]—doing it in the aircraft in a storm is the only way . . . ." Price's first mission for NOAA was in 2008 and involved a 2 a.m. takeoff to fly into a storm in pitch darkness. He also flew into the strongest hurricane on record, Hurricane Patricia in 2015.

A Hurricane Hunter flight crew usually gets 48 hours notice before having to leave on a mission. The preflight check takes 2-3 hours. The crew discusses the flight path, the purpose of the mission, current conditions and expected developments, and any anticipated hazards. The crew begins the mission with the best preparation possible, but Price says, "The only sure things about your trip through [the hurricane's eyewall] are the bumps behind you." A mission usually lasts about eight hours, and then a second crew embarks on a follow-on mission. Crews commonly fly six days in a row into a slow-moving storm.

The crew for one of these flights includes the pilot and co-pilot, flight engineers, the navigator, the flight director who runs the scientific mission, a data system operator, an operator who drops sensors into the storm, and other specialists. The information that the crew collects from inside the storm helps weather experts know when and where people should evacuate and businesses should shutter their facilities. This dangerous work can save lives and millions of dollars in property.

The Science of Hurricanes

The official name of storms that most Americans call hurricanes is tropical cyclones. People call these storms hurricanes in the North Atlantic and in the eastern North Pacific. In the western North Pacific, around China, Japan, and the Philippines, they are known as typhoons. In the western South Pacific and in the Indian Ocean the common term is tropical cyclones or simply cyclones. All of these terms refer to the same kind of storms. The prime season for tropical cyclones in the Northern Hemisphere is July through September, while in the Southern Hemisphere it is January through March.

Weather systems develop and move as a result of winds, moisture, the geography of the earth, and the movement of the earth. These monster storms in the Atlantic usually begin as low atmospheric pressure disturbances near the western coast of Africa at least 300 miles from the equator. The storms develop over a warm tropical ocean where the water temperature is 80° or higher as far down as 165 feet below the surface. Clouds form and begin to circulate around the low pressure center at the water's surface. As the air circulates, it draws the energy of water vapor and heat from the surface of the sea; and the warm, moist air begins to rise. As it rises, it cools and forms clouds. The clouds build and thicken as they circulate.

The circulation almost always results in a dense, circular storm. The low pressure center is called the eye. The winds in the eye are relatively calm, and the sky directly above the eye is fairly clear. The friction of the wind blowing against the ocean surface lowers the wind speed and causes the wind to turn

Commander Scott Price

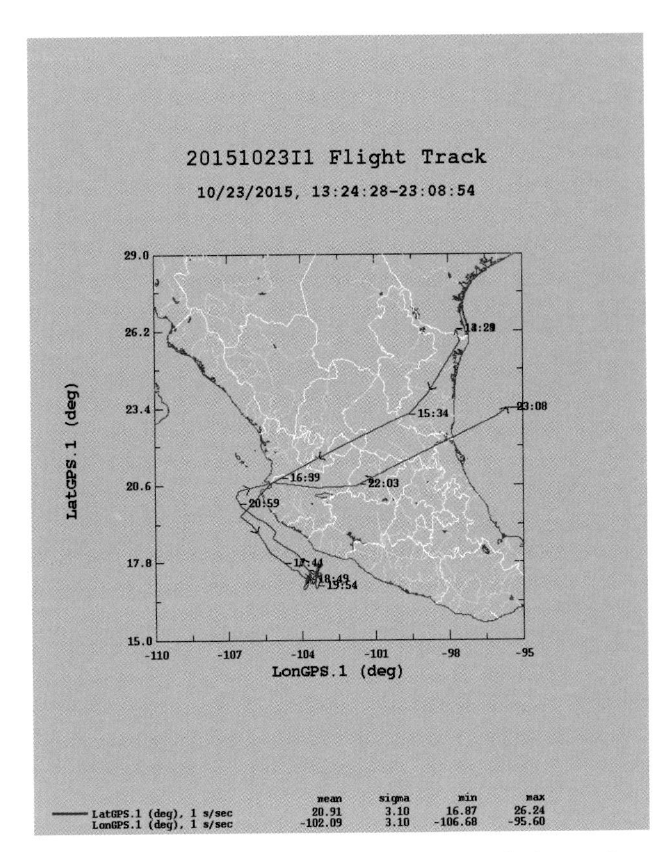

This NOAA chart shows the track of the flight made to observe Hurricane Patricia in the Pacific Ocean off the coast of Mexico.

inward toward the area of lower pressure. However, as the storm builds, the movement of wind around the calm center creates the eyewall, which comes to have the strongest winds in the storm. Winds in the eyewall can build to 150 to 200 miles per hour. Updrafts of wind in the eyewall result in towering cloud formations and the heaviest rainfall. Rainbands spiral out from the eyewall, creating a storm that can spread over 1,000 miles in diameter and rise as high as ten miles.

The earliest stage of this kind of storm is called a tropical disturbance. When sustained winds (as opposed to gusts) in the storm reach an average of 23 miles per hour, meteorologists designate it as a tropical depression. When the winds reach 39 miles per hour, it becomes a tropical storm. At a wind speed of 74 miles per hour, the storm is designated a tropical cyclone, typhoon, or hurricane. In the Northern Hemisphere, a tropical cyclone rotates in a counterclockwise or cyclonic direction. In the Southern Hemisphere, the storm rotates in a clockwise or anticyclonic direction. The rotation of the earth influences this circulation in what scientists call the Coriolis effect. In the Atlantic, the Caribbean, and the Gulf of Mexico, trade winds push the storm from east to west, but the rotation of the storm can cause it to curve northward.

When a tropical cyclone makes landfall, it causes destruction and loss of life in three ways: the high winds, the torrential rainfall, and the storm surge. This surge is the elevation of the surface of the sea because of the storm picking up water from the ocean. A storm surge can have a water surface up to 20 feet higher than the normal sea level. Once the storm makes landfall, it loses strength because it is not drawing energy from the warm ocean water. However, even a declining storm can cause severe flooding and spawn tornadoes before it dies out completely. A storm that moves northward over the ocean can lose strength as it passes over colder water. If an Atlantic hurricane has enough strength to remain formed after passing over a peninsula, such as Florida, it can regain more strength and cause additional destruction along the Gulf Coast.

Severe Tropical Cyclones in History

Every tropical cyclone that makes landfall creates its own tragic story of destruction and loss of life. The storms that strike the United States are the ones Americans know best, but cyclones and typhoons in Asia often cause greater loss of life.

The most devastating tropical cyclone in recorded history is the Bhola Cyclone that struck a low-lying delta region in what was then East Pakistan on November 12, 1970. The storm had winds estimated at 140 miles per hour and created a 20-foot-high storm surge. Estimates of fatalities range from 300,000 to 500,000. Frustration with the response of the Pakistani government increased the desire of many people there for the independence of East Pakistan and contributed to the outbreak of a war for liberation and the creation of the country of Bangladesh in 1971.

The deadliest weather disaster in United States history was the hurricane that hit Galveston, Texas, on September 8, 1900. The storm resulted in around 8,000 deaths, though estimates of the loss of life vary between 6,000 and 12,000. The storm turned north and continued (with reduced wind speeds) through Oklahoma Territory and Kansas.

The Saffir-Simpson Hurricane Wind Scale categorizes hurricanes by sustained wind speed and the damage that can be expected with each level. Any hurricane that is Category 3 or higher is considered major because of the greater potential for property damage and loss of life. The damage expected with each category involves ever greater numbers of homes damaged, trees uprooted or shattered, power poles downed, and power outages occurring. Recovery may take days after a Category 1 storm and up to weeks or months after a Category 5. In the western North Pacific, the term super typhoon describes tropical cyclones that have sustained winds of 150 miles per hour and greater.

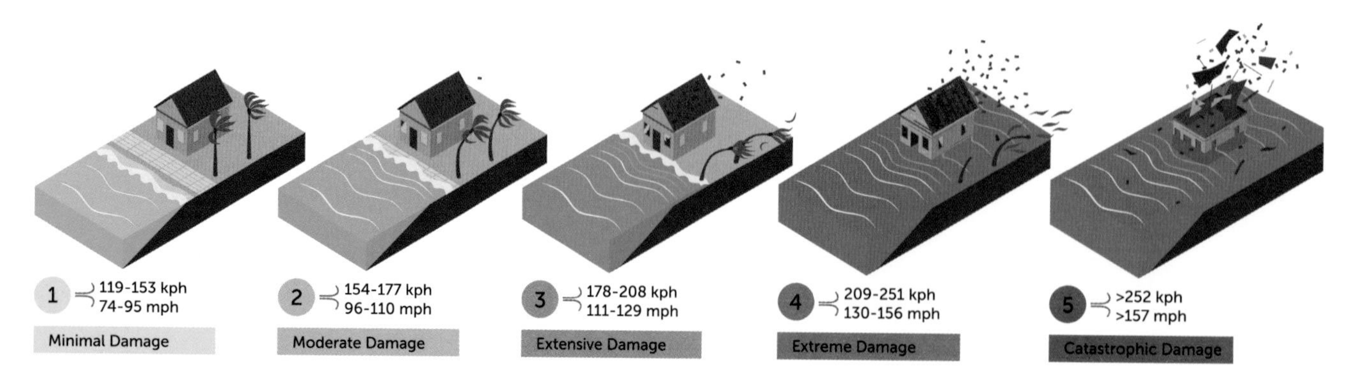

This image is from a NOAA flight into the eyewall of Hurricane Katrina in 2005.

It then crossed the Great Lakes, New England, and southeastern Canada, before finally dissipating over the North Atlantic.

The costliest tropical cyclone in U.S. history was Hurricane Katrina in 2005. About 1,200 people died in Louisiana and Mississippi as a result of the storm. Property damage was estimated at $75 billion.

The strongest tropical cyclone to hit the islands in the Caribbean struck Dominica, Saint Croix, and Puerto Rico in September 2017. Hurricane Maria was a Category 5 storm with sustained winds measured at 160 miles per hour. The region was already reeling from Hurricane Irma that had struck two weeks earlier. The estimates of fatalities reported by different sources vary from 3,000 to 5,000. Property damage in all of the affected islands exceeded $91 billion. Their governments were not prepared to undertake the huge recovery efforts required, and the islands suffered for a long time before they saw significant progress.

Predicting Hurricanes

Mankind has come a long way in our ability to predict hurricanes. The people who lived on the Caribbean islands centuries ago could tell by wind changes, the behavior of birds and fish, and the color of the sun when a major storm was coming, sometimes two to four days before it struck. The phenomenon was new to European explorers, who had to learn this aspect of weather forecasting from the indigenous people.

Christopher Columbus encountered hurricanes on two different voyages. In 1502 he perceived a tropical cyclone approaching and warned the Spanish governor of Hispaniola not to send a fleet of thirty ships to Spain as the governor was planning to do. Columbus sheltered his ships in a safe harbor, but the governor ignored the warning and lost all but one of his fleet.

Flooding in San Juan, Puerto Rico, persisted for weeks after Hurricane Maria in 2017.

When the giant storm was approaching Galveston in 1900, people knew a storm was headed their way but they did not know the intensity of it and thus did not take any precautionary action. The result was devastation and thousands of lives lost.

Weather planes have flown into the eyes of hurricanes for decades, but it was only after NASA began launching weather satellites in 1960 that meteorologists were able to have a better idea of hurricane intensity and movement. That progress has continued with improved satellites and better computer modeling from the data gathered by Hurricane Hunters like Scott Price. As a result, even though the population of coastal areas has increased, the number of fatalities resulting from hurricanes has decreased. Forecasters are now able to give up to 36 hours notice regarding where they expect a hurricane to make landfall, within an 80-mile range of accuracy.

Forecasters have become proficient at tracking storms, but they still lack the ability to determine how strong a storm will be. In other words, they are still learning. For instance, only in recent years did they learn that an eyewall can deteriorate and another eyewall can form around it. The need to continue gathering information is the reason why Price and other Hurricane Hunters still fly into the storms, even with accurate satellites and increased computer technology. Storm forecasters utilize information about a storm as well as historical records regarding the location, time of year, and behavior of previous storms in that place. The computer models that scientists generate utilize different variables, which is a way of saying, "If this happens, the hurricane will likely do this." This means that hurricane forecasting still has an element of making an educated guess, although the work of forecasters is much more accurate than in years past.

Naming of Hurricanes

Before World War II, weather observers identified tropical cyclones in various ways. The San Mateo Hurricane of 1565 struck Florida the day after the feast of St. Matthew (San Mateo in Spanish) in the Roman Catholic calendar. The hurricane that hit Galveston was known simply as the Great Galveston Hurricane of 1900. In the early 1900s, meteorologists often identified developing storms with latitude and longitude data.

During World War II, weather scientists began giving female names to storms. This practice helped forecasters identify storms more accurately, especially if more than one storm were taking place at the same time. The names also allowed the media to identify storms more easily for the public. Starting in 1953, the National Weather Service (NWS) assembled a list of female names each year that were given to the storms in turn as they developed during the hurricane season. In 1979 the NWS began using men's names, alternating with women's names.

The World Meteorological Organization (WMO) of the United Nations now oversees the lists of names that are used in various parts of the world. For tropical cyclones that originate in the Atlantic-Caribbean region, there are six lists that rotate annually. In other words, the list that will be used in 2021 will be employed again in 2027. However, if a storm is particularly destructive, the WMO retires that name from the list and replaces it with another out of sensitivity to the victims of the earlier storm.

Severe storms can have a cleansing effect on the earth's environment, but they also exact a heavy price in the loss of lives and property. God is still sovereign even during such terrible catastrophes that can characterize life on this amazing planet that groans for redemption. The inspired writers of Scripture sometimes used storms to illustrate their teaching.

I would hasten to my place of refuge
From the stormy wind and tempest.
Psalm 55:8

Assignments for Lesson 127

Gazetteer Read the entries for the Dominican Republic, Grenada, Haiti, Jamaica, and Montserrat (pages 217, 218, and 219).

Worldview Copy this question in your notebook and write your answer: How did the resurrection of Jesus change the worldview of the apostles?

Project Continue working on your project.

Student Review Answer the questions for Lesson 127.

Harvesting Sugarcane, Jamaica

128 How Sweet It Is—And Wasn't

First, there was honey. Honey satisfied the human desire for something sweet. To obtain it, however, people had to deal with bees, so honey was not widely available. It was a delicacy generally reserved for the wealthy and powerful.

At some point, European explorers and traders who visited the islands in the Pacific discovered sugarcane. This perennial tall grass grows to be two inches thick and 12 to 15 feet tall. It can be planted, harvested, processed, and refined to produce another kind of sweetener: sugar—and with no bees! People transplanted sugarcane to India and other parts of the world. The Portuguese transplanted sugarcane to the Canary Islands and other places near their home country. Muslim traders brought sugar to the Middle East. In the 1400s, it was introduced to the Mediterranean world. Columbus brought sugarcane to the New World on his second voyage in 1493 and found that it grew well there. Sugarcane fields and processing mills sprang up on Hispaniola, the island that later became the Dominican Republic and Haiti.

When the Portuguese began colonizing the New World, they brought the process of growing sugarcane to Brazil. The next issue to be addressed was who was going to grow the crop and process the sugar. Sugarcane only grows in tropical and subtropical climates. It requires significant heat and water and takes 12 to 18 months of fertilizing, weeding, and other work to be ready for harvest. Few Europeans wanted to do that kind of difficult farm work in such a setting. After the arrival of Europeans, many indigenous people died because of warfare and disease. The Portuguese decided to deal with the shortage of labor by importing slaves from Africa to work in the sugarcane fields and the sugar processing mills of Brazil.

Meanwhile, Europeans were discovering sugar and couldn't get enough of it. At first, it was, like honey, a delicacy only for the rich. Royal banquets often featured elaborate sugar sculptures and sugar-sweetened desserts. The growth of sugar production in the New World made sugar more widely available. Its use filtered down to the masses, especially to sweeten their daily cups of tea. People could do more with sugar than they could with honey. Besides having more recipe options using sugar, it's hard to create a decorative sculpture for the royal dining table out of honey.

Promoters put forth all kinds of ideas about how people could use sugar. Not all of these ideas were practical or accurate, but that has not always stopped promoters. Sugar was suggested as a preservative, a decoration, a medicine, a condiment, a sweetener, and a food.

As European demand for sugar grew, production in Brazil couldn't meet the demand, so sugarcane growing spread to the islands of the Caribbean that the Portuguese, French, and English settled. These islands had the geographic advantage of being closer to Europe than Brazil is, so the cost of shipping processed sugar from the Caribbean was lower. The labor situation played out largely the same on the Caribbean island sugar plantations: Europeans didn't want to do the work, so they imported slaves from Africa.

During the period in which Britain expanded its North American empire, it also developed its manufacturing strength. A triangular trade system developed. Britain sent manufactured goods (such as guns and rum) to Africa to trade for slaves. British slave traders took slaves from Africa to the New World in ships that had horrible conditions along a route that came to be called the Middle Passage (this practice says volumes about the worldview of the British and other Europeans regarding Africans). Then British traders transported processed sugar and other agricultural and craft products to sell in Britain. The sugarcane plantations in the British West Indies eventually produced 90% of the sugar that people in Western Europe consumed.

Dutch Sugar Plantation in Suriname (c. 1850)

As opposed to sugarcane, sugar beets are a root crop that was grown in Europe for centuries as a garden vegetable and as fodder for livestock. A sugar beet grows to be two to four pounds in weight. A German chemist produced sugar from sugar beets in 1747, and the first sugar beet processing facility was built in Poland in 1802. During the Napoleonic Wars of the early 1800s, France could not obtain sugar from the West Indies because of British interference with French trading ships. Napoleon encouraged the development of the sugar beet industry in France, and 40 processing plants were constructed there. Other countries in Europe followed suit. By 1880, more sugar was produced for European consumption from beets than from sugarcane. This demonstrated that Europe could have sugar without slavery.

Thus the development of an international trading system, the expansion of slavery, and the transformation of the human diet (with all of these factors having geographic aspects) came about in great measure because of one item: sugar.

The Making of Sugar

The term sugar applies to a number of different compounds that make up the simplest group of carbohydrates. Two places where sugar is found are in the sap of seed plants and the milk of mammals. The most common form is sucrose. Almost all plants contain sugar, but only sugarcane and sugar beets have concentrations high enough to make extraction practical and cost-effective.

Harvested sugar beets can be stored for relatively long periods, and they are generally processed to produce white sugar. On the other hand, harvested sugarcane cannot be stored for more than one or two days, so the timing of the harvest and the location of processing facilities near sugarcane fields are more important issues. Sugarcane is generally harvested in cooler months; but in some places, such as Cuba, Colombia, and the Philippines, harvesting occurs year-round.

Processing sugar from sugarcane involves extracting the juice from the cane; clarifying the juice by the addition of heat, lime, and additional compounds; concentrating the juice by evaporation; crystallizing the resulting syrup by further evaporation; and drying it to produce raw cane sugar. Refining this raw cane sugar produces high quality sugar that is used to make granulated sugar, sugar cubes, and brown sugar, which has some molasses content. Molasses, a concentrated syrup, is a byproduct of the processing of both sugarcane and sugar beets. Besides human consumption, molasses is a major ingredient in animal feed. The liquor drink rum is made from molasses.

Le Marron Inconnu ("The Unknown Maroon") *is a statue completed in 1967 by Haitian architect Albert Mangonès. The man represents the formerly enslaved people of Haiti who gained their independence. This 2012 photo shows the statute in front of the Haitian National Palace, which suffered extensive damage during a 2010 earthquake there.*

Slavery Was (And Is) a Bad Idea

The sugarcane industry on the Caribbean islands depended on slave labor. You might think that slavery would make slaveowners very wealthy. After all, labor is a major part of the cost of the production of goods and services. After the initial investment, a slaveowner avoids having to pay salary, provide benefits, and cover other expenses. Some slaveowners in the Caribbean and the American South did enjoy large profits.

However, the practice of slavery was actually a bad idea in economic terms. Enslaved persons had no incentive to work more efficiently. They had no motivation to learn better skills, and slaveowners had no motivation to pay for training. As a result, areas that did invest in such training passed the slave economy in terms of productivity. Slaveowners incurred costs for housing, food, clothing, medical care, and old age care, so even slavery was not expense-free. Most slaveowners sent their profits back to their home country instead of investing them in the Caribbean. Related industries such as shipbuilding and the manufacture of sugar processing and refining machinery grew, but that profited the West Indies very little. Slaves could not buy the products of their own labor or the labor of others, so they contributed little to economic growth.

Most important, of course, is the negative effect of slavery on human beings. Slave ships took an estimated 13 million people from Africa to the New World, and as many as two million of them died. Six million of these slaves worked on sugar plantations. Children of slaves commonly had to go to work when they were five or six years old. Conditions were terrible, and life expectancy was short. Slavery robbed people of the opportunity to grow, learn, and develop as they otherwise could have. How many slaves in the United States or the Caribbean islands had the chance to become doctors, scientists, entrepreneurs, professors, authors, or others who could have contributed to the economic or cultural wealth of society? Very few. Moreover, the slave trade did immeasurable damage to Africa, taking from the continent millions of laborers and many who possibly could have helped the continent in innumerable ways. Finally, slavery debased slaveowners because it negatively affected their worldview and how they saw their fellow human beings. Exploitation of human beings does not pay, however you measure it.

Many of those who owned sugar plantations in the Caribbean continued to live in Europe. Absentee ownership is not a good economic practice. It was not unusual that by the third generation of ownership, a plantation was no longer profitable. When owners did go to the plantations, they would sometimes be gone for years at a time, which was hard on the families they left behind. France paid a severe price for its slave policy when slaves and people of mixed race in Saint-Domingue on the island of Hispaniola rebelled against the French government. The rebellion began in 1791 and continued until 1804 when France granted independence to the country that became known as Haiti.

The Sweet World Has Passed It By

With the end of slavery, Caribbean sugar production and island economies had to undergo a painful but necessary transition. Today, workers on sugarcane plantations in the Caribbean still work hard for low wages. Two-thirds of the harvesting of sugar cane in the world takes place by hand with workers who use long machetes. It's hard for that to compete with mechanized sugar beet farming. Some Caribbean plantations have closed. Caribbean island economies have increasingly turned to tourism and petroleum and now import much of the sugar that the people of those islands consume.

The world has not lost its sweet tooth. People still consume vast quantities of sugar. The average annual per capita consumption in the United States of added sugar (not what occurs naturally in foodstuffs but what is added to cereals and other processed foods) is 57 pounds, or 17 teaspoons per day (a soft drink and a candy bar will do it). The American Heart Association daily recommendation is 3-6 teaspoons for children, six teaspoons for women, and 9 teaspoons for men. We like our sugar; unfortunately for our collective health, we like it too much.

People still love sugar, but we get much of it from different sources now. Brazil, an early source of sugar, has the highest level of production of any country in the world. In the United States, the leading sugar-producing state is Minnesota, which is not exactly a tropical setting but where sugar beets grow in abundance. Louisiana, Florida, and Texas lead in sugarcane growing. Over half of the American sugar production now comes from beets. Sugar beets have enabled many countries of the world to become sugar producers to try to capitalize on the world's appetite for sugar.

The sweetness of sugar has brought people much joy. Sadly, the way that the sugar industry offered that enjoyment to us exacted a bitter price for those

The image on the left depicts slaves harvesting sugarcane on the Caribbean island of Antigua in 1823. The photo on the right shows workers harvesting sugarcane in Madhya Pradesh, India, in 2011.

who were enslaved and those who have suffered physically because of sugar. There is no more precious commodity in our world than human beings. This advice from Proverbs refers to honey, but certainly we can apply it to sugar as well:

Have you found honey? Eat only what you need,
That you not have it in excess and vomit it. . . .
It is not good to eat much honey,
Nor is it glory to search out one's own glory.
Proverbs 25:16, 27

Assignments for Lesson 128

Gazetteer Read the entries for Puerto Rico, Saint Barthélemy, Saint Kitts and Nevis, Saint Lucia, and Saint Martin (pages 219-221)

Worldview Copy the following question in your notebook and write your answer: Why is it often hard to change your worldview when you encounter something that challenges your thinking?

Project Continue working on your project.

Student Review Answer the questions for Lesson 128.

129 So Close: Cuba

It is so close, just 90 miles from the United States. It is also so close to prosperity, in that it has the resources to have a thriving economy. But its economy struggles because of the direction that the government has taken, a government that is very different from the one 90 miles away.

Ninety miles is the distance between the southernmost point of the United States—Key West, Florida—and the island of Cuba. Those 90 miles not only separate two forms of government, two economic systems, and two worldviews; they also separate two sets of hopes and possibilities, and they separate many families.

As close as the U.S. and Cuba are in the geographic terms we have described, in one place United States territory and Cuban territory are even closer. In that place, only a fence separates them.

The Geography of Cuba

Cuba is a beautiful island. Pleasant trade winds moderate the tropical climate which features fairly constant temperatures year-round. Many rivers and streams flow on the island, although none is navigable for any great distance. Four mountain ranges rise from its mostly flat-to-rolling terrain that provides good farmland. Cuba has significant oil and gas reserves on land and offshore, and it has mineral and timber resources as well.

The island is home to more plant and animal species than anywhere else in the Caribbean. The thousands of flowering plant species include over 40 species of orchids. Cuba's national tree, the royal palm, grows to 50 to 75 feet in height and is common in rural areas. The national flower is the mariposa. Its fragrant white petals stand atop stems that often grow over five feet high.

Perhaps most distinctive among the 300 bird species found on the island (about two-thirds of which are migratory) is the bee hummingbird. This is the smallest bird in the world, averaging just over two inches in length and weighing less than one ounce. Unfortunately, several other bird species have been hunted to extinction.

Slightly smaller than Pennsylvania, Cuba is the largest island in the Caribbean. It is the westernmost island of the Greater Antilles. Cuba is located where the Atlantic Ocean, the Gulf of Mexico, and the Caribbean Sea come together. Cuba averages 60 miles in width and measures 750 miles from east to west. About 1,600 smaller islands, islets, and cays are also part of the country.

The island's eleven million people are about two-thirds white, one-fourth mixed, and one-tenth black. When Fidel Castro seized power in Cuba in 1959, about 85% of the population was Roman Catholic. Having grown up under atheistic Communism, most of the younger generations are not religious and have not received religious training. Some churches are still open, and some Protestant and evangelical groups have grown, but the government closely monitors their activities and official persecution of Christians does take place.

Probably between 50 to 70 percent of Cubans practice the Santeria religion to some degree. The Yoruba people of West Africa brought this religion when Spanish landowners brought Yorubans to Cuba as slaves. The religion is based on an identification that people made between Yoruban deities and Roman Catholic saints. Followers of this religion make appeals and sacrifices to their deities to obtain good things and to ward off troubles.

The Taino and the Spanish

We don't know for sure how the first humans arrived in Cuba. The Taino, an Arawak group, were the most numerous people on Cuba and other Caribbean islands when Europeans arrived. They developed an extensive agriculture and lived in a complex society.

We do know how Europeans arrived in Cuba. Christopher Columbus and his crew landed on the island during Columbus' first voyage in 1492. Primarily because of warfare and disease, within a hundred years the Taino had been almost completely wiped out on Cuba. The Spanish later imported slaves from Africa to perform farming work for them. Eventually about one-fourth of the population on Cuba were slaves. Spanish agriculture emphasized tobacco production early on and sugar growing later. These two crops became important Cuban exports.

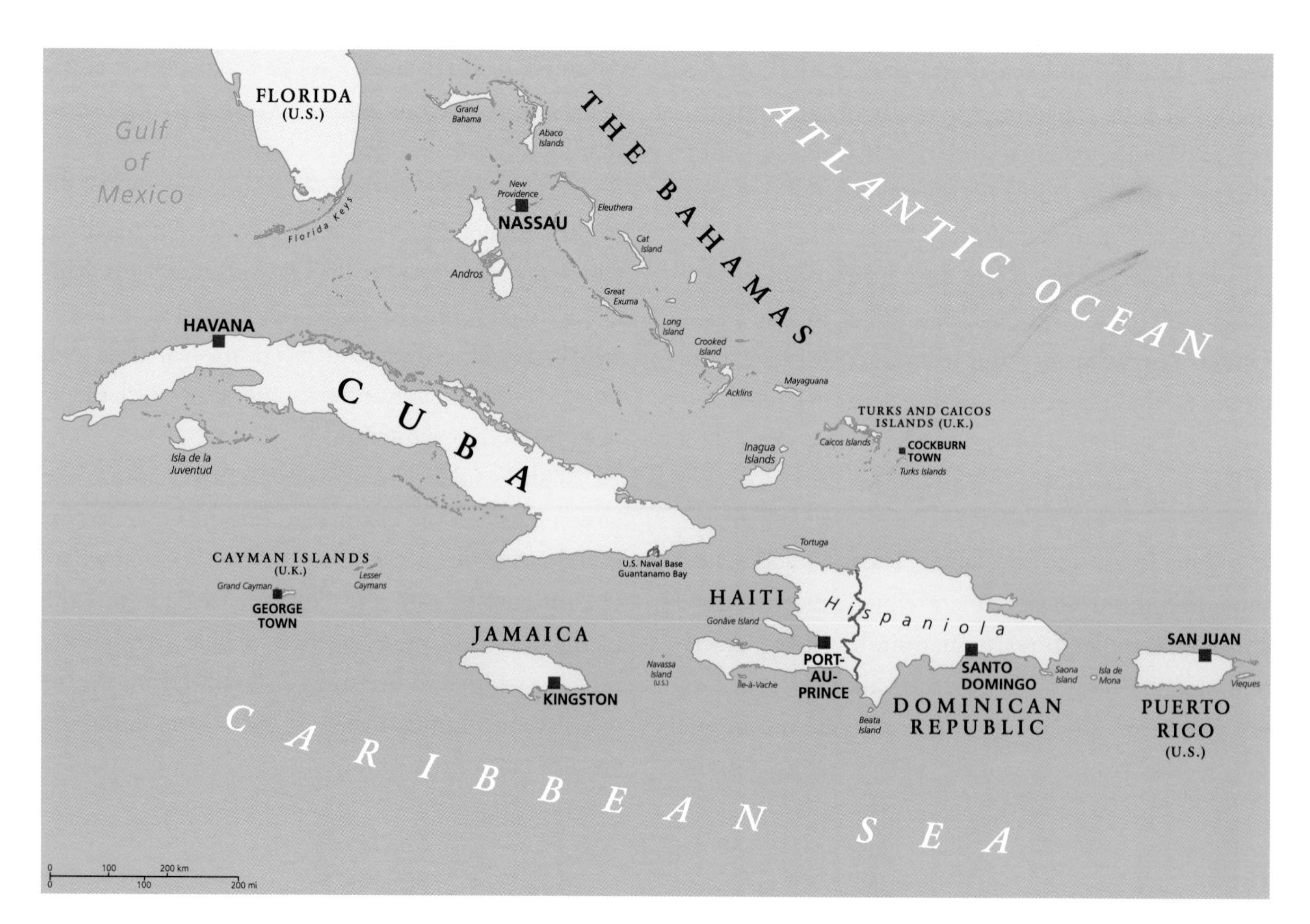

So Close to the United States

Cuba's geographic proximity to the United States has affected its national story for over two hundred years. Spain controlled Cuba for centuries. By the 1800s, however, Spanish rule in Cuba, though still imperialistic, had become weak.

Presidents as early as Thomas Jefferson as well as other political leaders and elements of the press expressed an interest in acquiring Cuba for several reasons. First, expansionists thought it was a logical place to extend American power. Second, slaveholding interests wanted to make it another state that practiced slavery. Third, many Americans wanted to insure that the island did not fall into the hands of another country, especially Great Britain.

Because of Cuba's proximity to the U.S., many Americans saw acquiring the island as a high priority. For instance, the Ostend Manifesto was a proposal circulated among a few members of the American government in 1854 giving reasons why the United States should acquire Cuba and suggesting the use of force if necessary to do so. European and antislavery reaction against it was so strong that the American government never acted upon the proposal.

Meanwhile, insurgents in Cuba began fighting their Spanish rulers for greater political rights and ultimately for independence. Spanish control over the island was so strong that it was not until 1886 that Spain abolished slavery in Cuba. The conflict between Spain and Cuban freedom-fighters became especially intense in the mid-1890s. The United States did not want to become involved in a civil war, but many Americans did support the cause of independence. In addition, Americans who had significant business investments in Cuba wanted either a continuation of the status quo or a peaceful transition to independence.

In 1897 the United States sent the battleship USS *Maine* to the Havana harbor to protect American citizens and interests and to be ready for whatever happened. On February 15, 1898, a huge explosion sank the battleship and resulted in 260 American lives being lost. The cause of the blast

This 1896 Spanish cartoon satirically suggests that Uncle Sam is reaching out to hold Cuba, "protecting the island so it won't get lost."

was never clearly determined, but many Americans blamed Spain. The United States Congress declared war on Spain in April.

The war was no contest. By July Spain sought an end to hostilities. In the peace treaty signed in December, Spain ceded Cuba, Puerto Rico, the Philippines, and Guam to the United States. In return the United States paid Spain $20 million for infrastructure that the Spanish had built in the Philippines. In one stroke the United States acquired a small but widely spread overseas empire and became a significant player on the world stage, while Spain's era of worldwide power ended. The U.S. finally possessed Cuba.

The United States oversaw the creation of a republic in Cuba. However, in 1901 the Platt Amendment to an Army appropriations bill in Congress stated that the United States had the right to oversee Cuba's internal and international affairs. It also claimed the right to lease land for a naval base at Guantanamo Bay on Cuba's southeastern coast. Congress repealed most of the amendment's provisions in 1934, but the U.S. maintained the base. The base can only be closed if the United States decides to do so.

The Twentieth Century

The Republic of Cuba began in 1902, but it had a troubled existence. Contested elections, attempted rebellions, routine corruption, and changing U.S. policies to try to deal with a difficult situation led to decades of instability. The United States helped Fulgencio Batista come to power in 1952. The Cuban economy grew, but the profits mainly went to a few wealthy Cubans and foreign investors. The majority of Cubans lived in poverty. Meanwhile, Batista's government continued the pattern of corruption.

During the 1950s, an opposition movement led by brothers Fidel and Raul Castro and Che Guevara from Argentina attracted supporters. On January 1, 1959, Batista fled the island. Fidel Castro led a takeover of the capital of Havana and declared a new government. It soon became apparent that Castro's movement was Communist oriented, as the government assumed control of the economy and implemented an authoritarian form of rule. The United States ended diplomatic relations with Cuba and imposed a trade embargo, but Cuba began a favorable trade arrangement with the Soviet Union.

Castro's government became the target of a group of anti-Castro Cubans living in the United States. The U.S. Central Intelligence Agency secretly assisted the group in preparing for an invasion to take over the island. An invasion took place at the Bay of Pigs in Cuba in April 1961, but it failed in part because of a lack of air support by the United States.

In 1962 the United States became aware of the Soviet Union installing nuclear missiles in Cuba that could reach the U.S. mainland. The U.S. imposed a naval blockade that prevented any further materials from reaching Cuba and insisted that the Soviets remove the weapons. For about two weeks in October, the world appeared to be on the verge of a nuclear conflict. The two sides reached a geographic tradeoff when the Russians agreed to remove the missiles from Cuba permanently, and the United States agreed to remove missiles in Turkey that were aimed at the Soviet Union.

This 1927 photo, taken from an airplane, shows ships of the U.S. Navy fleet at Guantanamo Bay during a training exercise.

Standoff and Restoration

Before Castro's revolution, business and travelers flowed freely between the United States and Cuba. Afterwards, trade was cut off. The only immigrants were those who escaped from Cuba secretly, often at the risk of their lives. The Cuban government and the American government eyed each other warily, suspicious of each other's intentions and actions. In 1987 the U.S. agreed to admit 20,000 Cubans per year, but many more came illegally. When the Soviet Union fell, Cuba lost its biggest trading partner; and the struggling Cuban economy suffered even more. Cuba's situation improved when it began taking small steps toward capitalism, and the island benefited from a trade agreement with Venezuela.

In 2014 President Barack Obama announced a policy of easing relations with the goal of eventually establishing full diplomatic relations, which took place the next year. Some thawing has taken place, but most of the benefits have gone to Cuba. The island has seen an increase in tourism, but the Cuban government still holds political prisoners and keeps repressive control over its people. For instance, Cubans cannot buy computers or have access to the Internet without special authorization from the government. This has led some Cubans to seek such purchases and access through black market transactions.

Fidel Castro stepped down in 2008 and died in 2016. His brother Raul assumed power upon Fidel's retirement. Raul retired in 2018, and Miguel Diaz-Canel Bermudez was his handpicked successor. Trade and cultural interchange have increased, but in 2017 U.S. President Donald Trump vowed to keep some economic sanctions in place until the Cuban government frees all political prisoners.

The Cactus Curtain

Naval Station Guantanamo Bay (nicknamed "Gitmo") is a 45-square-mile U.S. Navy installation that plays a key role in the U.S. presence in the Caribbean region. The facility has fulfilled several purposes in its history, but most recently it is best known as the detention site for terror suspects.

A fence surrounds the base. Just after Castro's revolution, some Cubans sought asylum by climbing the fence onto the base. Castro ordered miles of cactus to be planted along the northeastern section of the fence to discourage these efforts. This became known as the Cactus Curtain, after the Iron Curtain in Eastern Europe. The U.S. and Cuba both planted landmines along either side of the perimeter of the base. The U.S. has replaced its landmines with lights and motion sensors, but the Cuban landmines are still in place. The Cuban government would like to see the United States close the base, but that is not likely to happen under current circumstances.

Classic Cars

One intriguing aspect of contemporary Cuban culture is the widespread love of classic American cars. Cubans own an estimated 60,000 American cars from the 1930s, '40s, and '50s, which they acquired before Castro seized power and the United States imposed an economic blockade that prevented the importation of newer models. Many of these cars are family heirlooms that get passed down from one generation to another. Some serve as taxis for foreign tourists.

The import blockade also means that replacement parts are difficult or impossible to find. Many cars are held together by what has been called imaginative engineering. It is not unusual to find a car with a classic American body, a Japanese or Russian engine, and other parts imported from several different countries or even homemade. You can see some Soviet-era cars on the streets; but these are plain, poorly built, and do not have the charm and appeal that classic "Yank Tanks" do.

Classic Cars in Havana

Los Pinos, Musical Group in Trinidad, Cuba

The United States has done much to help Cuba, but it has not always been a good neighbor to the island. On the other hand, the governments of Cuba have not led the island as they should have, and on at least one occasion threatened the United States homeland. At one time, Cuba and the United States were close to nuclear confrontation. Today the tense relationship has improved in small ways, but the opportunity for freedom and prosperity in Cuba still seems far away.

Proverbs teaches the right way for neighbors to live with each other.

Do not devise harm against your neighbor,
While he lives securely beside you.
Proverbs 3:29

Assignments for Lesson 129

Gazetteer Read the entries for the British Virgin Islands, the Cayman Islands, Cuba, Curacao, and Dominica (pages 214-216).
Read the excerpts from "To Cuba and Back" (pages 325-329) and answer the questions on it in the *Student Review Book*.

Geography Complete the map skills assignment for Unit 26 in the *Student Review Book*.

Project Continue working on your project.

Student Review Answer the questions for Lesson 129.

Protesters Outside the White House (1965)

130 Changing Your Worldview

In one of my first political science classes in college, in 1970 or 1971, the subject of states rights came up (this is sometimes spelled as states', possessive, but I'll keep it simpler). This college was in the South, and I was from the South. States rights had been the mantra of most Southerners for the previous 25 years or so, let alone its role in the Civil War. Whenever national politicians talked about a stronger role for the federal government or a new program that seemed to threaten state sovereignty, Southerners would cry foul and appeal to states rights—the right of states to govern themselves as they pleased without any involvement by the federal government.

True enough, the Constitution created a limited federal government with only enumerated powers. The Tenth Amendment stated that powers not expressly granted to the federal government nor denied the states were reserved to the states or to the people. For instance, the Constitution did not include education or voting and elections as areas in which the federal government could exercise power. Nevertheless, in previous years, the federal government had become involved in these areas and had passed laws that states had to obey.

So, in the discussion in my political science class, one student voiced the typical and comfortable appeal to states rights to say that the encroachment of federal power in certain areas was wrong and unconstitutional. Right, I thought. That ought to settle the matter.

Then the professor asked, "But what if the states aren't doing their job?"

Oh. Wait a minute. I had never thought of that before. Did the professor mean that something trumped states rights, namely the rights of American citizens? For years some Southern states had routinely denied blacks the right to vote. We had seen states all over the country fail to protect the civil rights of blacks. Did the doctrine of states rights mean that if the state of Mississippi, or Iowa, or Massachusetts denied equal protection under the law to some of its citizens, as guaranteed by the Fourteenth and Fifteenth Amendments, that was just too bad? Those people had just picked the wrong state to live in, and there was nothing the federal government could do to enforce the application of the fundamental law of the land? That didn't seem right. As correct and powerful as the doctrine of states rights was, something else was even more basic: human rights, the rights of American citizens, for all Americans regardless of the state in which they lived. Black American citizens who were also citizens of Mississippi deserved to be able to vote because they were Americans, regardless of the traditional laws and attitudes in Mississippi.

That professor's question challenged my assumptions, my worldview. No, I didn't become a flaming liberal, and yes, I do believe the federal government can overstep and has overstepped its Constitutional limits in some areas. But the question caused me to rethink the issue, to see that there was more to the debate than just two sides standing their ground in defiance of each other while American citizens who had the same rights I had were denied those rights. This discussion in a college classroom changed my worldview.

Education, travel, books, interaction with people who don't share your background, and other life experiences are liable to change your worldview. This doesn't mean you have to toss out everything you believe and have been taught. Some life experiences will confirm and strengthen some of your beliefs. But life will challenge you to rethink your worldview; and that reassessment, while it can be painful, can lead you to a better, more mature,

Library in Governor's Harbour, Bahamas

Port-au-Prince, Haiti

and—dare I say it—more correct understanding of the world.

My father enlisted in the U.S. Army in 1941 as American involvement in World War II was getting closer. He told the story of a man in our small Southern hometown who wanted his son to be ready and to look well-dressed when it was his turn to go off and join the Army. Therefore, this thoughtful and conscientious father went out and bought his son two new pairs of—overalls. This is what the man's worldview led him to do, but his action based on that worldview was totally inadequate for preparing his son for serving in the military. Those overalls would be useless and not a little embarrassing at basic training camp, or if he were stationed in New York City as my father was for a time, or when he was shipped overseas to Europe or the Pacific.

I took part in a mission trip to Mexico in 1996. One day our group went to visit a Christian woman and her son who lived in a small house that had walls of made of flattened cardboard shipping boxes extending from dirt floor to tin roof. And the woman was happy! She did not mutter or complain about her lot to us visiting Americans. She was content in the situation she was in, reflecting Paul's outlook that he expressed in Philippians 4:11. For me, who had always lived in a comfortable frame or brick house with heat, air conditioning, electricity, and indoor plumbing, this was a revelation. I had always heard about poor people, and I had occasionally driven by a poor family's house—in an automobile that my family owned. Now I was seeing poverty first hand, in third world living conditions, and I realized that real human beings, just as valuable to God as I was, lived this way. That encounter—indeed, that entire mission trip—affected my worldview.

We take many things in our world for granted until something happens to threaten them. For me such eye-opening experiences have included reading a book or seeing a documentary that described a different way of life, meeting someone who had smuggled Bibles into the Soviet Union, hearing a physician describe an appointment with a pregnant female patient who wanted him to perform an abortion, and experiencing tragedies in my family that I never thought would happen to us. People and events don't stay put. Life has a way of challenging our safe and comfortable world and helping us—forcing us—to grow and to develop a broader perspective on the world; in other words, to have a different worldview.

A common temptation is for a person to try to force a new experience or idea into the limited mold of his or her pre-existing worldview, thus trying to redefine reality to fit that person's preconceived notions. This person might think, "Oh, that doesn't matter because . . ." or "This doesn't change my basic belief that . . ." or "That person simply doesn't understand that . . . ." Instead, you and I have to be willing to learn and accept a broader perspective, to see the eternal principles at stake, and to recognize and admit our limited and often immature beliefs. As I indicated earlier, I don't need to let new (to me) realities and information knock all the props out from under my belief system and lead me totally to reject what I have come to believe is true. But you and I have to stretch our thinking to fit situations we had not previously considered, to reconfirm

Festival of San Sebastian, San Juan, Puerto Rico (2016)

what is rock solid even in a new situation, to adjust what has been our limited thinking, and to replace what was wrong with a new and more Christ-like and realistic understanding that better fits the real world in which we live. Without flexibility and the willingness to grow and to learn to discern, the only options are to reject the old completely or reject the new completely. Neither option accurately takes reality into account.

The terrorist attacks on the United States on September 11, 2001, were an experience that created a new reality, which in turn challenged people to change their worldview. Any books or articles written before 9/11 regarding American security and international relations now sound simplistic because they didn't—they couldn't—describe the world as it is today. In the same way, the COVID-19 pandemic changed our world and thus our view of it. Any analysis of our world that doesn't incorporate the new reality of our world doesn't seem adequate or realistic.

Before the spring of 2020, billions of people around the world got up every morning and went to work in their offices and factories. From August to May, millions of school children headed off to classes in the mornings. When you needed something from a store, you simply went there, paid for it at the checkout, and returned home. If you needed a break, you went to the beach, the pool, or the park. On Sundays millions of people went to their churches, often exchanging hugs and handshakes. Millions of people routinely went out to eat at restaurants, sometimes standing in line and crowding in to occupy every table and booth. People went to see movies in theaters. Fans by the thousands filled arenas to watch sporting events and concerts. The goal of every event was to be a sellout, filling every available seat. All this and much more was considered normal.

Face Masks for Sale in Mexico City (2020)

Not any more. Many of these familiar ways of going about life have been altered, and some of them may never go back to the way they were before. Every person with a cough or fever became suspect. Masks became common in public places. Crowding into anywhere—buses, subways, elevators, sports events—became life-threatening.

The coronavirus changed our worldview. It made people rethink some core questions, such as: What is important? What is essential? What is fear, and what is wisdom? Can we ever see going to church or going shopping or going to a sporting event or family reunion the way we did before? We looked at education, retail sales, and interactions with other people and with other countries in an entirely new way. We had to confront a new reality, and as a result our worldview changed in significant ways.

The common worldview when polio was an ominous threat is different from now, when we hardly ever think about it. Those of us who grew up during the Cold War with the constant possibility of nuclear destruction think differently from those who came of age after 1989 with the fall of Communism in Europe. Children born after September 11, 2001—your generation—don't know first-hand what life was like for people who had grown up before that date. Children born after the coronavirus pandemic of 2020 won't have the same worldview as those who lived through it.

Another Worldview-Changing Situation

I grew up in Tennessee. My great-great-grandfather was a Confederate soldier who died during the war. My grandfather was an unreconstructed collector of Civil War memorabilia and had a room in his home dedicated to his collection. I always thought it was neat to have this personal connection to history, although I certainly didn't buy into the cause for which the Confederacy fought.

The author's grandfather, Earl Notgrass, collected guns, swords, bullets, and cannonballs from Civil War battlefields in the early 1900s.

A few years ago I was giving a talk to 15 or 20 people about how to make history interesting to students, a presentation I had made many times. One point I made was how we need to emphasize the personal connection that we have to historical events, places, and people. To illustrate that point I told about my great-great-grandfather being in the Confederate army. As I was using that illustration, I realized that at least a third of the audience was African American. This had never been the case in my previous presentations of this material. I thought I could see on every one of their faces a look that said, "Where is he going with THIS?"

I made it through the talk and no one spoke to me about it afterward, but that evening I rewrote my talk for my next presentation and took out that part. I had always thought that the story was an interesting illustration. I had not considered how painful it could be for descendants of people who had been enslaved. The uncomfortable situation caused me to change my worldview.

The Constants Amid the Change

Despite the changes that we experience, some things remain constant. God is still in charge, and He is still loving. People still matter. Living well to serve God and others is still our calling. But after the coronavirus, even these beliefs, perspectives, and commitments took on a new meaning. We see that they must apply to situations we never expected to face. Our world has gotten more complex. Our worldview has matured. As God gives us time in our lives, we can expect further challenges and opportunities to grow. What will be the next Pearl Harbor, polio cure, fall of the Berlin Wall, 9/11, or coronavirus pandemic that will change our world and our worldviews forever?

Therefore from now on we recognize no one according to the flesh; even though we have known Christ according to the flesh, yet now we know Him in this way no longer. Therefore if anyone is in Christ, he is a new creature; the old things passed away; behold, new things have come.
2 Corinthians 5:16-17

Assignments for Lesson 130

Gazetteer Read the entries for Saint Vincent and the Grenadines, Sint Maarten, Trinidad and Tobago, Turks and Caicos Islands, and the Virgin Islands (pages 222-224).

Worldview Recite or write the memory verse for this unit.

Project Finish your project for this unit.

Student Review Answer the questions for Lesson 130.
Take the quiz for Unit 26.

Colonia Independencia, Paraguay

27 South America Part 1

This is the first of two units on South America. The huge Amazon geographic region dominates the continent and impacts many people. The Pampas region is home to the legendary gauchos. Lake Titicaca is the highest navigable lake in the world and the Uru people live on it. The War of the Pacific that took place in the late 1800s is still a live issue for some South Americans even today. The worldview lesson examines the remarkable growth of Christianity in the Global South.

Lesson 131 - The Rubber Soldiers of the Amazon
Lesson 132 - Gauchos of the Pampas
Lesson 133 - Living on Lake Titicaca
Lesson 134 - "The Sea Belongs to Us by Right": The War of the Pacific
Lesson 135 - Evangelism in the Global South

Memory Verse

Memorize Habakkuk 3:19 by the end of the unit.

Books Used

The Bible
Exploring World Geography Gazetteer
Tales from Silver Lands

Project (Choose One)

1) Write a 250-300 word essay on one of the following topics:
   - Decide on one aspect of the Amazon River basin and write a report about it. (See Lesson 131.)
   - Considering the continuing resentment of Bolivia regarding their loss of coastal territory as a result of the War of the Pacific, write an essay about the importance of forgiveness, reconciliation, and moving forward. (See Lesson 134 and Philippians 3:13-14.)
2) Draw a picture of a floating Uru island in Lake Titicaca. (See Lesson 133.)
3) Write a story that tells about an individual who learns the gospel, turns from his or her former worldview, and becomes a Christian. (See Lesson 135.)

Literature

Tales from Silver Lands won the Newbery Medal in 1925. Charles J. Finger collected the folktales firsthand as he traveled through Central and South America. Like folk and fairy tales throughout the world, the stories include talking animals, wizards, giants, and other fantastic creatures interacting with ordinary people. The tales cleverly illustrate the importance of traits such as loyalty, courage, generosity, and fidelity. They also serve as a window into the culture and geography of the regions where they were created.

Finger was a dynamic adventurer who lived many places and held many jobs during his interesting life. He was born in England in 1867. His parents emigrated to the United States in 1887, and Finger eventually followed in 1896, later becoming a citizen of the United States. His many stages of employment included sailor, shepherd, tour guide for an ornithological expedition at the southern tip of South America, music teacher and concert tour arranger, railroad foreman, auditor, and manager, magazine editor, and writer. He settled in Fayetteville, Arkansas, and wrote thirty-six books during the last twenty years of his life. He died there in 1941.

Plan to finish *Tales from Silver Lands* by the end of Unit 28.

Collecting Latex, Brazil

131 The Rubber Soldiers of the Amazon

You might think of rubber as just a fun product that people use in insignificant items such as pencil erasers, rubber bands, and children's bouncy balls. But rubber has many diverse and important applications. Take the military, for instance. Without rubber, jeeps don't move, planes can't land, and tanks don't roll. In other words, without rubber even the mightiest military machine grinds to a halt. These uses are especially urgent in wartime.

Where Does Rubber Come From?

Natural rubber comes from the sap we call latex, which is derived from the rubberwood tree. The scientific name for the rubberwood is *hevea brasiliensis*. As the name suggests, the tree is indigenous to the Amazon region that covers 60% of Brazil. Before Spanish and Portuguese explorers began roaming through Brazil, indigenous people had already learned to make game balls and other products from latex.

The rubberwood tree can grow to 100-130 feet in height and live for one hundred years. When it is six years old, a sapper can remove a sliver of bark and collect the sap that flows out into a bowl. When the fruit of the rubberwood tree ripens and bursts, seeds can scatter up to 100 feet away.

How Great Britain Became King of Rubber

In the mid-1800s, Great Britain was developing a worldwide empire; but that empire did not include Brazil. The British wanted easy access to latex to use in many of the products that its mighty industries were producing. In 1876 the British explorer Henry Wickham didn't exactly follow Brazil's laws for exporting goods out of the country. In other words, he stole 70,000 rubber tree seeds and took them to England.

Not long afterwards, British merchants took some of the seedlings that the seeds had produced to Ceylon and Malaysia in south and southeast Asia, to a climate similar to that in Brazil. Cultivating the trees in that part of the British Empire proved to be easier than it was in Brazil, and the Asian rubber industry was born. In 1920 King George V knighted Wickham for his thievery—that is, for his contribution to the growth and development of the British Empire.

U.S. President Franklin Roosevelt (right) traveled to Brazil to meet with Brazilian President Getúlio Vargas in January 1943.

Then Came War

This worked out well for Britain and the world, and Malaysia became the world's leading producer of rubber. But then in the early 1900s, the Japanese military began taking over various areas of the Pacific region. Just after the Japanese attacked Pearl Harbor, they seized Malaysia. The United States and its allies faced the fact that 95% of the world's rubber production was in the hands of the enemy. Scientists in several countries had experimented with making synthetic rubber for years, but what they produced was not of the quality or amount that the war effort needed. So the Allies returned to the original source for rubber, Brazil. U.S. President Franklin Roosevelt arranged a deal by which Brazil provided all the rubber it could produce to help the war effort in return for a payment of $145 million. Brazil had been neutral about the growing war, but this deal pulled the country firmly onto the Allies' side.

The Brazilian government gave the men of Brazil a choice: they could either serve in the armed forces, or they could go into the Amazon and obtain latex from rubber trees to help the Allies. About 55,000 Brazilian men volunteered to go sapping. They came to be called rubber soldiers, fighting the jungle instead of the Germans and Japanese. Some of the men moved their families with them into the Amazon region.

Unfortunately, the rubber soldiers made the worse choice. Almost half of the total number, about 26,000 men, died from illness, snakebite, accidents, and other causes. What is more, the men never got paid.

Somewhere between the United States' payment of the money to Brazil and the time when the men expected to be paid, the money apparently wound up in the pockets of Brazilian government officials. Several years later the Brazilian government began paying the men a small pension, but it was much less than they had expected to receive. It was only in 2014, when many of the surviving rubber soldiers were in their eighties and nineties, that the Brazilian Congress agreed to make larger payments to the men and their survivors.

After the war, the development of synthetic rubber continued and improved; but natural rubber is still a component of many products we use today.

Amazonia

In 1541-1542, Spanish explorer Francisco de Orellana explored the length of a large river in South America. He started from its headwaters in the Andes and moved east. Orellana reported fighting tribes of female warriors, whom he compared to the Amazon warriors of Greek mythology (some historians believe that female warriors really did exist in some ancient civilizations). Thus the river came to be called the Amazon. The region surrounding the river is sometimes called Amazonia.

Everything about the Amazon basin is not just big; it is huge. It covers 1.5 million square miles, 60% of Brazil and parts of Peru, Colombia, Venezuela, Ecuador, Bolivia, Guyana, Suriname, and French Guiana. Two-thirds of the Amazon basin is rainforest. This single rainforest is twice the size of India and is larger than all the other rainforests in the world combined. The equator runs through the Amazon region, which means that it is a tropical rainforest, hot and wet. The average rainfall for the entire Amazon basin is 120 inches per year, but some places receive up to 400 inches per year.

The Amazon region is home to 80,000 plant species, which are half of the plant species in the Western Hemisphere. It contains one-third of the world's flowering plants. Orchids are just one of the many kinds of flowering plants found there. There are 16,000 species of trees in the rainforest; the tallest of which is the *dinizia excelsa*, which can reach almost 300 feet in height. Other plants in the Amazon include cocoa trees, banana trees, Brazil nut trees (which can live to be 1,000 years old), mango trees, and sugarcane.

One-fifth of the world's bird species live in the Amazon region, including 300 species of hummingbirds. Amazonia contains 2.5 million species of insects. In sum, the Amazon basin holds ten percent of the world's species of living things—we think, but scientists have not identified all the species that live in Amazonia.

Layers of a Rainforest

A rainforest has four main layers. The *emergent* layer is the highest, where trees grow to 200 feet tall and more. These giants tend to grow far apart. Most leaves are at the top, and they have large bases at ground level that supply the nutrients they need and give them support in strong winds. The animals that live in this layer are either birds or small animals that the slender branches at the top can support. The rubberwood tree, the Brazil nut tree, and the kapok tree are emergent layer trees.

The next highest layer is the *canopy*, an area of vegetation about twenty feet thick. The canopy is home to most of the rainforest's species. It blocks much of the light from getting to the lower layers. Fruit-bearing trees in this layer attract many animals. The thousands of insect species that live here attract reptiles.

Next comes the *understory*, which has less light and is more still and humid. Here palms and philodendrons grow alongside other plants that are generally less than twelve feet in height. The plants in this layer often have large blooms and large leaves to receive what little light gets through. Animals take advantage of the colors and low light for camouflage.

The forest *floor* is the darkest and lowest level. This is the layer where decomposition of fallen leaves takes place (a process which provides nutrients for the trees). It's where foraging animals eat insects and roots, and where we find slugs, worms, ferns, moss, and fungi. The capybara, the world's largest rodent (which can grow to 170 pounds) lives here.

Other animals that live in the Amazon rainforest are several species of monkeys, sloths, jaguars, ocelots, bats, and the 30-foot long anaconda snake.

Capybara and White-Throated Kingbird, Brazil

Medicinal Plants

A fourth of all pharmaceuticals are made from rainforest plants. For centuries people in the Amazon region have cited the healing properties of plants that grow there, although medical science has not confirmed all of these claims. Among these plants are:

- achiote, with spiky red fruit and seeds, the juice of which makes an insect repellent. People boil the leaves to make a medicine that reduces fever and heals wounds.

- cordoncillo, which produces an anesthetic. If you chew it, your mouth goes numb. Applied to a wound, it deadens the pain. It is also used as a disinfectant and to treat gallstones.

- shapumvilla, which contains a coagulant that stops bleeding.

- quinine, which was the first effective medicine to treat malaria.

- cat's claw, so named because it has claw-shaped thorns and attaches to tree trunks, which helps relieve the pain of arthritis and rheumatism.

Seventy percent of plants that are reported to have anti-cancer properties exist only in the Amazon. These include tawari tree bark.

The River

Even though we know a great deal about the Amazon River, some things about it are still unknown. For instance, we know where it ends, but we're not exactly sure which of its many tributaries (some of which are large themselves) is the farthest from its mouth; in other words, experts have suggested more than one possibility for where the Amazon actually begins. We are pretty sure that it is the second longest river in the world after the Nile. The Amazon is over 4,200 miles long, which makes it longer than the distance from New York City to Rome, Italy.

In one remarkable confluence of tributaries, the Rio Solimões—which carries the equivalent of six Mississippi Rivers and is coffee-colored and rich with sediment—converges with the Rio Negro, which carries the equivalent of two Mississippi Rivers and is black with decayed leaf and plant matter. The Solimões is cooler, denser, and faster. The two streams flow side by side in the same channel for some distance. Their boundary in the water, marked by an obvious difference in color, is visible from space. The two eventually mix and form the Lower Amazon River.

Confluence of the Rios Negro and Solimões

The mouth of the Amazon is 40 miles wide. The river pours 1.3 million tons of sediment into the sea every day, but ocean currents sweep it away so that there is no delta as there is with the Nile. The river water dilutes the saltiness of the ocean for about 100 miles out to sea. The tidal bore that comes into the river with the incoming tide moves at 10-15 miles per hour and creates a wall of water from 5 to 12 feet high.

The Amazon River holds one-fifth of the world's fresh water and between 2,000 and 2,500 species of freshwater fish (compared to 300 fish species in all of the streams and lakes of Europe). The leaves of Amazon water lilies can grow to be over six feet in diameter. The river is home to the boto or pink river dolphin, one of the few freshwater dolphin species in the world.

The People

And yes, people also live in the Amazon region. Hundreds of tribes there have different languages, cultures, conflicts, and beliefs. It's hard to know exactly how many people live there because some tribes have not yet been contacted or even identified by outsiders.

Chief Akanawa of the Pataxó Hã-Hã-Hãe people uses a computer in 2011.

We do know that the indigenous people of the Amazon region invented blowguns and the hammock, discovered the use of quinine to treat malaria, learned how to tap rubberwood trees for the sap and make products from it, and were experts at using dugout canoes and sailing rafts.

Exploration and Exploitation

People have been exploring the Amazon ever since Orellana did so almost 500 years ago. The richness, complexity, and mystery of the place draw them to it. They come to map the region, to collect scientific data, and simply for adventure. In recent years the main interest people have had in the Amazon has been to extract as much of the natural resources as they could in order to enhance the lives of people elsewhere. The rubber soldiers were only one of many extraction projects that have taken place there and continue to take place. A major emphasis has been the cutting of trees in the Amazon region for several purposes: to take the lumber for many different projects, to clear land for cattle grazing, and to open routes for new roads.

The Brazilian government has moved large numbers of people into the Amazon region to help alleviate overcrowding in other parts of the country. Large hydroelectric projects provide power for many people throughout the country. Petroleum and mining companies have found rich deposits of gold, oil, and other minerals. A gold rush in the 1980s brought about a half million people into the Amazon region.

Unfortunately, these changes do not take place without a cost. Trees absorb carbon dioxide and give off oxygen, and they are an important part of the water cycle. Fewer trees means less carbon dioxide absorbed and greater water runoff. The reservoirs behind the hydroelectric dams change the ecosystem.

Miners used large amounts of mercury to extract the gold from the land, and that mercury made its way into the waterways of the Amazon basin. Deforestation also changed the environment. We may not even yet know some of the environmental effects of these activities.

Clearing Amazon Rainforest (2010)

The need in the Amazon, as in many parts of the world, is to achieve a balance between man's needs and the need to maintain the health of the Amazon rainforest for the world and for future generations. Different people have different ideas about what that balance is. Both developers and environmentalists can go to extremes in their positions; but since God commanded mankind to rule over the earth and subdue it, we can trust that people can achieve the right balance. The oversight of the unique geography of the Amazon basin is a stewardship that people—especially Brazilians— must exercise wisely for this generation and for generations to come. We must remember that if a species of any living thing becomes extinct, it would require another Creation for the world to have that species again.

The psalmist says that even the created world will rejoice at the coming of the Lord:

Let the heavens be glad, and let the earth rejoice;
Let the sea roar, and all it contains;
Let the field exult, and all that is in it.
Then all the trees of the forest will sing for joy. . . .
Psalm 96:11-12

Assignments for Lesson 131

Gazetteer	Study the maps of South America and read the entry for Brazil (pages 225-226 and 229). Read the excerpt from *Through the Brazilian Wilderness* (pages 330-331) and answer the questions in the *Student Review Book*.
Worldview	Copy these questions in your notebook and write your answers: What does the gospel provide for the person who does not have much material wealth? What does the gospel provide for the person who has significant material wealth?
Project	Choose your project for this unit and start working on it. Plan to finish it by the end of this unit.
Literature	Begin reading *Tales from Silver Lands*. Plan to finish it by the end of Unit 28.
Student Review	Answer the questions for Lesson 131.

Cattle and Power Lines on the Pampas, Argentina

132 Gauchos of the Pampas

It is a legendary scene. The lone gaucho, the cowboy of Latin America, sits astride his horse, surveying thousands of cattle as they graze on the vast, flat, grassy Pampas of Argentina.

The gaucho wears a colorful woolen poncho. He has on loose bombachas trousers and a large boina hat tilted slightly to one side to keep off the sun. Around his neck is a bright kerchief. Next to a woolen faja sash around his waist is a rastra, a stiff leather belt that he has decorated with coins and which holds his long facon knife. He also has on spurs and a vest.

He is skilled in throwing a boleadora or bola. This implement consists of the ends of three ropes tied together, with the free ends weighted with rocks or metal balls. The skilled gaucho throws the boleadora in such a way that it entangles the legs of a calf or cow and stops it from running away.

The gaucho spends days alone on horseback on the Pampas, using his poncho as a blanket at night. When gauchos do occasionally gather in a town, some of them live rough and are quick to fight. Occasionally they compete with each other in games that display their talents at handling horses. They have developed a style of dancing called malambo, which involves amazingly fast and fancy footwork.

The life of the gaucho is the stuff of legend, much like that of the cowboy in the United States: tough, independent, wanting to live on his own, embodying traits that many Argentinians, especially men, idealize. The gaucho has been the subject of much literature, poetry, and music (can't you hear the harmonica playing, the guitar strumming, and the campfire crackling in the background?). Perhaps the best known work is the long poem *El Gaucho Martin Fierro* by Jose Hernandez, published in 1872.

As with most stereotypes and legends, the image of the gaucho is at least partly based on truth and partly idealized. The lifestyle he embodies, however, has mostly disappeared.

The Pampas

The term Pampas comes from a Quechua word that means level plain. In scientific terms the Pampas is a temperate grassland biome. Except for a few jagged sierra mountain ranges in the northwest and south, the Pampas is mostly flat or gently rolling land and it is usually windy there. Slightly larger than Texas, the Pampas dominates central Argentina. It stretches from desert areas near the Andes in the west to the humid Atlantic coast, also covering Urugauy and part of southern Brazil. The

Pampas slopes gently downward from northwest to southeast, declining from 1,640 feet above sea level until it reaches the sea.

In the northeastern part of the Pampas are wetlands, including Esteros del Ibera, the second largest wetland in the world. In that region gauchos raise cattle on the land that rises above the water and manage to get around by boat, canoe, and horse.

The most recognizable plant growth on the Pampas is the silver pampas grass, a flowering plant that grows in bunches with blooms that stand tall above the ground. Few trees grow on the Pampas. The ombu tree is actually a bush, although it often grows 40 to 60 feet tall and its canopy of leafy, flowering branches that develop crimson berries can spread 40 to 50 feet in diameter. Gauchos and cattle often rest in its shade. It is immune to locusts and other pests, and it requires little water. This is good in a region with only 10 to 20 inches of rainfall in a typical year.

Animals found on the Pampas include the guanaco, a relative of the llama; the maned wolf, which is a member of the canine family but is neither a wolf nor a fox; the gray fox; and Geoffroy's cat, a nocturnal feline about the size of a domestic cat. The greater rhea is a flightless bird that grows to about four feet in height. It is related to the ostrich and emu and is the largest bird found in South America.

Gauchos in Argentina (c. 1890)

The range of the guanaco extends into Patagonia, south of the Pampas. There they mingle with penguins on the coast of Argentina.

The Story of the Pampas and the Gaucho

Beginning in the 1500s, the Spanish began settling what became Argentina. Some claimed huge estates called *estancias*. The Spanish introduced horses and cattle to South America, but they did not do a good job of managing the animals. Many escaped into the wide open Pampas. Gauchos, who were mostly mestizos or mixed race, took on the difficult job of catching and training horses so they could then do the difficult job of capturing the cattle. The horsemanship skills of the gauchos became legendary.

Argentina chafed under Spanish rule for many years, and in 1810 Argentine patriots began a war for independence. The gauchos generally sided with the patriots and served as scouts and intelligence gatherers. When the war ended in 1818 with Argentine independence, the gauchos received honor for their contribution to the effort. The gauchos began serving the owners of estancias as something on the order of independent contractors, rounding up cattle and doing other work for them but not thinking of themselves as employees.

During the last half of the 1800s, life in Argentina changed in many of the same ways that it changed in the United States. People began seeing the rural lifestyle as uncivilized compared to the appeal of urban life. Owners of estancias began fencing in their land to control their livestock and devoting more of their land to farming crops. The need for the work that gauchos had done declined. Many went to work in the cities, while others became employees on the estancias as herders, shearers, and handymen. Owners hired immigrants to cultivate crops. Railroads ran through the Pampas to carry cattle to the coast and from there to Europe and other markets.

The poem *El Gaucho Martin Fierro* was not just a celebration of gaucho life but a lament for what was a disappearing way of life even in 1872. A few gauchos, however, still lived and worked on the open Pampas. During the 1900s, landowners introduced new equipment such as tractors to accomplish the work they wanted done on their land. They began growing crops such as wheat, alfalfa, and soybeans. In the last few years property owners have found vineyards to be a profitable venture.

One surprising group that took part in the transition of gaucho life were Jewish immigrants from Europe. As persecution of Jews was increasing in Russia and other countries in Europe in the late 1800s, the Jewish German railroad magnate and philanthropist Baron Maurice de Hirsch formed the Jewish Colonization Association. The group bought huge tracts of land in the United States, Canada, and Brazil to give Jews in Europe a place to go beginning around 1881. Hirsch was most successful in Argentina, where his group bought land equivalent to the size of Delaware and arranged for some 50,000 Jews to settle there over many years. The last group of European Jews who emigrated to Argentina came from Germany in 1936, as even worse persecution was taking shape there.

Jews established Moisés Ville, Argentina, in 1889. This is one of three synagogues they eventually built in the town. This one is known as the Baron Hirsch Synagogue.

Cooking Beef in Argentina

The subculture of Jewish gauchos and their families thrived for many years. They built strong communities on the Pampas, where community meeting halls and adobe synagogues were filled with activity. Younger generations, however, have not embraced the rural lifestyle, choosing instead to seek life in the cities. The Jewish communities on the Pampas are dying out, and most of the synagogues have closed.

Argentine agriculture has diversified since the heyday of the gaucho, but the beef industry is still vitally important to the country and the region. Latin America and the Caribbean account for 13.5% of the world population but produce 23% of the world's beef and buffalo. Latin America is the greatest beef exporting region in the world.

Argentinians consume significant amounts of beef every year. One long-standing tradition is for friends and families to gather on Sunday afternoons for a *parrillada* or barbecue.

Another strong Argentine tradition is La Rural, an agricultural and livestock festival held in Buenos Aires every July since 1886. Sponsored by the Argentine Rural Society, the event draws hundreds of thousands of attendees. It features livestock showing, horsemanship competition, artwork exhibitions, and much more. Other towns and cities have similar but smaller annual festivals. The city of San Antonio de Areco has a gaucho museum and hosts the Festival of Tradition for gauchos every November.

The New Gaucho

The day of stalwart, independent gauchos overseeing herds with thousands of cattle on the vast Pampas is gone. Instead, many gauchos today have embraced, however reluctantly, a growing industry of the twenty-first century: tourism. Argentine and international tourists pay good money to stay on an estancia and interact with the gauchos they have heard about.

The modern gauchos prepare meals, play music, guide horseback rides, and explain their work in order to keep the story of the gauchos alive. A small but increasing number of these modern gauchos are women, performing tasks that were once considered difficult for men and impossible for women. In between hosting tour groups, however, you can still find these modern gauchos herding cattle, shearing sheep, and branding horses, much like they did centuries ago.

A geographic setting impacts the lifestyle of the people who live there, but other factors influence their lives as well. The lives of the people who live on the Argentine Pampas today still bear traces of earlier times, even as the modern age surrounds them.

On the Argentine Pampas, the Lord provides the grass for the livestock to graze and the crops that people plant and harvest.

Festival of Tradition in San Antonio de Areco

He causes the grass to grow for the cattle,
And vegetation for the labor of man,
So that he may bring forth food from the earth. . . .
Psalm 104:14

Assignments for Lesson 132

Gazetteer Read the entry for Argentina (page 227).

Worldview Copy this question in your notebook and write your answer: Why do you think interest in Christianity has declined in Europe and the United States from what was once the case in those regions?

Project Continue working on your project.

Literature Continue reading *Tales from Silver Lands.*

Student Review Answer the questions for Lesson 132.

Cattle in Argentina

Lake Titicaca

133 Living on Lake Titicaca

These people live on a lake.

Not by a lake, but on a lake. Not on natural islands in a lake, but on a lake.

It's a unique lake, and their living arrangement is very unusual.

The lake is the highest navigable lake in the world and the largest lake by volume in South America.

The islands on which they live are ones that the people make themselves, out of reeds.

The Lake

We are talking about high altitude country. Some of the highest peaks in the Andes are nearby, reaching heights of over 21,000 feet in elevation. La Paz, the capital of Bolivia and highest national capital in the world, is thirty miles away in a river canyon—a canyon, mind you—that bottoms out at 10,650 feet above sea level. Parts of the city are at 13,250 feet in altitude. It is in this part of the world that Lake Titicaca sits on the border between Peru and Bolivia on the Altiplano (High Plain) plateau. The lake's surface is 12,500 feet above sea level.

Lake Titicaca has a surface area of about 3,200 square miles, 120 miles long by 50 miles across at its widest point. Lake Maracaibo in Venezuela has a larger surface area, about 5,100 square miles, but Maracaibo is actually a bay or inlet off the Caribbean Sea. Titicaca is the largest lake by volume in South America. It has an average depth of 460-600 feet, with its deepest point about 900 feet. Water temperature averages about 55°F.

Titicaca is the eighteenth largest lake in the world. It is about the size of Delaware and Rhode Island combined. By way of comparison, Lake Ontario, the smallest of the Great Lakes, is over twice the size of Titicaca, at 7,340 square miles. A narrow strait divides Titicaca into a larger part and a smaller part. Some 41 natural islands dot the lake.

Twenty-five rivers empty into Titicaca, but only one river drains out of it. The Rio Desaguadero drains about five percent of what the lake loses; the sun and wind evaporate the rest. The level of the lake rises and falls with the wet and dry seasons.

The Habitat

Lake Titicaca has about 530 aquatic species and water birds. Both resident and migratory birds find a home there. The primary fish species include 23 species of killifish and two species of catfish. The Titicaca water frog grows to a length of almost one foot.

In the 1930s Peru and Bolivia both saw the lake as an economic opportunity that could help them in a difficult time. They appealed to the United States for assistance. The U.S. sent a representative of the U.S. Fish and Wildlife Service, who studied the lake during 1935 and 1936. He suggested stocking the lake with North American fish to enhance the fishing industry. The U.S. sent a half million trout eggs and two million whitefish eggs. The whitefish eggs did not survive, but the trout flourished. Then in the 1950s, Peru and Argentina introduced Argentinian silverside fish into the lake. The imported species turned out to be good for fishing but hard on native species as they competed for food.

Agriculture is a thriving industry around the lake. Terraced fields have been a long-standing practice since ancient times, but wide open fields are more common now. The terraces take more work to be able to begin crop production, but they tend to hold moisture better than fields which drain more rapidly into the lake. Quinoa and potatoes are common crops. Scientists believe that the potato originated here. The highest known cultivated plot in the world, a field of barley growing at 15,420 feet above sea level, is near the lake. Livestock raised here include alpacas, llamas, sheep, and cows.

Ruins on Isla del Sol

Uru Women with Woven Rugs

The People

The lake and its environs were important to the Incas and other ancient civilizations. It was so important to the Incas that they believed it to be the birthplace of mankind. Many ruins of ancient structures lie near the lake. Isla del Sol, the largest of the natural islands in the lake, has some 80 ruins on it. In 2000 archaeologists discovered a structure they believed to be a temple at the bottom of the lake.

About two million people live in the lake region. Some islands are densely populated. Isla del Sol has about 800 families that live on it, even though no motorized traffic is allowed on the island.

The Urus

The Urus are an ancient people who don't live near Lake Titicaca or on its natural islands. They live on the lake on islands they make themselves.

We don't know why the Urus took up this unusual living arrangement. Perhaps they were threatened by the Incas or some other nation. They used to live about nine miles out from shore, but a storm in 1986 devastated them, so now they live close to shore near Puno, Peru.

The floating islands they create are about 50 feet by 50 feet. The base of an island is about twelve feet thick. The base is made of clods of earth tethered together and anchored to the lake bottom. On top of this base are several layers of totora reeds, which grow along the shore and which Americans know as California bulrushes. The reeds also grow in Central America. Of course, reeds decompose, so the people must constantly repair and occasionally rebuild their island bases. The Urus also build their houses and boats out of the totora reeds. An island is home to about 10-15 persons, usually members of an extended family. There are about one hundred of these islands, and the total Uru population is about 1,200.

The Urus live in a primitive way, but they have not missed out on modern times. The Peruvian government has installed solar panels on most of the houses, which enable the residents to have electric lights, charge their cellphones, and use electricity in other ways. Many of them cook on open fires on top of a layer of stones, using more reeds for fuel. They have also established an FM radio station which provides them with news and entertainment.

The Urus have become a tourist attraction. They paddle their boats to shore and pick up about 200,000 paying tourists per year, whom they take to their islands. Once there, they dress the visitors with traditional garments. Some residents allow the tourists to spend the night with them—for an extra fee. The locals also have handmade trinkets and garments for sale. The Uru families alternate the days on which they receive tourists. A family will play host to guests one day and live as regular people the next.

The Problems

The people of Lake Titicaca haven't avoided environmental problems. Those who fish for a living have overfished the capacity of the lake, and some species are endangered. The water has become contaminated with farm fertilizer because of runoff from the fields, waste from nearby industry, and mercury and other substances related to mining. On the other hand, eco-toilets have helped the problem of raw sewage being dumped in the lake.

Uru Boats

Above is an aerial view of the floating islands built by the Uru people. Below is a close-up view of one island with solar panels installed.

Why did people come here in ancient times? Did they believe that the elevation of the land allowed them to be closer to the gods? How did they quarry and transport the stones for their structures? What changed that resulted in one structure being underwater? How did the Urus learn to live by the totora reed?

How does the geography where you live shape your life?

The prophet Habakkuk compared the sureness of a deer climbing on high places to his confidence in the Lord.

The Lord God is my strength,
And He has made my feet like hinds' feet,
And makes me walk on my high places.
Habakkuk 3:19

Assignments for Lesson 133

Gazetteer	Read the entry for Peru (page 237).
Worldview	Copy this question in your notebook and write your answer: What do you think would help to spark a spiritual revival in the United States (trusting that God is in control, of course)?
Project	Continue working on your project.
Literature	Continue reading *Tales from Silver Lands*.
Student Review	Answer the questions for Lesson 133.

Museum About the War in La Paz, Bolivia

134 "The Sea Belongs to Us by Right"

The men and women of this navy have a clear mission. It is a mission they have heard about since childhood. It is a mission that defines who they are as a navy and as a country. The importance of this mission is made more dramatic because the country has a navy but no coastline. This navy does not go to sea. Their mission, their purpose, their dream is to regain access to the sea.

How do you think Americans would feel if the United States lost a war with Canada, and in the treaty that ended the war the U.S. had to give the coastline of Washington, Oregon, and California to Canada? What if that were the only coastline that the U.S. possessed? Do you think Americans might have any lingering resentment? Might Americans think that the lost region still rightfully belonged to them and not really Canada? What might this loss of territory mean for America's access to the world and the world's access to America?

A piece of geography can have a strong hold on a nation's consciousness. Americans are proud of the Grand Canyon, the Grand Tetons, the Chesapeake Bay, and numerous other special places. They are part of the national identity. To lose one would be a national tragedy. This is what happened to Bolivia when they lost their coastline.

On a map of South America, along the western coast you will see where Peru, Chile, and Bolivia come together. There you will also see a strip of land belonging to Chile that separates Bolivia from

In this map of South America made before the war, you can see a narrow strip of blue touching the Pacific Ocean as part of Bolivia.

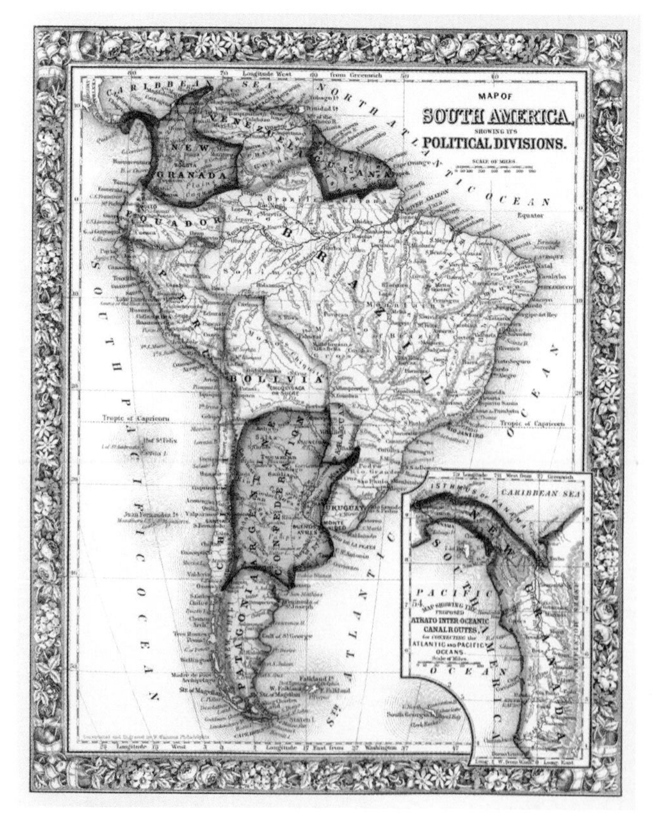

the Pacific Ocean. That strip of land did not always belong to Chile. The northern part once belonged to Peru, and the southern part once belonged to Bolivia. Those two countries lost the region in a war with Chile. Bolivians have not forgotten. Even today, 140 years later, they want it back.

The Disputed Area

In the 1800s national borders were sometimes ill-defined. As long as disputes did not erupt over contested areas, the countries involved usually got along. In the area of west-central South America that we are discussing in this lesson, Peru claimed the Tarapaca Region to the north and Bolivia claimed the Antofagasta Region to the south. In 1866 Chile and Bolivia agreed to divide the Atacama Desert area at 24 degrees south latitude, just south of the port city of Antofagasta. The agreement allowed citizens of both countries to access and develop mineral deposits there, and both countries agreed to share the revenue that came from the mining activity. The primary mineral known to be in the region was nitrate, an important ingredient of fertilizer and explosives. Copper had been discovered there also. Chilean and European (mostly British) companies owned the most lucrative mines.

Soon thereafter, however, Bolivia repudiated the agreement and began levying taxes on a Chilean mining company that was operating there. Chile, Bolivia, and Peru began acquiring arms to prepare for conflict. In 1873 Bolivia and Peru entered into a secret mutual defense treaty to ward off any Chilean aggression.

In 1879 Chilean troops moved north and invaded the Atacama region. Chile's superior forces overwhelmed the troops of the other two countries. Chile's army occupied Peru's capital of Lima on January 17, 1881, a humiliation that Peruvians still resent. Peru surrendered with the Treaty of Ancon in 1883 and gave up its Tarapaca Region, although Peru regained some of the territory through a 1929 agreement with Chile. Bolivia fought on until 1884 but then surrendered and gave up the 46,000-square-mile Antofagasta Region and its 240 miles of coastline. Some 35,000 soldiers were killed or wounded. Peru had other coastal areas but Antofagasta was Bolivia's one access to the sea. Bolivia was now landlocked. A 1904 treaty between the two countries reaffirmed the arrangement.

Eduardo Abaroa led a small group of Bolivian civilians who fought to the death against Chilean soldiers at the Battle of Topáter, the first battle of the War of the Pacific. Abaroa is remembered as a national hero in Bolivia. This monument to him stands in La Paz.

Chile profited greatly from the sale of nitrates from the newly acquired region. For the next 40 years, over half of the Chilean government's revenue came from export taxes on nitrate. Some Chileans became wealthy, a rising middle class formed, and a number of immigrants from Europe contributed to the country's economic growth.

Meanwhile, Bolivia suffered national humiliation and increasing economic difficulty. Today, Bolivia is the poorest country in South America in terms of per capita gross domestic product. Many Bolivians blame their troubles on the lack of a coastal port of their own.

Bolivian Naval Base at Lake Titicaca

Still Not Over It

Each of the three countries developed a different explanation for what took place. Chileans believe that Bolivia was the aggressor nation and that they had no other option but to fight. Bolivia and Peru say that the war resulted from Chilean expansionism. Peru has mostly gotten over its loss of what Peruvians call their "captive" territory, although the seizure of Lima still stings. In Bolivia the War of the Pacific is an open wound, even after 140 years. There people refer to the defeat and loss of territory as an "injustice" and a "stabbing."

Stymied by the loss of its coastal territory, many years later Bolivia looked in the opposite direction for relief. The Bolivian government tried to devise a way to reach the Atlantic coast by the Rio de la Plata river system that crosses the barren Gran Chaco region, which is part of both Bolivia and Paraguay and which was thought to contain large oil reserves. Bolivia's first step in trying to reach the Atlantic was to attempt to take control of Gran Chaco. Skirmishes began occurring in 1928. Conflict heightened in 1932, and Paraguay officially declared war on Bolivia in 1933.

A truce was arranged in 1935 and signed in 1938 under the oversight of the leaders of Argentina, Brazil, Chile, Peru, Uruguay, and the United States. This Chaco War cost about 100,000 lives. Paraguay gained clear title to most of the disputed area, but Bolivia did receive a corridor along the Paraguay River and access to the river port of Puerto Casado in Paraguay. This was better than nothing, but it still

This 2008 parade in Cochabamba, Bolivia, included a group of Bolivian Marines and a ship pulled on a float.

left Bolivia subject to the rules and taxes of other countries.

In 2009 the presidents of Bolivia and Paraguay signed an accord that resolved the border dispute and blamed the conflict on foreign oil interests that fomented the dissension in hope of gaining access to whatever oil deposits might be found in Gran Chaco.

Meanwhile, there was still Chile. In 1963 Bolivia formed the Armada Boliviana, which consisted of four used boats from the United States. The name was changed in 1966 to the Bolivian Naval Force. The force now has 5,000 members. Its vessels consist of speedboats, tankers, and castoff ships from China. They practice maneuvers on Lake Titicaca. Their duties take them along the rivers of Bolivia delivering medical supplies to remote villages, hunting down drug traffickers, and giving aid during natural disasters. By law they are not allowed to enter any ocean. Yet.

Every May 23 the country celebrates the Day of the Sea. Children march in parades and wave Bolivian flags, and the people sing songs such as "We Will Recover the Sea." On the outside of the naval base at Lake Titicaca is the motto of the Bolivian Naval Force: "The Sea Belongs to Us By Right, Recovering It Is Our Duty."

In 2010 Peru granted Bolivia a 99-year lease to a 1.3-square-mile port facility. Through this port Bolivia now ships two-thirds of its foreign trade.

This was still not enough for Bolivia. Lease agreements and mere access to another country's port left Bolivia subject to the actions of those other governments. In addition, there is still the factor of national pride, which led Bolivians to believe that the lost region is rightfully theirs. Diplomatic relations between Bolivia and Chile were broken off in 1962, restored in 1975, but severed once more in 1978. Their current strained relations now take place below the ambassadorial level.

In 2013 Bolivia brought suit against Chile in the International Court of Justice (ICJ) in The Hague in the Netherlands. Their demand was not for the

Southwestern Bolivia has fascinating geological formations. The Siloli Desert is also called the Desert of Dali because the vast landscape and the unusual shapes of rocks there are reminiscent of the paintings of artist Salvador Dali. The Salar de Uyuni is the world's largest salt flat. In some seasons the surface looks like a giant mirror because of the way it reflects the heat of the sun like a mirage. At other times it looks like a huge puzzle with hexagonal pieces. The area is a breeding ground for pink flamingoes.

Argentina and Chile resolved a separate border dispute in 1902. To honor that agreement, the two countries erected a 23-foot statue, "Christ the Redeemer of the Andes," on the border between the two countries in 1904.

return of territory but only for Chile to enter into good faith negotiations over the matter. However, in 2018 the ICJ ruled that Chile had no obligation to enter into such negotiations. The decision pleased Chile and infuriated Bolivia.

Before this lesson you may have never heard of the War of the Pacific or the treaty arrangements that came about as a result of it. Bolivians, however, live with a consciousness of their loss every day and defiantly proclaim, "What once was ours, will be ours once more."

Jesus urged adversaries to settle their differences quickly and put their dispute behind them for their own good.

Make friends quickly with your opponent at law while you are with him on the way, so that your opponent may not hand you over to the judge, and the judge to the officer, and you be thrown into prison.
Matthew 5:25

Assignments for Lesson 134

Gazetteer	Read the entries for Bolivia, Chile, and Paraguay (pages 228, 230, and 236).
Geography	Complete the map skills assignment for Unit 27 in the *Student Review Book*.
Project	Continue working on your project.
Literature	Continue reading *Tales from Silver Lands*.
Student Review	Answer the questions for Lesson 134.

Assembly of God Church in Búzios, Brazil

135 Evangelism in the Global South

For several centuries, the common perception of the typical Christian, at least among people who lived in Western culture, was that of someone who was white, either European or American, and who lived in a moderate-sized town or city. This typical Christian was probably a member of a mainline denomination.

Not anymore. Today, the most typical Christian—if there is such—would be a black woman who lives in a village in either Brazil or sub-Saharan Africa. Today's typical believer is also likely to be charismatic.

This change exemplifies the explosive growth in the number of followers of Jesus in the geographic regions of the world that make up what many call the Global South.

What Is the Global South?

People generally think of the countries in Central and South America, Africa, and southern Asia as making up the Global South. As you can see, a great many of these countries are in the southern hemisphere. The Global South includes many economically developing nations. Generally speaking, these countries are not as wealthy nor are they as industrialized as are European and North American countries. People in these countries have relatively limited educational and economic opportunities.

Opinions differ about whether to include some countries in the Global South. For instance, some say China is part of the Global South because it is still developing economically. Given its current trend, however, at some point most economic experts will see China as developed. Some people include the countries of the Middle East in the Global South, even though they are north of the equator, but many countries in the Middle East are well developed economically. Most people do not consider Australia and New Zealand to be in the Global South because, even though they lie in the southern hemisphere, they are largely Westernized and are economically developed.

Two things we can keep in mind in thinking about the Global South: a majority of the population is non-white, and the countries are economically disadvantaged to a significant degree.

The Changing Religious Makeup of the World

People in the fields of economics, politics, and cultural studies speak of the Global South in terms of what the phrase means in those areas of study; but our emphasis will be on the religious beliefs of the people in the regions we have described.

Christianity began in what is now called Palestine in the Middle East. We have some records of it spreading into northeastern Africa (such as Egypt and Ethiopia) and into India during the early centuries of the church, but the greatest growth in its first millennium took place in Europe. What had been a pagan region became a majority Christian region. Then Christianity spread where Europeans spread: into North America (largely Protestant) and Central and South America (largely Roman Catholic), where again the people there had been pagan.

From the 1700s on, British and American missionaries took the gospel into Africa, India, and Asia with varying levels of success. Some people did decide to follow Jesus, but large numbers remained committed to the Buddhism, Hinduism, and folk religions they had known.

Starting in the last half of the 1800s, the widespread acceptance of Darwinian evolution and liberal theology (defined here as the rejection of Scripture as the authoritative and inspired Word of God) led to many people abandoning their faith. In the twentieth century, a large number of people reacted to the calamities of two devastating world wars by walking away from Christianity. Today it is

A leader of the Igreja Batista Renovada do Espírito Santo (Renewed Baptist Church of the Holy Spirit) performs baptisms in São Paulo, Brazil, in 2013.

not at all unusual for the majestic cathedrals all over Europe to be largely empty on Sunday mornings.

The trend is the same in the United States and Canada, although the trend is not as far advanced in the U.S. as it is in Europe. Church membership and attendance have been declining for decades while the percentage of the population who expresses no religious preference (called "nones") has been increasing. On the other hand, over the last several decades the number of people who profess faith in Christ in sub-Saharan Africa, India, and China has been increasing.

Thus, in a series of ironies, the area of the world where the Christian faith began has become dominated by Islam, and in Israel itself Judaism is the majority religion. At the same time, the long-time bastion of Christendom, Europe, has seen declining numbers of churchgoers, as has North America. On the other hand, Africa, where for centuries relatively few Christians lived, has experienced dramatic growth in churches. South America, which was culturally and nominally Catholic for centuries, has seen a revitalization of the Christian faith, largely among Pentecostal and charismatic churches.

The Numbers Tell the Story

In 1910 about two-thirds of the Christian believers in the world lived in Europe, which had been the center of Christianity for over fifteen hundred years. An estimated four-fifths of the Christian population of the world lived in the Global North (mostly Europe and North America), while only one-fifth lived in the Global South.

A century later, in 2010, only about one-fourth of the world's Christian population lived in Europe. That year one-fourth of the world's confessing Christians lived in sub-Saharan Africa and another thirteen percent lived in Asia and the Pacific Islands. Slightly less than 40% lived in the Global North and over 60% lived in the Global South.

This photo is from a Congregação Cristã no Brasil meeting place in the 1950s. The Christian Congregation in Brazil was established in 1910 by an Italian American missionary. A brass band often accompanies hymn singing in these churches.

Specifically in sub-Saharan Africa, in 1910 the ten million people there who confessed faith in Christ accounted for ten percent of the population. In 2010 an estimated 360 million believers lived in sub-Saharan Africa; and these people made up half the population. Experts expect these trends to continue.

Some Christian growth is taking place in the Global North among people who are from the Global South. For instance, some refugees from the Middle East who have fled to Europe are Christians or are becoming Christians. In the United States, the numbers of those from Hispanic, Asian, and African ethnic groups are growing; and many of these people are Christians.

Today there are about 2.2 billion of what demographers call adherents of the Christian faith around the world, or about one-third of the world population. These believers are so widespread that no one continent is the center of Christianity as Europe was for centuries.

The nature of Christian expression is changing as the location of the majority of believers is changing. Christianity in the Global South is generally more theologically conservative than what one commonly finds in Europe and North America. These Global South believers are strongly evangelistic. They also are more likely to be Pentecostal or charismatic than are Christians in the Global North. (Note: Pentecostal and charismatic believers commonly share many beliefs, including a belief in miraculous works such as healing and speaking in tongues. Pentecostalism is a recognized denomination, while charismatic believers can be members of any number of denominations. In other words, pretty much all Pentecostals are charismatics, but not all charismatics are Pentecostal.) Many believers in the Global South, especially in South and Central America, are former Catholics. Pentecostalism is growing three times as fast as Catholicism in Latin America. Brazil has more Catholics than any other country, but also more Pentecostals than any other country.

What we know about Asia is that, over the last century, Christianity grew twice as fast as the population. What we don't know for sure is the growth of Christianity in China. Because China is largely a closed society and because of the risk of government persecution that Christians face there, house churches try to avoid detection. As far as any reliable figures show, the United States is the country with the most confessing Christians in the world. However, Christianity is definitely growing in China, and some observers think that church attendance in China might soon surpass that in the United States if it has not done so already.

Why Are These Changes Happening?

Two factors in the Global North have contributed to the decline of the Christian faith there. As indicated above, a large number of people in the Global North have turned away from Christianity. In addition, the rate of population growth in the Global North has been declining in recent decades, to the point of approximately zero growth in some European countries.

As the influence of Pentecostalism grew in Latin America, some Catholic leaders began to embrace charismatic beliefs and practices. One of these leaders was Rafael Garcia Herreros. He founded Minuto de Dios (Minute of God), a Catholic organization in Colombia that hosts radio and television programs and organizes efforts to support the poor through housing and education.

On the other hand, Christians in the Global South have been evangelistic in bringing friends to Christ, and population growth has been dramatic in many parts of the Global South.

Many people in Africa, Asia, and South America who are poor have become followers of Jesus. These people do not have access to wealth, and so they probably feel a need for something more in their lives than what they have—a desire that Jesus can fulfill. The Christian population in the Global North tends to be better off financially. While it is possible for the wealthy to be faithful Christians, it is difficult, as Jesus said, for a rich man to enter the kingdom of heaven (Matthew 19:23).

The poor in the Global South can identify personally with many parts of the Biblical narrative. They may know what it feels like to be robbed while on the road, as Jesus described in the parable of the Good Samaritan. They can understand the desperate search by the widow who had lost a coin, as in the parable Jesus told in Luke 15. They can identify with the famine that struck in the days of Joseph, the one that Ruth experienced, as well as the famine that the early church endured (see Acts 11:28). When they read in James 4:14 that life is a vapor, they understand.

Many Christians in the Global South have a worldview that accepts as reality the supernatural spiritual warfare that is taking place. Many have claimed to have witnessed what they believe are physical healings. They have quite possibly endured persecution for their faith, often at the hands of extreme Muslims. In other words, they have many reasons to believe the story of the Bible on the basis of their own experience.

In a small way, the evangelistic dynamic of the world has begun moving in a direction that is opposite of what it was for many years. Churches in the Philippines and Brazil, for instance, are sending missionaries to other countries, and some of those missionaries are coming to the United States. Some Christians in China have made a commitment to move to the Middle East to share the gospel there.

Methodist Church in Paysandy, Uruguay

In parts of the Global North, people seem to be less interested in spiritual matters than in previous centuries. From the perspective of the world as a whole, however, interest in Christianity is increasing.

In the book of Acts, Jewish Christians scattered from Jerusalem because of persecution. Some of those Christians went to Antioch and preached to Gentiles.

And the hand of the Lord was with them, and a large number who believed turned to the Lord. The news about them reached the ears of the church at Jerusalem, and they sent Barnabas off to Antioch. Then when he arrived and witnessed the grace of God, he rejoiced and began to encourage them all with resolute heart to remain true to the Lord. . . .
Acts 11:21-23

Assignments for Lesson 135

Worldview	Recite or write the memory verse for this unit.
Project	Finish your project for this unit.
Literature	Continue reading *Tales from Silver Lands.*
Student Review	Answer the questions for Lesson 135. Take the quiz for Unit 27.

Wax Palm Trees, Cocora Valley, Colombia

28 South America Part 2

We discuss Jimmie Angel's discovery of "the cascade from the sky," a beautiful geographic setting known as Angel Falls. We describe a war over a little piece of geography known as the Falkland Islands. In the third lesson we go north to south, following the American Cordillera mountain ranges and the Pan-American Highway. Then we look at the Patagonia region in southern South America, focusing on the story of an exodus movement that went there from Wales. The worldview lesson discusses the meaning and purpose of life and how you can know and live up to the meaning and purpose of your life.

Memory Verse Memorize Colossians 3:17 by the end of the unit.

Books Used

The Bible

Exploring World Geography Gazetteer

Tales from Silver Lands

Project (Choose One)

1) Write a 250-300 word essay on one of the following topics:
   - What is an exceptionally beautiful geographic feature you have seen? Use vivid descriptive language to tell about it. (See Lesson 136.)
   - What is the meaning and purpose of your life? How do you know? How do you live it out? (See Lesson 140.)
2) Imagine that you have driven the length of the Pan-American Highway. (See Lesson 138.) Create a fictitious interview in which an interviewer asks you questions about the experience. What would you like to hear from someone who had done this? What would you like to tell about it? Perhaps you can get a friend or family member to ask you the questions.
3) Imagine that you are part of the exodus movement that moved from Wales to Patagonia in South America. Write a series of letters to a friend in Wales about the experience. Talk about adjustments you have had to make, the homesickness you feel, and unexpected blessings you have found. (See Lesson 139.)

136 "The Cascade from the Sky": Angel Falls

I have been practically speechless ever since and even now I just have a sort of sinking, hopeless feeling at trying to record even a bare hint of what all this is like. It is grand, awesome, awful, beautiful, marvelous, and terrible. It seems impossible that such things can exist on our earth. It makes the greatest famous scenery of the world seem puny. Having seen it makes one feel that he can never be the same again. A man who had been to the moon and explored its craters might feel similarly exalted and cut off by the depths of his experience from his fellows and from all that he knew before.

—George Gaylord Simpson, upon seeing Angel Falls for the first time while flying in a small plane piloted by Jimmie Angel

The pilot of a small plane can see a grand sweep of geography. He can view a breathtaking combination of mountains, valleys, rivers, lakes, grasslands, forests, seashores, and more. When a pilot sees something amazing that is also unexpected, it adds to the wonder he feels at the beauty in front of the plane.

This happened to Jimmie Angel.

The Place

La Gran Sabana or the Great Savannah is a vast region in southeastern Venezuela, part grassland and part rainforest. Rising above the surface in this region are just over one hundred *tepuis*, tabletop mountains that reach as high as 9,000 feet above sea level. These mountains are not barren like the mesas in the western United States. Instead, they teem with plant and animal life, much of which is found only there, as well as rivers and water pools. The area surrounding a tepui is tropical, but at the top is a microclimate that is much cooler. Clouds sometimes form below the summit. The tepuis are isolated from each other and do not form a connected range.

The word *tepui* means "house of the gods." Surely they must seem to be such to the Pemon people who live below them until they learn about the one true Maker of heaven and earth and of all the wonders He has made which command man's attention.

Simply getting to the tepuis in this remote region is difficult enough. Getting to the top is even more difficult. The sides of the tepuis are sheer or nearly sheer escarpments. Climbing one is difficult if not impossible, so people have thoroughly explored less than half of them. Modern tourists can get there by helicopter or small plane, if a large enough landing site exists on the top. Mt. Roraima, one of the tallest at some 9,200 feet elevation, was unexplored until 1884. This mountain marks the *punto triple* or triple point, where Venezuela, Brazil, and Guyana meet.

The wonders of these remarkable geographic features do not cease on the surface. Beautiful caves open beneath the surface of the tepuis, with quartz walls that have a pinkish hue. Acids in the water running through the caves turn the streams in them red. The speleothems—stalactites and stalagmites—take unusual shapes, sometimes appearing as clouds or spray. Unique species of living things are in the caves as well. The only way into many of these caves is down from the top of the tepui.

Another feature of the tepuis are the waterfalls that plummet down the sides of some of them. The sighting of one special falls was the thrill that a gifted, daring pilot experienced in 1933.

The Pilot

James Crawford ("Jimmie") Angel was born in Missouri in 1899. As a young man he was fascinated with the new invention of airplanes and wanted to fly them. Following World War I, Angel worked as a barnstormer (going from place to place giving daring flight performances), a test pilot, a movie stunt pilot, and a flight instructor.

During the 1920s he began flying into unexplored regions of Mexico and Central and South America. He worked for mineral exploration companies that were searching for oil, diamonds, and gold; government agencies; and scientific exploration teams. Jimmie loved flying and he loved that area of the world.

In 1933 Angel was an aviator and guide for the Santa Ana Mining Company of Tulsa, Oklahoma. On November 18 he was flying solo over La Gran Sabana and approached Auyántepui, the largest (though not the tallest) of the tabletop mountains with an area of 348 square miles. In his journal he later wrote: "I saw a waterfall that almost made me lose control of the plane. The cascade from the sky!" It seemed to him that the waterfall was "at least a mile high." The falls emerged from the side of the tepui about a hundred feet below the top and descended to a lower level above the grassland. The falls were so tall that by the time the water reached the lower level it was mostly mist. At that level it reformed into a stream and soon descended again.

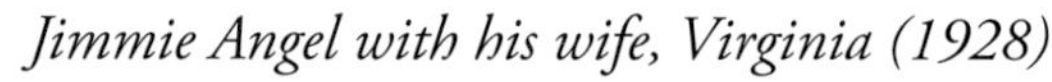

Jimmie Angel with his wife, Virginia (1928)

Angel Falls

Few people believed Angel's report of the waterfall because Auyántepui is so remote. Some even doubted the existence of the mountain. It was only by taking others with him that he was able to convince the world that it really existed. Jimmie Angel was the first non-indigenous person we know of to have seen the falls, and by 1937 people were referring to it as Angel Falls in his honor.

The Falls

Jimmie Angel missed his guess about the height of the falls on his first sighting of it, but we can understand his excitement. The beautiful and impressive Horseshoe Falls of Niagara are 167 feet high. The majestic Victoria Falls in Africa descend 355 feet. Journalist and photographer Ruth Robertson led an expedition to Angel Falls in 1949. They measured the first descent of the falls to be 2,648 feet. Combined with the second cataract, Angel Falls is a total of 3,212 feet high. It's not a mile high, but it is over a half mile! It is the world's tallest waterfall.

The Rest of the Story

Angel's missions and subsequent publicity helped to create international interest in La Gran Sabana and the tepuis there. Scientists have since explored, mapped, and studied the Gran Sabana extensively; and yet there is still much more that can be done. Auyántepui is now in the Canaima National Park.

Jimmie Angel's plane, El Rio Caroni, *remained on top of Auyántepui until 1970 when members of the Venezuelan Air Force recovered it. Many of the original parts were used to build a restored airplane that was put on display at the airport in Ciudad Bolivar, Venezuela.*

Angel became obsessed with finding gold in the region after hearing reports of a river of gold. He believed that Auyántepui was the location of that river. Angel flew many missions in search of such a river, but he never found it. Two years after his first sighting, Angel landed on top of Auyántepui; but his wheels got stuck in the mud. He, his wife, and two other travel companions had to climb and walk for eleven days to get back to their base camp.

Angel also continued to fly missions sponsored by private companies and the Venezuelan government. Jimmie Angel died in 1956 from complications stemming from injuries he sustained while making a landing in Panama.

In 2009 the president of Venezuela at the time, socialist Hugo Chavez, declared that the falls should have an indigenous name rather than be named for an American adventurer. Chavez decreed that the falls should be called Kerepakupai-Merú, Pemon for "waterfall of the deepest place." But for most people around the world, the beautiful and inspiring phenomenon will always be known in Spanish as Salto Angel, and in English as Angel Falls.

God's works are so amazing that sometimes all we can do is ponder them in wonder and amazement.

On the glorious splendor of Your majesty
And on Your wonderful works, I will meditate.
Psalm 145:5

Assignments for Lesson 136

Gazetteer Read the entries for French Guiana, Guyana, Suriname, and Venezuela (pages 234, 235, 239, and 241).

Worldview Copy this question in your notebook and write your answer: What event, book, speaker, or other influence has helped you see your meaning and purpose in life more clearly?

Project Choose your project for this unit and start working on it. Plan to finish it by the end of this unit.

Literature Continue reading *Tales from Silver Lands*. Plan to finish it by the end of this unit.

Student Review Answer the questions for Lesson 136.

Stanley, Falkland Islands

137 Fighting Over the Little Sisters: The Falklands War

At first glance, they might not seem like much to go to war over. The islands are mostly rocks and hills, sheep and penguins. Trees don't grow there naturally. The highest point is only about 2,300 feet above sea level. It rains over half of the days in the year. The total land area is less than the state of Connecticut, and the population is about 3,200.

But two countries—two democracies—both claim the place, and in 1982 those two countries fought a war over the islands that cost about nine hundred lives. In this story we find a lesson about the importance of location and the power of patriotism.

Falkland Islands/Islas Malvinas

The British call them the Falkland Islands. East Falkland and West Falkland are the two largest islands, although the archipelago includes about two hundred islands in all. The government of this overseas territory of the United Kingdom also administers South Georgia Island and the South Sandwich Islands, which lie hundreds of miles away. The Falklands are about three hundred miles east of the southern coast of Argentina (See the map of Argentina in the *Gazetteer*, page 227).

An Englishman first sighted the uninhabited islands in 1592. An Englishman first landed on the islands in 1690. The French established the first settlement there in 1764. Spain assumed control of the islands from the French in 1766 and called them Islas Malvinas. But Britain never formally gave up its claim to the islands, and in 1833 the British established a naval garrison there to assert what the British government believed was its prior sovereignty. Argentina, on the other hand, believes that it gained sovereignty over the islands at the time of its independence from Spain.

This is the basis of the conflict.

It Began with Scrap Metal

Britain did not show great interest in the islands for many years. The British government sponsored few improvements and kept only a small defense force on the islands. Argentina showed more interest in the islands than the British government did, but the residents of the islands overwhelmingly wanted to stay with Great Britain. Argentina and Great Britain had engaged in on-again, off-again negotiations over the Falklands for fifteen years prior to 1982.

In March of 1982 an Argentine scrap metal company with a legitimate contract with the British government sent workers to South Georgia Island to remove an old whaling station. However, Argentine marines disguised as company workers were part of the landing party. In an attempt to establish an Argentine presence on land held by the British, the workers raised the Argentine flag. After a series of inconclusive diplomatic moves failed to resolve the standoff between Argentina and Great Britain on South Georgia Island, the Argentine government decided to invade the Falklands. The leader of the military junta, General Leopoldo Galtieri, was being accused by his opponents of mishandling the economy and violating human rights. He hoped that taking the Falklands would push these issues to the background and unite Argentina under his leadership in a swell of patriotic fervor.

On April 2, 1982, Argentine forces took Stanley, the capital of the Falklands, as well as South Georgia Island two days later. U.S. President Ronald Reagan implored Galtieri not to continue the invasion, but the Argentine leader ignored the request. British prime minister Margaret Thatcher resolved not to surrender any sovereign territory of Great Britain, and the British military mobilized. While British army, navy, and air forces hastened to cover the 7,900 miles from Britain to the Falklands, fruitless diplomatic efforts took place to resolve the conflict. Argentinians turned out in demonstrations supporting the invasion, while the British people supported Thatcher's policy. Most European countries and the United States supported Britain (and the U.S. provided military aid), while most Latin American countries backed Argentina's claim.

Argentine Soldiers in Stanley

By the end of April, Argentina had stationed 10,000 troops in the Falklands, although most of them were draftees who were poorly supplied. Argentine ships and planes were mostly old American and French hardware that the country had bought after the end of World War II. Also by the end of April, British forces began arriving en masse. On April 25 British forces retook South Georgia Island.

The air and naval war was intense. On May 2 a British submarine sank the Argentine cruiser *General Belgano*, with the loss of 323 lives. This loss sidelined the Argentine navy for the rest of the war. On May 4, Argentine planes sank the destroyer H.M.S. *Sheffield*; and Argentine attacks also destroyed other British ships during the conflict. Argentina lost several planes in the fighting.

British troops landed on East Falkland north of Stanley and headed for the Argentine-held capital. Despite the Argentinians' brave stand and some intense hand-to-hand combat, the British retook the city on June 14 to end the ten-week conflict. British forces took over 11,000 prisoners but soon released them all. Argentina lost 649 men while the British lost 255.

Argentine troops laid thousands of mines to defend against British attacks. The minefields were fenced off after the war, and they became havens for wildlife. A 1997 international treaty required the UK to begin removing the mines. Teams of workers, primarily from Zimbabwe, were nearing completion in 2020.

Since the War

The ruling Argentine military junta lost favor with the people over its failure to hold the islands, and civilian rule was restored in the country the next year. By contrast, in Great Britain Margaret Thatcher's Conservative Party government won a landslide victory in the 1983 elections.

Great Britain has increased its involvement with and commitment to the Falkland Islands since the war. Britain has granted full British citizenship to residents of the islands. It has stationed about 1,200 troops there who have access to modern equipment.

The Falklands economy is growing as a result of this greater investment. The chief exports used to be wool and mutton, but now the main export is squid. Tourism has become a significant industry on the Falklands as well, as people are drawn to see what all the fuss was about. Offshore drilling for oil and natural gas has begun and developers are optimistic. Reindeer were brought to South Georgia Island in the early 1900s to provide food for whalers. Some were moved to the Falklands in 2001 to provide meat for the locals and tourists as well as to be an export.

Great Britain and Argentina restored diplomatic relations in 1989. In 1998 the Argentine government declared that it would only use peaceful means to assert its claim to the islands. Argentinians refer to the islands as the little lost sisters; but Great Britain demonstrates no interest in abandoning them. A March 2013 referendum of the residents of the Falklands (almost all of them British subjects) indicated that 99.8% of Falkland Islanders wanted to remain with Great Britain. And so they remain, but Argentina has neither forgotten nor given up the desire for the little sisters to come home. In 2020 Argentina created the National Council of Affairs Relating to the Malvinas, South Georgia, South Sandwich Islands, and the surrounding maritime spaces. Its goal is to take control of the Falklands/ Malvinas.

Reindeer and Penguins on South Georgia Island

This 2004 photo shows graffiti in Buenos Aires, Argentina. It says "English out of the Malvinas! We will be back."

The story of the Falklands reminds us that history and geography can be important to a nation's identity. Patriotism can be a powerful motivator. And the location of a given piece of geography—even a relatively small piece of geography—can be a matter of strategic importance from the perspective of a nation's worldview. Nine hundred people gave their lives in a demonstration that all of these points are true.

People have often resorted to war to settle their disagreements. The psalmist wanted peace but found only those who wanted war.

Too long has my soul had its dwelling
With those who hate peace.
I am for peace, but when I speak,
They are for war.
Psalm 120:6-7

Assignments for Lesson 137

Gazetteer	Read the entries for the Falkland Islands and for South Georgia Island and the South Sandwich Islands (pages 233 and 238).
Worldview	Copy this question in your notebook and write your answer: Who is someone who seems to you to have a clear sense of his or her purpose in life?
Project	Continue working on your project.
Literature	Continue reading *Tales from Silver Lands.*
Student Review	Answer the questions for Lesson 137.

Andes Mountains, Peru

138 The Longest Line on the Map

You can get in a vehicle near Prudhoe Bay on the Arctic coast of Alaska and drive—over a period of months—to Ushuaia, Argentina, the southernmost city in South America. This fantastic and memorable trip is possible because of the Pan-American Highway, a 19,000-mile series of connecting roads that runs from the top of North America to the bottom of South America. The Highway is a fascinating route; and its creation is a feat of international political cooperation, highway planning, and civil engineering. We talk about the highway in this lesson, but first we need to mention the fact that God took things even further.

The American Cordillera

If you look on a map of the Western Hemisphere, you see that when God created the world, He created a series of mountain ranges that run, once again, from the top of North America to the tip of South America. Moreover, most scientists agree that these ranges, the "backbone" of the Americas, actually go further and run under the Southern Ocean onto the continent of Antarctica. This chain of ranges is the American Cordillera. The name comes from the Spanish word for cord and refers to a series of parallel and interconnecting mountain ranges.

The northernmost part of the Cordillera consists of the Alaska Range and the Brooks Range in Alaska. It then continues into the Yukon Territory and British Columbia in Canada. The Cordillera becomes the Rocky Mountains of Canada and the United States as well as the mountains on Vancouver Island and the parallel Pacific coast ranges that include the Cascades and the Sierra Nevada.

In Mexico, the Cordillera continues as the Sierra Madre Occidental (western), the Sierra Madre Oriental (eastern) and the mountains on the Baja California peninsula. The mountain ranges of Central America continue the chain to South America, where it becomes the Andes and parallel chains that run to the southern tip. The undersea South Georgia Ridge connects to the Graham Land mountains on the Antarctic Peninsula. The Cordillera forms the eastern side of the Ring of Fire that encircles the Pacific Ocean.

The Andes

The Andes Mountains, the southern portion of the American Cordillera, is the longest mountain chain in the world. It extends about 5,000 miles in length but is never more than about 500 miles wide. The chain runs through Venezuela, Colombia,

Ecuador, Peru, Bolivia, Chile, and Argentina. Mount Aconcagua in Argentina is the highest peak in the Western Hemisphere at 22,841 feet above sea level.

The Andes have a wide variation in climate. The northern section is warm and wet with tropical forests and cloud forests. The central section has seasonal variation in weather patterns, while the southern range is mostly dry and icy. The mountains affect the weather to the east of it as they block Pacific winds and moisture from the rest of the continent.

Remarkable animals make their home in these mountains. The Andean or spectacled bear is the only bear species in South America. The Andean condor is the world's largest raptor and largest flying bird, weighing up to 33 pounds and having a wingspan of up to ten feet.

The people of the Andes have learned to cope with their challenging environment. For instance, every year the Quechua people of four villages work together for three days to rebuild the 120-foot Q'eswachaka rope bridge that spans the 60-foot deep Rio Apurímac gorge. They gather straw and fashion ropes that are as thick as a man's thigh. They carry one cord across the old bridge, cut loose the old bridge and let it fall into the gorge, then carefully weave a new bridge in its place. The bridge was once part of the Great Inca Road that connected the Incan empire. The Quechua used to build many such bridges, but with modern bridges being put in place this is the last one of its kind. The people of these villages have kept this bridge building tradition for five centuries.

The Longest Road in the World

The longest road in the world didn't start as a road. In fact, it didn't start on a road at all. When Hinton R. Helper, U.S. consul to Argentina, was traveling by ship from Buenos Aires to New York in 1866 and became seasick, he realized the need for a means of land travel between North and South America. His idea was not for a road, however, but for a railroad. At the time railroads had begun transforming American life and were opening up new possibilities for travel. The first east-to-west transcontinental railroad in the U.S. would be completed just three years later. However, Helper's idea, pardon the expression, didn't go anywhere. It was too expensive and difficult at the time to build a railroad that connected North and South America.

Instead, the next big transportation concept for the Americas involved digging a canal across Panama. A French company tried and failed to build it in the late 1800s, but the U.S. government took up the project and completed it in 1914.

Q'eswachaka Bridge, Peru

The Idea of a Road

During the 1920s, when the United States generally withdrew from world affairs after its costly involvement in World War I, the U.S. increased its contact with other countries in the Western Hemisphere. In 1881 the United States had initiated the first Pan-American Conference for the countries in the Western Hemisphere. These meetings became regular during the 1920s.

As cars and trucks proliferated, the conference took up the call for the construction of a highway to connect the continents. Although an intercontinental highway was first discussed at the 1923 conference, it did not gain approval until 1928. Most countries agreed that it was a good idea, but funding for the project was slow in coming. Some work took place in Central America in the 1930s, financed by New Deal funding since President Franklin Roosevelt was a supporter of the idea.

Mexico completed the first official portion of the highway in 1950 and was able to fund its part of the project itself. The route is not one single highway and no one administrative unit oversees the entire length of the road. Instead, the route often uses long-established roads and each country oversees the route within its own borders. The Pan-American Highway Congress, with representatives of participating countries, meets every four years. Financial support from the United States was crucial to the project's success, but the U.S. does not have any official segment of it. In fact, the United States does not officially recognize it.

How to Get from Here to There

The route that people call the Pan-American Highway runs through fourteen countries: the United States, Canada, Mexico, Guatemala, El Salvador, Honduras, Nicaragua, Costa Rica, Panama, Colombia, Ecuador, Peru, Chile, and Argentina.

Let's start at the northern end of the route, in the charmingly named community of Deadhorse, Alaska, near the Prudhoe Bay oil fields. The first 414 miles are the Dalton Highway, which runs to Fairbanks. This road was built in 1974 to support the Trans-Alaska Pipeline. It's pretty desolate here; there are no services on one 240-mile stretch. At Fairbanks we can take a 94-mile road to Delta Junction, where the Alaska Highway begins.

Work on the Alaska Highway (1942)

During World War II, Japan seized two islands in the Aleutian chain of Alaska and threatened to take more. The U.S. Army Corps of Engineers built the Alaska Highway in 1942 with the labor of 10,000 soldiers. The road linked Alaska with the 48 existing states to the south and helped soldiers and supplies reach Alaska. The Alaska Highway is 1,387 miles long and is completely paved. (The U.S. retook the islands from Japan in 1943.)

The Alaska Highway ends at Dawson Creek, British Columbia. A highway runs from there to Edmonton, Alberta. From Edmonton the traveler can take one of two routes. The first runs to Minneapolis, Minnesota, and follows Interstate 35 to Laredo, Texas. The other route goes through Calgary, Alberta; Billings, Montana; and then joins with Interstate 25 through Denver, Colorado, to Las Cruces, New Mexico. From there the route follows Interstate 10 to San Antonio, Texas, and then connects with the I-35 route. Of course, the traveler in the United States can follow any number of routes to get to Laredo.

Across the Rio Grande from Laredo is the Mexican city of Nuevo Laredo. Here begins the first official section of the Pan-American Highway. The road runs across Mexico to Guatemala and through all of the countries in Central America except Belize. Other than in Honduras, the road goes through the capital of each country: Guatemala City, Guatemala; San Salvador, El Salvador; Managua, Nicaragua; San Jose, Costa Rica; and Panama City, Panama. In Costa Rica the road climbs to 10,942 feet elevation at the Cerro de la Muerte, which translates to Summit of Death. This is the highest point on the highway.

You Can't Get There Through Here

In southern Panama the highway stops at the Darien Gap, a dangerous 60-mile stretch of swampland, rainforest, and mountains. The place has never treated outsiders kindly. Spanish explorers built their first settlement on the American continent here in 1510, but indigenous people destroyed it 14 years later. A few attempts have been made to extend the Pan-American Highway through this forbidding area, but none has succeeded. Environmentalists have expressed concerns that such a construction project would ruin the habitat. Indigenous people fear that a road would threaten their traditional cultures.

In addition to the physical dangers, the Gap is a common route for drug traffickers. Despite the dangers, tens of thousands of people cross the Gap every year as illegal immigrants on their way usually to Mexico or the United States. Some do not survive. A handful of legitimate adventurers have made it through the Gap, but most people take a ferry to Colombia or Ecuador to resume their drive on the Pan-American Highway. The route through Colombia leads into Ecuador.

In South America

The highway runs through Quito, the capital of Ecuador. At 9,350 feet elevation, Quito is the second-highest capital city in the world after La Paz, Bolivia. Then we go from Ecuador to Peru, and from Peru to Chile, where the road runs through the Atacama Desert, the driest non-polar location in the world. Some parts of the desert have never seen rain (as long as people have been keeping records).

If you are beginning to feel lonely on the Atacama, get ready for a big greeting. South of the city of Antofagasta, in 1992 sculptor Mario Irarrázabal crafted a 36-foot Hand of the Desert sculpture out of concrete and iron rising out of the ground. This is a left hand. Ten years earlier, the same sculptor created out of concrete, plastic, and steel a right hand coming out of the sand on Brava Beach in Punta del Este, Uruguay.

Then wave goodbye to the hand and proceed on to Valparaiso, where an unofficial leg (pardon the expression) of the highway continues down the coast of Chile to Quellon. The official route turns

Chilean artist Mario Irarrázabal created these two sculptures in western and eastern South America. The left hand is located in the Atacama Desert of Chile. The right hand is located on the beach in Punta del Este, Uruguay.

east at Valparaiso, crosses the Andes, and proceeds to Buenos Aires, Argentina. Before 1980, the highway crossed the Andes at Uspallata Pass, which at 12,572 feet elevation used to be the highest point on the highway. In 1980 the Christ the Redeemer Tunnel through the Andes opened at 10,499 feet.

The Pan-American Highway officially ends in Buenos Aires, but an unofficial leg continues south along the Argentine coast. Near its end the road dips back into Chile, crosses the Straits of Magellan via a ferry, re-enters Argentina and ends at Ushuaia, on Isla Grande de Tierra del Fuego, the largest of the Tierra del Fuego island chain.

By the way, Argentina says that Ushuaia is the southernmost city in the world, while Chile claims that distinction for Puerto Williams. Puerto Williams is farther south, but the question becomes, what is a city. Ushuaia has a population of 71,000 while Puerto Williams only has about 3,000. What do you think?

How Long Does It Take?

So how long does it take to travel the Pan-American Highway? If you drove eight hours per day every day, you could do it in three months; but no one does it that way. There is so much to see and so many side roads to explore that most people take nine months to two years to travel the entire length of the road.

There are faster ways and slower ways. In 2018 Michael Strasser of Austria made it on a bicycle in 84 days, 11 hours, and 50 minutes, although he took a plane over the Darien Gap. On the other hand, George Meegan of the United Kingdom walked the length of the highway in 2,426 days, from January 26, 1977 to September 18, 1983—over six and a half years. Meegan started at Ushuaia and went north, detoured through the eastern United States, did not take extended breaks, and walked through the Darien Gap.

George Meegan

Pan American Highway in Costa Rica

Portions of the Pan-American Highway see significant traffic, but relatively few people travel the entire length of the route. Perhaps the main benefit of the highway is as a symbol of inter-American cooperation and a means for furthering that cooperation in practical terms.

The prophet Isaiah spoke of a highway in the desert in a prophecy of the coming revolutionary Messiah. Matthew tells us that the preaching of John the Baptist regarding Christ fulfilled Isaiah's prophecy.

A voice is calling,
"Clear the way for the Lord in the wilderness;
Make smooth in the desert a highway for our God.
Let every valley be lifted up,
And every mountain and hill be made low;
And let the rough ground become a plain,
And the rugged terrain a broad valley;
Then the glory of the Lord will be revealed,
And all flesh will see it together;
For the mouth of the Lord has spoken."
Isaiah 40:3-5

Assignments for Lesson 138

Gazetteer Read the entries for Colombia, Ecuador, and Uruguay (pages 231, 232, and 240).

Worldview Copy this question in your notebook and write your answer: For what things in your life are you most grateful?

Project Continue working on your project.

Literature Continue reading *Tales from Silver Lands.*

Student Review Answer the questions for Lesson 138.

Gorsedd Ceremony, Gaiman, Argentina (2019)

139 The Welsh at the Bottom of the World

The people in this community speak Welsh. The street signs are in Welsh. Many of the residents have roots that run deep in Wales. Their community is located in . . . southern Argentina!

To use a geographic expression, how in the world did this happen? To find out, we have to go back over 150 years.

The Migration

In the mid-1800s, the Industrial Revolution was transforming Great Britain. Factories needed coal, iron, and slate, and people to work in them. One source for all of these was Wales, the small country to the west of England on the island of Great Britain. The Welsh are a Celtic people whom the Anglo-Saxons pushed into this region when they invaded the island. The Welsh had tried to hold on to their language and culture, but England had relentlessly absorbed them over a period of several centuries by military conquest, by extending English legal jurisdiction into Wales, and finally in 1707 by incorporating Wales into the Kingdom of Great Britain.

At the time of the Industrial Revolution, many Welshmen were leaving their farms and mines and moving to England. A movement developed among some of the Welsh people to look for a new start somewhere else where they could maintain their identity. Some Welshmen had set up colonies where they could do this. The most successful were in Utica, New York, and Scranton, Pennsylvania. However, these Welsh settlers felt pressure to learn English and to adapt to the ways of American culture.

In 1861 Michael Jones called a meeting at his home in northern Wales. Jones was the principal of Bala College and a strong Welsh nationalist. The leader and meeting place were appropriate since about 15 percent of Welsh people have the surname Jones. The group wanted to identify a place other than the United States for a new settlement. One possible location was Patagonia, the southern region of Argentina and Chile in South America.

Patagonia

The region of Patagonia makes up most of the southern cone of South America. It lies mostly in Argentina, south of the Rio Negro, but extends into Chile south of Puerto Montt. The Atlantic coastal plain ends in tall cliffs that give way to a scrubby desert and semi-desert steppe marked by gradually rising plains to the stark and breathtaking beauty of the Andes. In Chile the region is marked by a rugged coast with numerous channels, fjords, and islands. In the mountains are icefields and glaciers that make up the largest expanse of ice in the Southern Hemisphere outside of Antarctica. Recent melting of the icefields is a cause of concern among some scientists. The climate is characterized by seasonal extremes and relatively little rainfall. Patagonia is one of the most sparsely populated regions in the world.

Patagonia has a rich variety of animal life. Species include armadillos and pumas; the guanaco, which is a member of the camel family and related to the alpaca and llama; and the rhea, a large flightless bird related to the ostrich. Rodent species include the rabbit-like cavy or mara and the burrowing vizcacha and tuco-tuco. The pudu is the smallest deer in the world, reaching a height of only about 13 to 17 inches at the shoudler.

In the Atlantic are elephant seals, fur seals, sea lions, and right whales. On the coast is the largest colony of Magellanic penguins in the world, along with other species of penguins.

Y DRAFOD.

LLONGAU.

CLYWEDION.

Meni Rhagoraf y Byd.

UNIG GYNRYCHIOLWYR TIRIOGAETH Y CAMWY

C. M. C.

Gwneuthuriad cryfach a gwell heb angen newid dim arnyat, am brisiau llai nag o'r blaen.

$ 175 $ 315 $ 360 $ 380 $ 520 $ 540

C. M. C.

LLOGAU.

Yn unol a phenderfyniad yr Hyrwyddwyr yn eisteddiad Rhagfyr 28, 1908, dymunir gwneud yn hysbys y codir llog o'r 1af o Ebrill, 1909, yn ol 12 y cant y flwyddyn

E. M. Morgan, Arolygydd.

Caballo Perdido.

Edictos Judiciales.

C. M. C.

INTERESES.

Desde el 1° de Abril 1909

COBRA

Por adelantos en cuent corriente

12 por ciento anual.

SE HA PERDIDO:

AVISO.

AR WERTH.

TY HARDD,

Pris $ 5,800.

Dr. LUIS A. VALLEJO, B.

CIRUJANO-DENTISTA.

Pascual Carubelli,

TRELEW.

COLLWYD.

CONSULTORIO:

HOTELES UNIDOS.

Dr. LUIS A. VALLEJO, B.

Llaw-feddyg Deintydd.

HOTELES UNIDOS.

GWAIR ALFALFA.

This is a page from the January 8, 1909, issue of the Y Drafod *newspaper. It has content in both Welsh and Spanish.*

The Migration

Michael Jones had been in touch with the government of Argentina about settling an area called Bahia Blanca. The Argentines had assured Jones that the landscape there would feel like home and that the Welsh could maintain their language and culture. The benefit to Argentina was that the area would grow in economic productivity and also that Argentina could control an area in western Patagonia that Chile claimed.

Some 165 people gathered in Liverpool, England, in late May 1865. The ocean voyage to their new home took eight weeks, and they landed at Puerto Madryn on the Argentine coast on July 27. They quickly learned that the Argentines either had not been honest or had not known the geography of Wales. In the Argentine winter, these Welshmen faced a barren, windswept, inhospitable semi-desert. There were few trees for building homes, and they had insufficient water and little food. Some of the people lived in caves they dug out of the soft rock in the cliffs by the bay where they had landed.

The New Community

The Welsh settlers received help from the Tehuelche people (whose ancestors Magellan had met in 1520) and welcomed mercy shipments from the people back home. They made their way to the Chubut Valley 40 miles inland where a river ran through the desert. Here they established the community of Rawson by the end of 1865, and they struggled through. One colonist, Rachel Jenkins, noticed that when the river flooded it enriched the land. She conceived the idea of building an irrigation system to water the fields, and it saved the colony.

New settlers came from Wales and Pennsylvania. In 1875 Argentina gave the settlers title to the land. More settlers arrived from Wales, which was undergoing a severe economic depression. Other groups migrated there over the coming decades. They transformed the scrub-filled semi-desert into a fertile and productive area. New settlements formed in the foothills of the Andes.

Their success attracted immigrants from other countries. In 1915 the Chubut region had a population of 20,000, half of whom were immigrants. At this point the Argentine government changed its policy and assumed direct rule over the settlements. The national government ended the use of the Welsh language in government and in schools.

Despite these setbacks, the Welsh settlers and their offspring persevered and thrived. In the Chubut region today there are numerous communities with Welsh origins, from Rawson and Trelew near the Atlantic coast to Trevelin and Esquel in the Andean region to the west. Some 50,000 people there claim Welsh heritage and 5,000 speak Welsh as their first language, although Spanish is by far the first language for most.

Today

In more recent years the residents of the region have been able to celebrate their Welsh roots publicly. A few people actively promote and teach Welsh to adults and children. Welcome signs are in Spanish and Welsh. Signs indicating streets named for "Miguel D. Jones," the founder of the exodus movement, are found in several communities, as are Welsh tea houses. Shops sell Welsh souvenirs and craft items. Such acceptance not only encourages Welsh identity but also brings in tourists.

Today tourists also travel in the opposite direction. Every year residents of Argentina make the 8,000-mile trip—by air—to the Eisteddfod, the annual celebration in Wales of music, poetry, and performance that takes place in a different city in Wales every year. These travelers have a hard time understanding English when they land in London, but when they get to Wales they communicate easily with their fellow Welsh-speakers.

Teapot at the Welsh Tea House in Gaiman

Ending the Way We Began

We have come to the last lesson about regions of the world. In this lesson we have learned how the geography of Wales in Great Britain has a human connection with the geography of Patagonia in Southern Argentina. People left Wales with its geographic setting of coal and iron and its geographic relation to England. They moved to Argentina, carved a life out of the semi-desert, and built communities that still exist today. Among these people the Welsh culture and the Spanish culture have merged and have enriched each other.

The bottom of the world proves to be like the rest of the world: people interacting with geography and geography affecting how people live. This is the world in which God has placed us, for our good and for His glory.

I will give thanks to You, O Lord, among the peoples;
I will sing praises to You among the nations.
For Your lovingkindness is great to the heavens
And Your truth to the clouds.
Be exalted above the heavens, O God;
Let Your glory be above all the earth.
Psalm 57:9-11

Assignments for Lesson 139

Worldview Read "Think Like Jesus: Put Others First" (pages 332-338) in the *Gazetteer* and answer the questions in the *Student Review Book*.

Project Continue working on your project.

Literature Continue reading *Tales from Silver Lands*.

Student Review Answer the questions for Lesson 139.

Perito Moreno Glacier, Argentina

140 What Is the Meaning and Purpose of Life?

When my wife and I got married, we received a set of Fostoria stemware glasses. They are beautiful glasses that hold a special place in our hearts and lives. My wife would be greatly distressed if I filled one with water, took it outside, and set it on the driveway while I shot basketball. That is not why people gave us those glasses. That is not their purpose.

Before my grandfather died, he gave instructions that I was to receive the flag that would drape his casket. The flag is a precious family heirloom. It would be terribly short-sighted and disrespectful of me to use it as a rag to clean the windshield of our car. That is not why my grandfather gave it to me. That is not its purpose.

You have been given a precious gift. You need to know the purpose of this gift because you have one chance to use it in keeping with its purpose. You don't want to turn it into a disposable water cup or a cleaning rag when its value is beyond estimating and its purpose is to honor God and to be a blessing to the world. This gift is your life.

The meaning and purpose of a ruler is to measure things. That is why its maker made it. The meaning and purpose of a math book is to teach math. That is why the author wrote it. Your meaning and purpose is to honor God by following the example of Jesus Christ in your life. That is why your Author and Maker made you. That is the meaning and purpose of your life.

Your understanding of the meaning and purpose of life in general and of your life in particular is an essential part of your worldview. You need to know where you came from, why you are here, and where you are going. It is a short-sighted and self-focused worldview to misuse expensive stemware glasses and a family heirloom. In the same way, it is short-sighted and self-focused to use your life for something less than its real value. You need to have an appropriate and accurate worldview concerning the meaning and purpose of your life.

Steps Toward Understanding The Meaning and Purpose of Your Life

1. Seek God to know His path.

It is easy to look inside yourself for your life's purpose. You might well ask, "What can I do? What should I seek to accomplish?" But it is too small a purpose for your life merely to acquire, achieve, impress, and become known. Seeking to gain the world, you risk losing your soul. Your life will be much richer if you find your meaning and purpose outside of yourself and in God. You can have no

Mountain Path, Peru

higher purpose than to honor God with all that you do. To live that way, you have to look to Jesus, see how He lived, and follow His example, not just do whatever "feels right" to you. You can know the joy of fulfillment with even the least that you give in the name of God. Life has great meaning when you know the importance of giving a cup of cold water.

Without knowing the right path, you will be aimless. If you don't know the right goal, you won't know what you are looking for or how to get there. How can you know goodness, or courage, or sacrifice, or loyalty, or love? You will drift from one cause or philosophy to another and never get anywhere. Loving God and loving others will help you know how to make a difference along the path. Where you put your faith, the path you decide to follow, is all-important.

2. Be grateful.

You received the gift of life without asking for it and before you knew how to live it well. You didn't ask for the shape of your life, and some elements of it you might wish were different. Nevertheless, you have many reasons to be thankful for your life. Count your blessings, and you will find that they are countless. Even the hardships can help you grow. A victorious worldview is one of gratitude.

3. Prepare yourself.

Opportunity comes to those who prepare for it. You need to be learning and growing with humility so that you will be better able to fulfill your purpose when you face a decision or have an opportunity to act or to learn something more deeply. If you try to come across as knowing it all, you will only be fooling yourself; most everyone else knows that you don't. Listen more than you talk; read more than you write. Some things that you learn you will lay aside as not helpful; other things that you learn through trial will be invaluable later on, even though you would rather not have to learn those lessons that way at all.

4. Live it now

Take small steps in living out the meaning and purpose of your life. You don't know everything, but you do know some things. Be faithful in these things, and the Lord will open more doors and give you a chance to learn more. A life is made up of some major decisions and many tiny, everyday ones. Don't derail yourself by making poor decisions. You can recover from your mistakes, but you will still have to face the consequences of them. The further you get off the right path, the harder it will be to get back on it. The more you make the right decisions, the more practice you will gain in living out your purpose.

5. Be a person of good character

We all have humdrum, everyday, routine things we must do. Living with a clear and worthy meaning and purpose does not eliminate these things. It is in living that we find life's meaning—in learning, serving, caring, connecting, and doing.

Do you want people to remember you as someone who could perform a beautiful piano concerto but who wouldn't lift a finger around the house? Audiences come and go, but the people you live with every day matter most. In this worldview, everyone matters. If the only things that matter in this life are political power, great wealth, and lasting fame, only a few people will achieve those things; but everyone can be a servant and honor God and love others. Since this is the victorious life, everyone can know victory. Directing your life in this way will help you to become a person of noble, upright character.

Welsh Chapel, Gaiman, Argentina

Some Questions to Ask About Meaning and Purpose

Who around you is living a godly life? How can you gain from their wisdom and experience?

Does the book, speaker, podcast, or philosophy that is before you honor God and help you to honor Him better? If not, what purpose does it serve?

Are you praying for a clearer sense of your purpose and direction?

Use the ten-year test. Ten years from now, what will you wish you had done—or not done?

Does This Seem Too Heavy?

All this talk about meaning and purpose, where you are headed, and choosing to live well sounds pretty deep, deeper than you might care about thinking. It's hard to think about meaning and purpose when you have a pile of laundry to fold or a sink full of dishes to wash or a yard to mow or an algebra lesson to get through.

But giving some thought to these issues will help you know how to handle those everyday tasks. Perhaps it will give you some perspective about how you spend your time and what you text about. Living with a clear meaning and purpose from God will bring true joy and fewer regrets. That will result in a happy life.

Where Are You Headed?

As much as you live in today and enjoy what you are doing right now, you have a whole life and even eternity to keep in mind. We all have various experiences, successes, and failures in life; but unless the Lord comes back first, everyone will experience the end of this life in death. Our wonderings and wanderings will be no more. Because what we do matters—to ourselves, to others, and to God—we need to live with an awareness that one day our life on the earth will end. We will leave behind a legacy of some kind. It might be a legacy of kindness, love, joy, learning, and service; or it might be a legacy of instability, anger, and hurt; or it might include any number of other ways one person can affect others.

What does all this mean for when you die? Does it all end at physical death and the only lasting impact will be the legacy you leave on the earth? Or will you face judgment based on what you have done and what you have believed? In other words, will the impact of your life extend into eternity? The teaching of Scripture and the logic of life tell us that our lives matter both here and hereafter.

Your worldview does not just include how you see this world. It also includes what you believe about the world to come. If a person's worldview has consequences, it seems reasonable to believe that a person's life does also. Some of our most precious possessions are things we did not ask for but for which we serve as faithful stewards. You did not ask for your life, but you have the responsibility to live it well. Honor God by doing so.

Paul taught the Christians at Colossae the meaning and purpose of their lives, and we can know this same meaning and purpose today.

For by Him all things were created, both in the heavens and on earth, visible and invisible, whether thrones or dominions or rulers or authorities—all things have been created through Him and for Him.

Whatever you do in word or deed, do all in the name of the Lord Jesus, giving thanks through Him to God the Father.
Colossians 1:16, 3:17

Assignments for Lesson 140

Worldview Recite or write the memory verse for this unit.

Project Finish your project for this unit.

Literature Finish reading *Tales from Silver Lands*. Read the literary analysis and answer the questions about the book in the *Student Review Book*.

Student Review Answer the questions for Lesson 140.
Take the quiz for Unit 28.

29 The People of the World

We have completed our trip around the world, looking at continents, countries, geographic highlights, and people groups. In the final two units we consider topics that relate to worldwide issues. We first look at demographic studies of the world population. Then we consider how and why populations change. In the third lesson, we contrast rural geography and urban geography and what they mean for the people who live in these places. We study the significance of land ownership in human life. The worldview lesson examines how we see ourselves and, as a result, how we see others.

Lesson 141 - All These People!
Lesson 142 - Population Changes, Too
Lesson 143 - City Mouse, Country Mouse
Lesson 144 - Who Owns the Land?
Lesson 145 - Peopleview — How We See Ourselves and Others

Memory Verse

Memorize Genesis 1:27-28 by the end of the unit.

Books Used

The Bible
Exploring World Geography Gazetteer

Project (Choose One)

1) Write a 250-300 word essay on one of the following topics:
   - Where would you like to live: a rural setting or an urban setting (or a small town setting)? Give reasons for your answer. (See Lesson 143.)
   - Develop your arguments for or against the institution of private property ownership. (See Lesson 144.)
2) Prepare and deliver a five-minute talk about God's love for all people. (See Lesson 145.)
3) Write a narrative comparing the lives of two people—one in a crowded city and one on an isolated ranch or farm. (See Lesson 143.)

141 All These People!

Jorg Junhold, director, Leipzig Zoo, Leipzig, Germany

Vishwanath Mahadeshwar, mayor, Mumbai, India

Asdis Magnusdottir, attorney, Reykjavik, Iceland

You

And a few billion other people

The world's population started with two people in the Garden of Eden. The number of people on the earth grew for some period of years in the area near the Garden. After the Lord scattered the people away from the Tower of Babel, people spread over the earth to the rest of the continents and to the islands of the world. As of June 2020, experts have estimated the human population to be about 7.8 billion people, including you. We've come a long way from a world population of two!

The study of the earth's human population is called demography. This word is from the Greek word *demos,* which means people. We get the word democracy from the same Greek word. Demographers have many ways by which they study the population of the world.

Population by Continents and Countries

Population estimates for some countries can vary widely as census counts in these countries are difficult to determine. The estimated population of the world by continents is as follows. (Except where noted, the figures in this unit are 2020 estimates.)

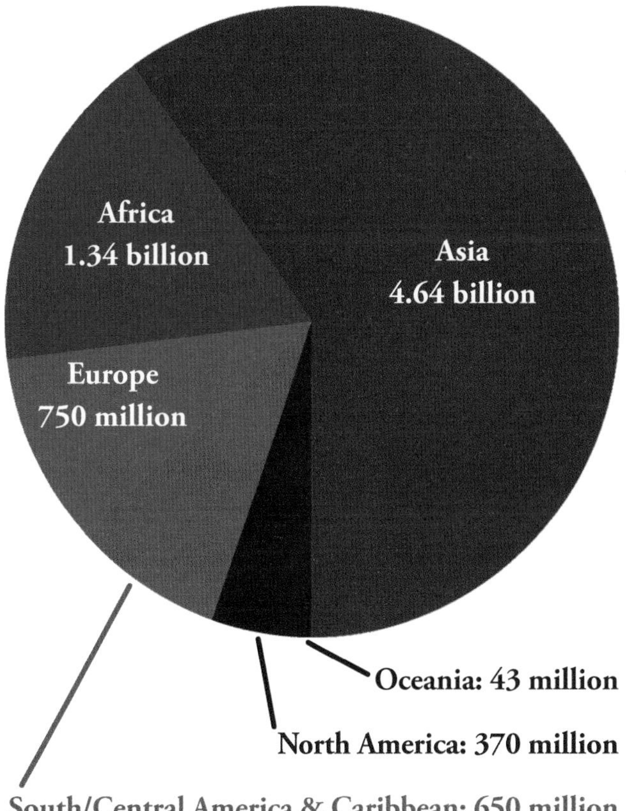

Top Ten Nations by Population (in Millions)

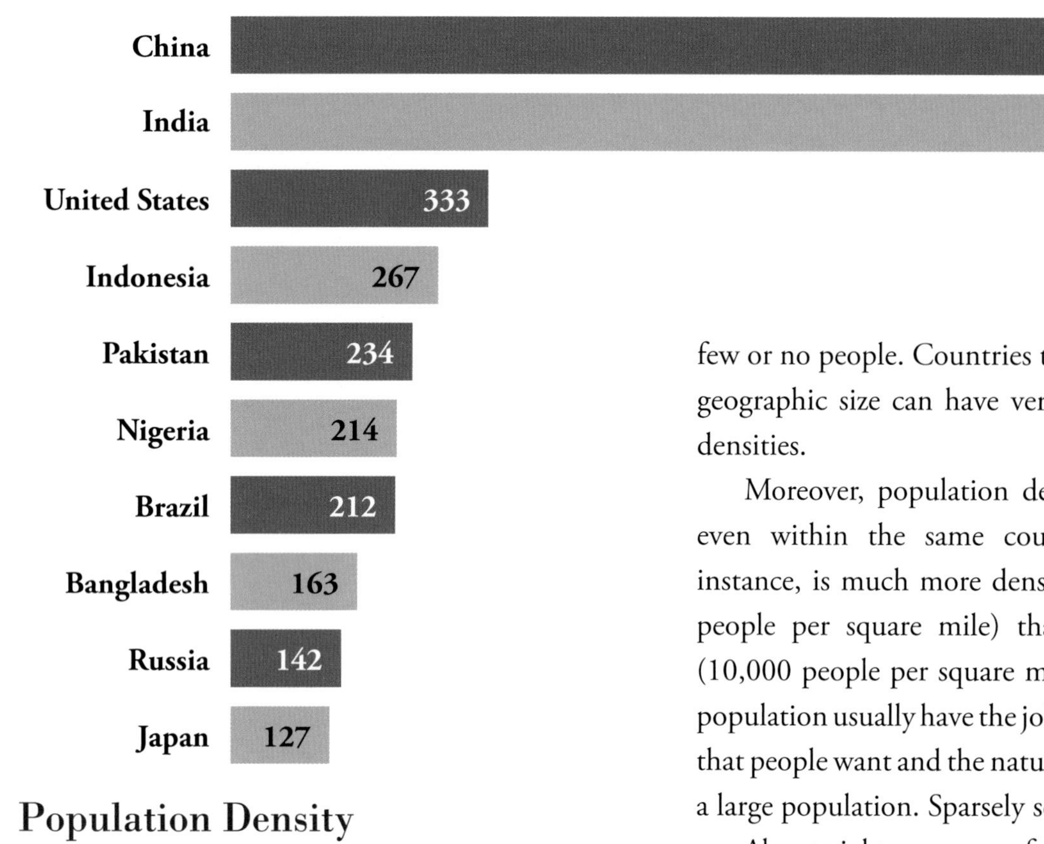

Population Density

The 7.8 billion people who live on the earth are not distributed evenly across the planet's land area. Some countries have huge numbers of people living in a relatively small area. For instance, Monaco has about 50,000 people per square mile, and Singapore has 20,000. Other places have large land areas with few or no people. Countries that are about the same geographic size can have very different population densities.

Moreover, population density can vary greatly even within the same country. Manhattan, for instance, is much more densely populated (70,000 people per square mile) than Washington, D.C. (10,000 people per square mile). Places with dense population usually have the jobs and other attractions that people want and the natural resources to support a large population. Sparsely settled areas do not.

About eighty percent of the world's population lives in what are considered less-developed countries. Less-developed countries do not have large, active economies; they do not have the level of social services that are available in developed parts of the world; and they have relatively limited educational and employment opportunities.

Population Density Per Square Mile of Select Countries

Bahrain	Bangladesh	India	Japan	Germany
5,136	3,051	1,066	862	603
Nigeria	Albania	Mexico	United States	Canada
563	258	171	87	10

Population Groups

The two most common ways of breaking down population numbers are by gender and by age group. As of 2019 there were slightly more men than women in the world. Demographers believe that about 105 males are born for every 100 females. Men are more likely to die from accidents, warfare, and violence than women are; and men tend not to live as long as women. Because of these factors, the world male and female populations tend to remain about equal. However, warfare can temporarily reduce the male population in specific areas. As we discussed in Lesson 91, the population control policies of China has thrown the male-female ratio out of balance there.

A common way to illustrate a nation's population breakdown by age is with a population pyramid. This shows in graphic form what age groups are largest and which are smallest in a country. It also shows the gender difference in each category. Examine the population pyramids of three countries at right. The bars show the percentage of the total population of males and females in each age bracket.

Kenya is a country with a huge number of young children because of the population boom occurring there now. The United States and Denmark have similar population distributions, though the U.S. has a slightly higher percentage of younger people and Denmark has more older people.

This information is more than just data that shows contrasts between countries. It suggests what priorities the people of a country will need to address both now and in the future. Kenya will want to provide health services for mothers and young children and education for their youth. As the population of the United States ages, a relatively smaller population of workers will have to support a relatively larger number of older citizens by providing care for older parents and by providing revenue for Social Security and Medicare. Denmark is in that situation now; and with fewer children being born, it will have to develop ways of caring for the elderly

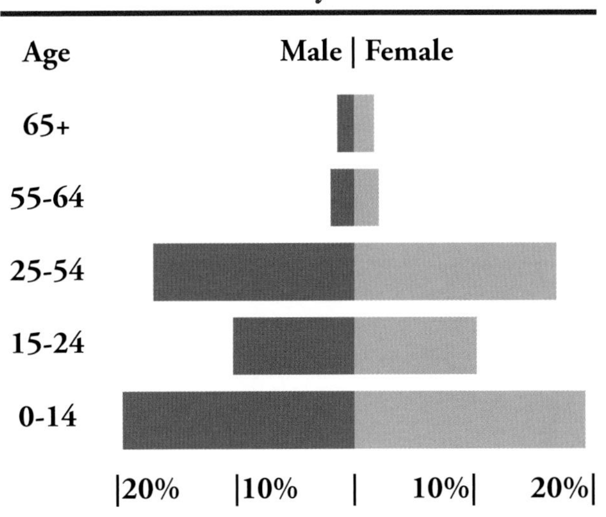

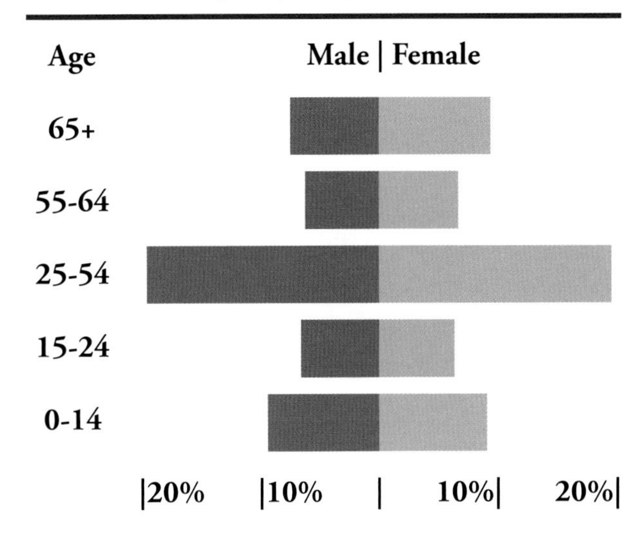

Denmark

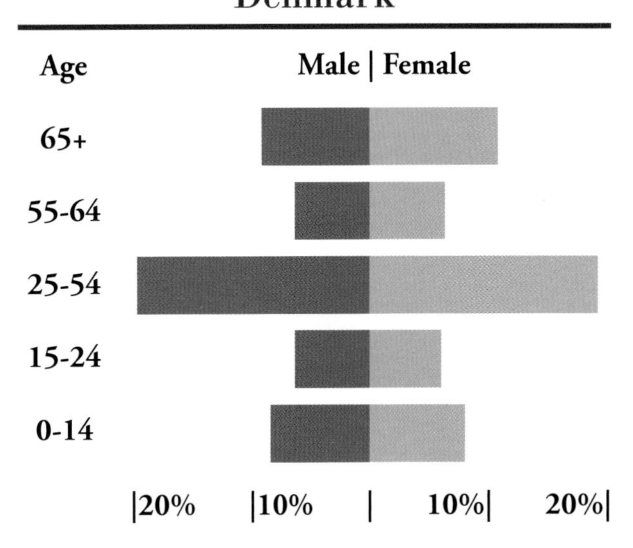

1925-1945	1946-1964	1965-1980	1981-2000	2000-2012
Traditionalists (or the Silent Generation)	Baby Boomers	Generation X (or Baby Busters)	Millennials	Gen Z

with an even smaller workforce and with fewer adult children of the elderly.

These realities do not necessarily mean that the government should step in to provide these services, but it does mean that the people of these countries will have to respond to these situations and trends to provide the services that their people need.

Another way to describe the age groups in a population is by describing cohorts, the people born between certain years. Except for the Baby Boomers, researchers and commentators differ on the precise years for each cohort. One breakdown of the population of the United States is shown in the chart above.

Because of their shared experiences with world events, technology, and other factors, large groups of people in each cohort may have similar priorities and expectations for work, church, family life, money, retirement, and other aspects of life. However, you will find many differences between members of each generation as well.

Life Expectancy

Life expectancy is the average number of years that a person born in a specific year can expect to live. Overall, life expectancy has increased over the last two centuries as health care has improved and as the infant mortality rate has decreased.

The World Health Organization says that the world average life expectancy for a child born in 2017 is 72.2 years. Life expectancy is highest in Monaco, 89.5 years, and lowest in Chad, 53 years.

The U.S. ranked 42nd in life expectancy among countries of the world at 79 years as of 2017. Life expectancy is increasing in the U.S. According to the Social Security Administration (SSA), an American man turning 65 today can expect to live, on average, to age 84.3; for a woman, it's 86.6. The SSA also says that one-fourth of Americans who are 65 today, having survived illnesses, accidents, and wars, will live past age 90, and one out of ten will live to be 95.

Ethnic Groups

An ethnic group is a group of people who share common traits such as race, language, nationality, or culture (including religion). Ethnic groups often form the basis of nations; however, a modern political nation can have several ethnic groups living in it. This can happen for several reasons. People from an ethnic group might migrate to a new country, as happened with Polish Jews and Italian Catholics coming to the United States. Members of a group (or their ancestors) might have been forcibly imported as slaves into a country or driven out of a country as victims of religious persecution. Years ago, those who held power in government—for example, those in the government of a nation that had overseas colonies—might have drawn the borders of other countries in such a way that ignored where people of certain ethnic groups lived. Alternatively, they might have drawn borders intentionally to divide ethnic groups among several nations so as to weaken that group's influence. The Kurds in the Middle East are one example of an ethnic group to which this was done (see Lesson 14).

The population of Norway is 94% Norwegian. China is 91% Han Chinese. Greenland is 88% Inuit. The majority of people who live in the Democratic Republic of the Congo are Bantu, but over two hundred other identified groups live there. Bangladesh recognizes 27 ethnic groups officially, but members of as many as 75 groups live there. Czechia (formally the Czech Republic) is about 64% Czech. The presence of many other groups in Czechia reflects the ethnic diversity and movement of people in Central Europe. Kuwait is 31.3% Kuwaiti, 27.9% other Arab, 37.8% Asian, 1.9% African, and 1.1% other, which includes European, North American, South American, and Australian. These percentages reflect the diverse business interests in Kuwait.

The United States is about 79.96% white or Caucasian, 12.85% black, 4.43% Asian, .97% Native American or Alaska Native, .18% Hawaiian or Pacific Islander, and 1.61% of two or more races. The U.S. Census Bureau does not consider Hispanic to be a race. It defines Hispanic as being "persons of Spanish/Hispanic/Latino origin including those of Mexican, Cuban, Puerto Rican, Dominican Republic, Spanish, and Central or South American origin living in the U.S. who may be of any race or ethnic group (white, black, Asian, etc.)." About 15.1% of the U.S. population is Hispanic.

Kuwait City, Kuwait

These people of Chinese ancestry are observing Chinese New Year at a Taoist temple in Malaysia.

Experts give widely varying numbers as to how many ethnic groups exist in the world, from 13,000 to as many as 24,000. This difference is because different experts define groups in different ways. For instance, are German-speaking Swiss a different ethnic group from French-speaking Swiss? A country in sub-Saharan Africa might have dozens of groups, each with a different language and belief system but many of whom share some traits in common.

Problems arise when ethnic groups come into conflict. The ruling majority ethnic group in a country might persecute or discriminate against minority ethnic groups. When the boundary lines defining political nations do not respect where an ethnic group lives, and the lines result in parts of the group living in two or more countries, the divided group might actively seek to become a separate nation.

In Malaysia, the minority ethnic Chinese (immigrants and descendants of immigrants from China) have overcome prejudice and discrimination toward them by the majority Malays and have made significant accomplishments in education and business.

Religion

Adherents of Christianity make up the largest religious group in the world, accounting for about 2.3 billion people. Christians predominate in areas with traditionally European roots, such as North and South America, Europe, Russia, South Africa, Australia, New Zealand, and the Philippines. The Vatican states that the Roman Catholic Church has about 1.2 billion members worldwide, making it the largest Christian denomination. Orthodoxy in its various forms (Eastern, Russian, etc.) totals about 260 million adherents and is the second largest Christian group. The broad category of Protestant totals about 800 million, but this single designation includes a large number of individual denominations.

Church in Sesfontein, Namibia

Christianity is growing in Africa and Asia. It is declining slightly in the United States and even more rapidly in Europe as older Christians are dying and many in the younger generations are not becoming members of a church.

Muslims are the second largest religious group in the world with 1.8 billion followers. Islam predominates in the Middle East (outside of Israel), North Africa, and several countries in Southeast Asia including Indonesia and Malaysia. Islam is growing more rapidly than Christianity in terms of the birth rate of women in those religions. If current trends continue, within a couple of generations Islam will overtake Christianity in terms of total number of adherents.

Population experts count about 1.1 billion followers of Hinduism. This religion predominates in India and Nepal. There are about one-half billion Buddhists. This is the majority religion in Mongolia and in several countries in Southeast Asia, such as Myanmar, Thailand, and Laos.

A group called Unaffiliated totals about 1.2 billion. Unaffiliated individuals, who claim to follow no religion, are the majority in China and North Korea.

Japan is a religious mixture. Over half of the population describe themselves as atheists. However, majorities also practice Shinto and Buddhism, with many of these observing practices in both religions. Only about 2% of Japanese claim to be Christians.

Language

Human beings speak over 7,000 languages. A few of these are spoken widely, while many are spoken by just a small number in a small area. The numbers in the next paragraph represent those whose native language or primary spoken language is the one indicated.

About 1.3 billion people speak Chinese. Of those, about one billion speak the Mandarin dialect. Spanish is the first language of about 437 million people. This is the result of the wide Spanish

Warning Sign in Singapore

influence in Central and South America centuries ago. English is the first language of about 372 million people, although English is the most widely spoken second language and is the most common language of business around the world. About 295 million speak Arabic as their first language, 260 million speak Hindi, and 242 million speak Bengali. Rounding out the ten most widely spoken languages are Portuguese (219 million), Russian (154 million), Japanese (128 million), and the Pakistani language of Lahnda with 119 million.

Factors of Physical Geography

One key factor in where people live and the lives they lead in those places is physical geography. For instance, the physical geography of an area influences the growth of a city. A city that is located where a large port can be built, whether it is on the seacoast or on a river, is likely to grow. A city depends on the natural resources of its region. Los Angeles, California, and Las Vegas, Nevada, for example, have enjoyed huge growth; but in recent years they have struggled to provide enough water for their residents because neither place is near a good source of abundant fresh water. The presence of fertile land near a city helps insure an ample supply of food for a city; otherwise, the city's population depends on businesses bringing food to them through trade.

The realities of world geography affect many factors in people's lives. A person is more likely to have access to sufficient food and health care in the United States than in North Korea. A resident of Geneva, Switzerland, is more likely to live in a healthy environment than someone who lives in Beijing, China. These factors affect a person's life expectancy and quality of life. Someone who grew up in Saudi Arabia is likely to be Muslim, whereas someone who grew up in Thailand is likely to be Buddhist.

The people of the world owe their existence to the God who created them, in whom they live and move and exist (Acts 17:28). Because of this, all of the diverse peoples of the world owe Him their praise, as the psalmist said:

Let the peoples praise You, O God;
Let all the peoples praise You.
Psalm 67:5

Assignments for Lesson 141

Gazetteer Read "World Population Growth" (pages 339-340). There are no questions in the *Student Review Book.*

Worldview Copy this question in your notebook and write your answer: What are some ways in which you see people devaluing others, and what can you do to show that you value others?

Project Choose your project for this unit and start working on it. Plan to finish it by the end of this unit.

Student Review Answer the questions for Lesson 141.

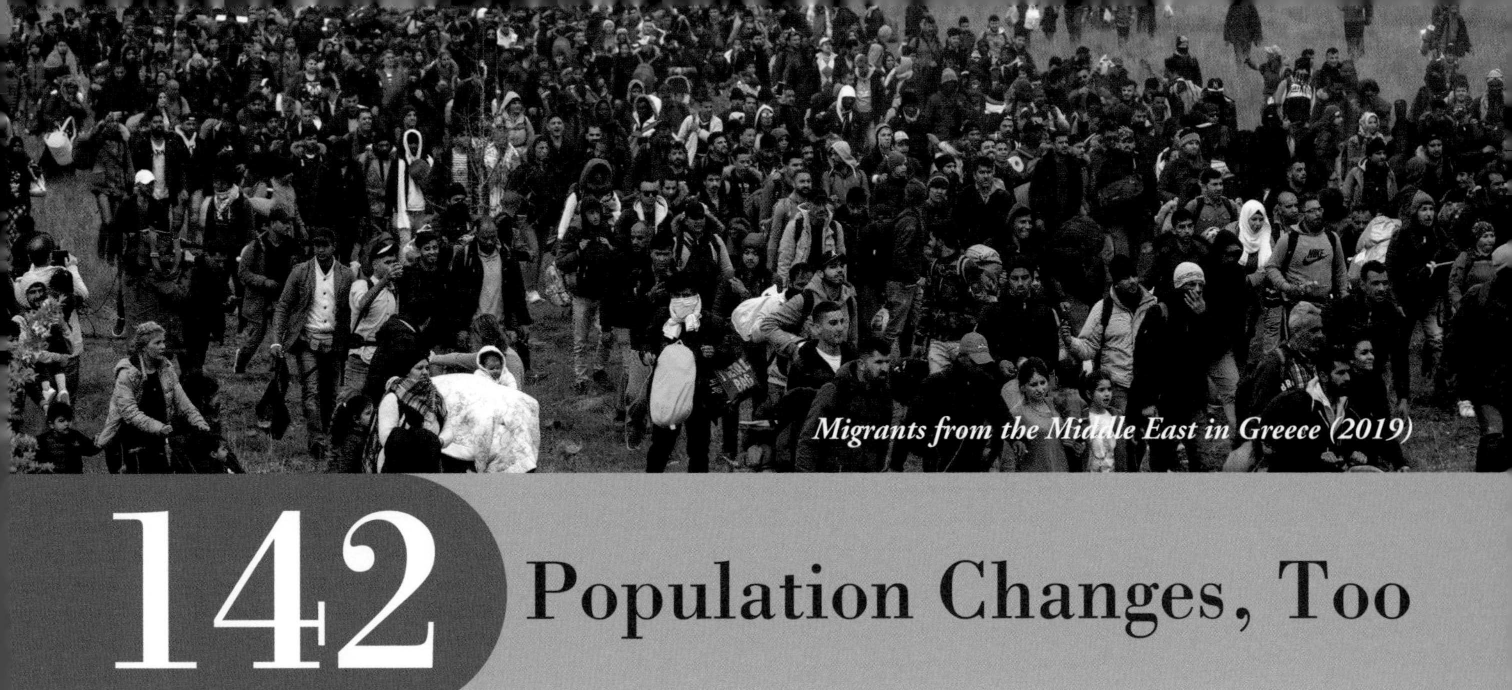
Migrants from the Middle East in Greece (2019)

142 Population Changes, Too

The previous lesson described various characteristics of the world's population. However, remember that human geography is a video, not a snapshot. The world's population statistics are constantly changing. The figures in the previous lesson will not be the same on the day you read this.

Population Growth

The total population of the earth increases as women have children and as those births outnumber the deaths that occur. For centuries, several key factors slowed population growth. One main factor was that most societies experienced a relatively high rate of infant mortality, which is the death of babies who are less than one year old. Another factor was the deaths of women that occurred as a result of the hardships of childbirth. Still another factor was limited healthcare in the face of deadly diseases. Additional factors, such as poor nutrition, warfare, and accidents contributed to slowing the rate of growth of the human population.

Progress in relevant areas of human life have enabled the world population to grow more rapidly. For instance, better health care has meant that fewer children die at a young age and fewer women die in childbirth. Improved health care and nutrition have meant that the impact of disease has decreased and life expectancy has increased, so more people are living longer. Wars are less frequent, and workplaces have become safer overall. Also, as more people live to adulthood, the number of couples having children increases.

The great majority of American soldiers who were involved in World War II came home to their wives or got married, and these families began having children. This resulted in the population growth period we call the Baby Boom.

The chart on the next page shows the estimates that demographers have made for how the world's population has grown over time. We cannot know the figures for sure, but experts generally believe that the population of the earth reached one billion around 1800. The population of the earth reached two billion much more quickly, around 1927. The total was four billion by 1974, six billion around the year 2000, and seven billion in 2011.

Population growth impacts the physical geography where that increase is occurring, such as in land use practices and the availability of water. The importation of food and housing supplies for growing populations increases the production of those items in other parts of the world. Increased oil

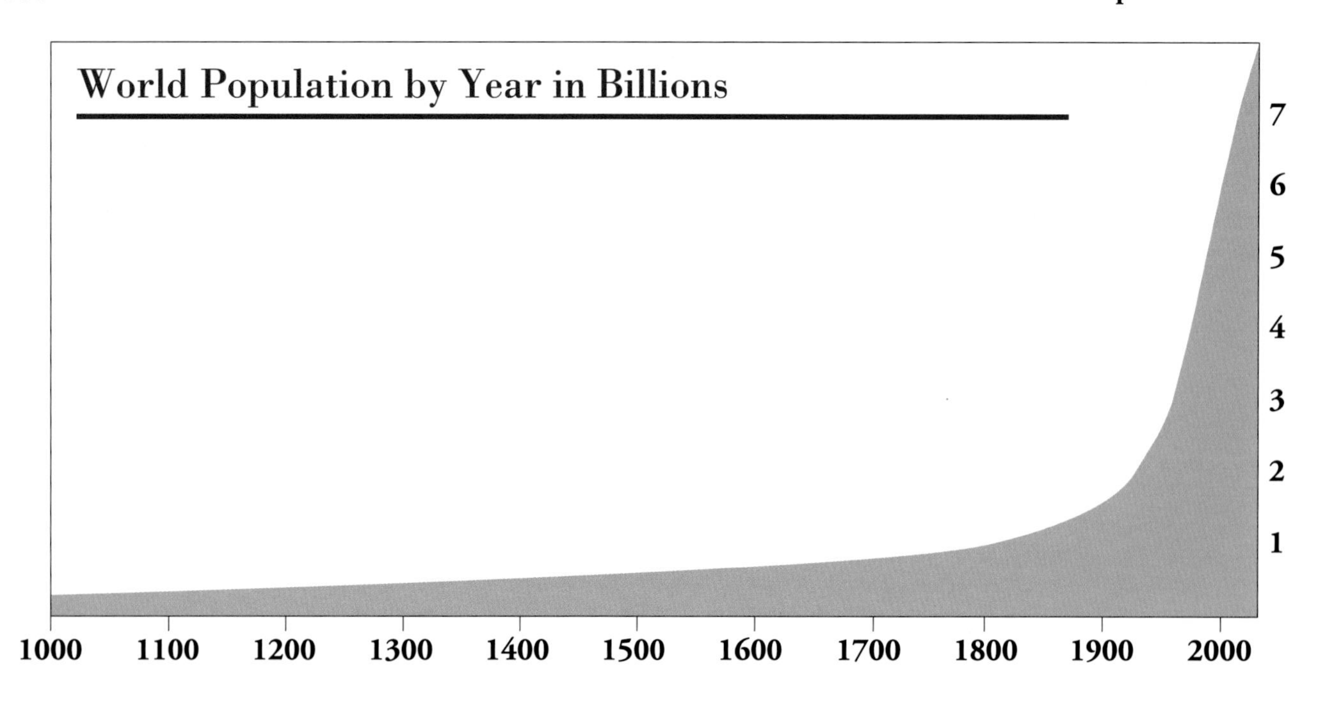

and coal production to meet rising demand affects the geography of the regions where those mining activities take place, such as the oil fields of North Dakota and the coal mines of China.

Population growth tends to take place in cities faster than in rural areas. This leads to changes in urban land use policies. For instance, population growth in cities in the midwestern United States has meant that owners of fertile and once productive farmland are selling that land to developers to build houses and businesses for the increasing number of people who live in those cities. This dynamic is repeated in many cities around the world. At the same time, agricultural technology has increased the productivity of farmland, with the result that farmland can feed a growing population more efficiently. The tradeoff is that a dependence on farm chemicals might be threatening the health of the growing population as well as causing environmental damage. Population growth can be complicated!

Periods of Population Decline

Although the general world trend has been one of population growth, certain periods have seen dramatic declines in population, at least in some areas of the world. For instance, the Black Plague that struck Europe in the 1300s is thought to have taken the lives of perhaps 25 million people, or about one-third of the population of Europe. Some have estimated that England's population in 1400 was half of what it was in 1300.

The population of indigenous Americans declined significantly after the arrival of European explorers and settlers for two main reasons. First, the Europeans brought diseases against which the local populations had no immunity. Second, when armed conflict occurred the Europeans used more powerful weapons to defeat indigenous warriors.

The population of Ireland decreased by about 20-25% in the mid-1800s because of severe potato famines. About one million people died and another million people left Ireland during that time, most of whom came to the United States.

The deaths that resulted from World War I and the world influenza epidemic that followed the war claimed the lives of some 29 million people. World War II resulted in the deaths of an estimated 50 million people from warfare, disease, lack of food, and genocide such as that practiced in Nazi Germany and the Communist Soviet Union. After the war, even though the United States experienced a postwar baby boom, in Germany and the Soviet Union the deaths of so many young men in the war resulted in fewer births taking place than would otherwise have probably occurred.

Growth Rate and Birth Rate

The annual growth rate of a country is the percentage that its population grows over the period of one year. This growth rate is made up of two main statistics: the net natural increase and the net immigration. The natural increase is the difference between the number of live births and the number of deaths that take place in a country. Net immigration rate is the number of immigrants into the country minus the number of people who emigrate out of the country. The total of these two figures, divided by the population at the beginning of the year, gives the annual growth rate.

For instance, assume that a country had a population of 1,000 on January 1. If that country has 100 live births and 90 deaths during the year, its natural increase rate is 1% (10 divided by 1,000). If this same country receives 20 immigrants and no one emigrates that year, its growth rate is 3% (total net increase of 30 people, divided by 1,000).

The estimated growth rate for the United States in 2020 was 0.72%. See the chart below for estimated growth rates in other selected countries. A negative growth rate means that the total population was lower at the end of the year than at the beginning.

The birth rate (also called the crude birth rate) is the number of live births per 1,000 people. The world average is 18. In 2020 in the United States, the birth rate was 12.4. It was 36 in countries such as Afghanistan and Cameroon and 7 in Japan and Andorra.

The percentages in the charts appear quite small, but even small numbers can result in big changes. If a country maintains a growth rate of 3%, the country's population will double in about 23 years. This happens because that same rate of 3% multiplies a larger beginning population each year, which results in larger numerical growth each year.

Another statistic that demographers consider is the fertility rate. This is the average number of children that a woman bears during her life. This number for a given country might be six or even ten, but a high infant mortality rate or a high maternal mortality rate (women dying in childbirth), lessens the impact of a high fertility rate on population growth. The rate of replacement for a human population is a fertility rate of 2.1 births per woman.

Growth rate, birth rate, and natural increase rate will usually be fairly similar; but it is important to be aware of the differences. For instance, people might advocate certain government policies based on the statistic that seems most to support their proposal, such as a low birth rate, when in fact the policy might not address the real situation, which might involve a high rate of immigration.

Given the trend of many countries having a lower birth rate and the trend toward longer life expectancy, many experts believe that one-fourth of the world's population will be 65 or older by 2100.

National Population Growth Rates

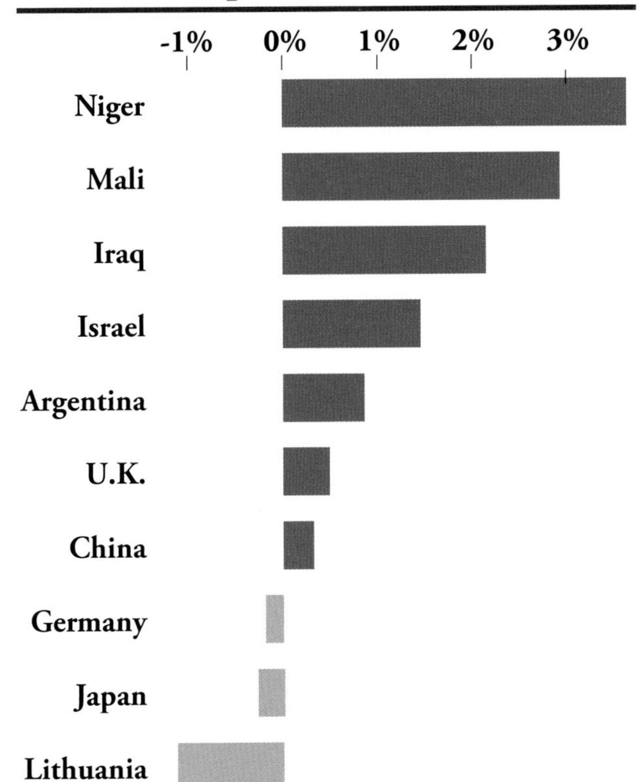

Migration

Population does not just increase or decrease. Population also shifts as a result of migration. Many factors contribute to migration. Demographers refer to these as push and pull factors. Push factors push people out of the country where they are living. These include war, famine, a lack of work opportunities, and religious and political persecution. Pull factors pull or attract people to move to another country. Pull factors include greater work opportunities, lifestyle improvements, and political and religious freedom. Chain migration describes people migrating to the same place where others from their country have moved. Emigration refers to out-migration, or people leaving a country and going to other countries; immigration refers to in-migration, or people coming into a country from other countries.

History tells the story of many people movements around the world, and these movements have had an impact on geography. For instance, the immigration of tens of thousands of English and Scots-Irish settlers led to changes in land use and population patterns in North America. Later, the policy of the United States government of forcing native nations to live on reservations, which were not established on the most productive land, had geographic impact.

We usually think of people choosing to move, as when millions of Europeans came to the United States, so we call this voluntary migration. Involuntary or forced migration happens when large numbers of people have no choice in the matter and move against their will. The transporting of Africans to other countries as slaves was forced migration.

Migration by itself merely shifts population. However, one result of migration is often that immigrants find more food, better healthcare, and a more abundant way of life, all of which can encourage and enable immigrants to have more children. For instance, Irish immigrants in the United States in the 1800s and early 1900s had more children to survive to adulthood than had been the case in Ireland before they left. On the other hand, migration can also lead to loss of life, such as occurred on the unhealthy slave ships traveling between Africa and the Western Hemisphere and on the Trail of Tears in the United States.

Another form of migration is internal migration, when people leave one area of a country for another part. In the decades after World War I, millions of African Americans left poverty and racial discrimination in the southern United States and moved to cities in the North. In more recent years, many people have left the northern states (called the Rust Belt) where industries were closing and have moved to the South and the Southwest (called the Sunbelt) for warmer temperatures and a lower cost of living.

People can also be temporary migrants, as when they move to another region or country to harvest farm products. When the harvest is completed, they might move to another region with a later harvest or return to their homes. Political unrest, as well as occasional natural disasters, have caused many people to leave their homelands and become refugees. These refugees create sudden demands on the people, land, and resources of the receiving countries.

During the 1930s, 40s, and 50s, Joseph Stalin ordered the forced migration of various groups of people within Soviet Union. This photo shows Romanian refugees displaced by the Soviets during World War II.

The Dadaab Refugee Complex in Kenya has hosted hundreds of thousands of refugees and asylum seekers since 1991. Most of them have been fleeing violence or famine in neighboring Somalia.

Population Growth in Less-Developed Countries

The population in less-developed nations (those with limited industrial development and economic opportunity) is growing faster than in developed countries. Better health care and lower infant mortality rates help a population to grow faster, and more-developed countries tend to have better health care and lower infant mortality. However, those factors are not the whole story.

Growth factors such as better health care and lower infant mortality are improving more rapidly in less-developed countries than they are in more-developed countries. In the developed countries, improvements in these factors are still taking place but are not as dramatic. Thus, the rate of change is greater in less-developed countries.

Worldview differences are probably at work as well in this comparison of less-developed and more-developed countries. The people in less-developed nations often value children as blessings from God, as potential workers who can contribute to the family's financial needs, and as care providers for aging parents. Developed nations are more likely to have numerous government programs that provide assistance from cradle to grave. In addition, many adults in more-developed nations value their autonomy more than they value having children. As a result, the need and desire for having children in more-developed countries will generally be less. In addition, couples tend to get married younger in less-developed nations, while people in more-developed countries tend to be older when they get married.

Is There a Limit to the Population the Earth Can Sustain?

In Genesis 1:28, God said to Adam and Eve, "Be fruitful and multiply, and fill the earth, and subdue it." Adam and Eve and their descendants have certainly multiplied, but is there a limit to the number of people the earth can sustain? The demographic term for this is the earth's carrying capacity. Many people are concerned that there is a limit.

In 1798, the English economist and demographer Thomas Malthus published "An Essay on the Principle of Population." In this work Malthus expressed concern that population growth would outrun the world's food supply, and as a result large numbers of people would starve or die from illness. He encouraged government programs to limit population growth.

Malthus was a pessimist and anticipated the worst. He failed to see significant truths. He could not foresee the development of agricultural technology that would enable a great increase in food production. He did not anticipate the widespread use of contraceptives. He also did not recognize, as many still fail to see today, that most famines are the result not of a lack of food but of government inefficiency and the unequal distribution of the food that is available. This is why so many people in the world die of starvation while the United States and other countries produce a surplus of food products.

Another pessimist who has had a wide influence is Paul Ehrlich, who published *The Population Bomb* in 1968. Ehrlich predicted that 100-200 million people worldwide would die every year from starvation during the 1970s and that 65 million Americans would die from a lack of food during the 1980s. Ehrlich urged much greater control of population growth.

Malthus and Ehrlich were not only wrong in their predictions; they were, as one writer put it, spectacularly wrong. Yes, some cities and some regions are very crowded. Yes, many people in the world do not have enough to eat. Yes, an increasing population brings the need for more jobs, more transportation and housing, more food, more services such as fire protection and sewer and water service, and a greater risk of pollution. But the earth's food production has continued to increase, health care and health care delivery have improved, and the standard of living for millions of people has risen. Politics get in the way. The developed nations of the West that practice greater birth control are not willing or able to get more food to those who are suffering under the governments of places such as North Korea and Somalia.

Long-term population predictions are notoriously unreliable, so policies based on those assumptions will likely be ill-directed or unnecessary.

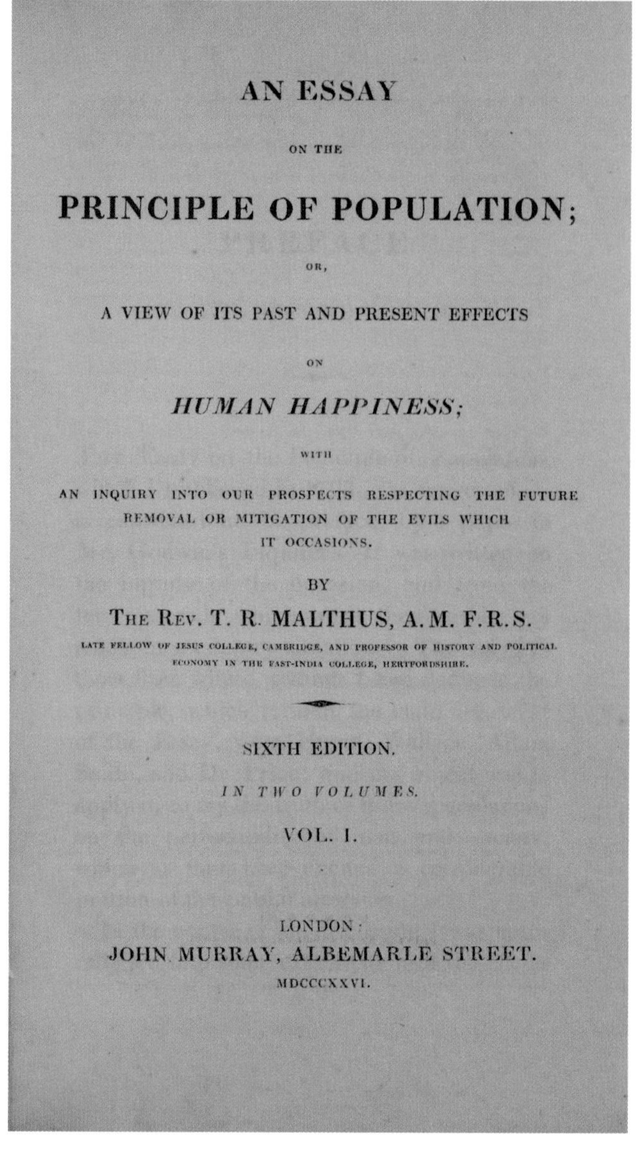

AN ESSAY

ON THE

PRINCIPLE OF POPULATION;

OR,

A VIEW OF ITS PAST AND PRESENT EFFECTS

ON

HUMAN HAPPINESS;

WITH

AN INQUIRY INTO OUR PROSPECTS RESPECTING THE FUTURE REMOVAL OR MITIGATION OF THE EVILS WHICH IT OCCASIONS.

BY

THE REV. T. R. MALTHUS, A. M. F. R. S.

LATE FELLOW OF JESUS COLLEGE, CAMBRIDGE, AND PROFESSOR OF HISTORY AND POLITICAL ECONOMY IN THE EAST-INDIA COLLEGE, HERTFORDSHIRE.

SIXTH EDITION.

IN TWO VOLUMES.

VOL. I.

LONDON:

JOHN MURRAY, ALBEMARLE STREET.

MDCCCXXVI.

The lengthy subtitle of the essay on population by Thomas Malthus is "A View of Its Past and Present Effects on Human Happiness; with an Inquiry Into Our Prospects Respecting the Future Removal or Mitigation of the Evils Which It Occasions."

We don't know what technology people will develop in the future. Some demographers have predicted that the world population will level off at about 9 billion or 12 billion, but those predictions are educated guesses. The Bible doesn't say anything about a limit to "filling the earth." Our efforts will be better directed toward solving the problems that we can see and not trying to solve assumed problems that we cannot see.

Each person, regardless of where they were born or where they live now, is created in God's image and is valuable in God's sight.

God created man in His own image, in the image of God He created him; male and female He created them. God blessed them; and God said to them, "Be fruitful and multiply, and fill the earth, and subdue it; and rule over the fish of the sea and over the birds of the sky and over every living thing that moves on the earth."
Genesis 1:27-28

Assignments for Lesson 142

Worldview	Copy this question in your notebook and write your answer: What does the parable of the Good Samaritan (Luke 10:25-37) say about the value of people?
Project	Continue working on your project.
Student Review	Answer the questions for Lesson 142.

Aerial View of Moscow, Russia, at Night

143 City Mouse, Country Mouse

Study the painting, *Paris Street; Rainy Day* by Gustave Caillebotte on the opposite page. The artist executed this work in 1877. The large painting, 7'10" wide and 6'1" tall, takes up an entire wall in the Art Institute of Chicago.

The painting shows people walking in various directions on a street in Paris at that time. The street is clean and the buildings are large and modern. The buildings contain apartments, offices, and shops. The couple in the center is well-to-do; notice their clothes and the woman's earring flashing in the light. In fact, most of the people we see are financially secure.

Caillebotte painted the picture a few years after the difficult Franco-Prussian War and the devastating Paris Commune. Before those events, Paris had begun a modernization project. At the time of the painting, Paris was enjoying an economic revival. The painting gives clues to this economic success. Notice the construction scaffolding in the very center of the painting. Caillebotte was from a wealthy family who owned property in this section of Paris. This intersection still exists.

Of course, not all the people of Paris were doing well economically. Notice the working woman behind the left shoulder of the central female figure and the worker carrying a ladder behind the man's left shoulder. Other Paris streets on that rainy day would have been experiencing poverty and unsanitary conditions. Caillebotte's painting showed one side of Paris and perhaps created an idealized impression of life in Paris at the time. In fact, the artist exaggerated the width of the boulevard, a decision which placed the people in the painting even further apart from each other than they would be in reality.

Think about the people in the scene. They are all in motion, headed somewhere. In another moment, the scene will be different because the figures will have moved on. More than likely, most of the people in the painting do not know each other. They are pretty much alone. (Why do you suppose the couple in the foreground are both looking to the side? Are they checking to see if they are being noticed?) All of these people just happen to be walking, riding, or working in the same street at the same time.

Now consider the rural scene in the Netherlands on the next page opening. The buildings are fewer and simpler—homes, barns, and churches. The roads are dirt. One group of people is enjoying a break from their harvesting work.

More than likely, the people in this setting know each other; many are probably related to each other. They have probably lived in the village all of their

Paris Street; Rainy Day *by Gustave Caillebotte (French, 1877)*

lives. Their families have probably lived in the village for generations.

The people in the village probably do not have as much money in the bank as people in the Paris painting do; in fact, the villagers might not have any money in a bank. But their shared experiences working together give them a powerful bond through relationships that are more valuable than money.

What thoughts and impressions do you have about these paintings?

These artistic works illustrate different ways in which people have interacted with their geographic settings. They are studies in urban geography and rural geography.

Urban Geography

People began building cities early in human history. Cain built a city and called it Enoch, after his son (Genesis 4:17). Part of the building project at Babel was the building of a city (Genesis 11:4). The war described in Genesis 14 involved the rulers of city-states (Genesis 14:1-2). Nevertheless, throughout most of history, the majority of people have lived in rural areas and small towns, farming their land and living in communities with only a few families.

As trade grew, cities became the main places where trade took place because that is where businesses found the most customers. A trader was more likely to stop in a city with thousands of people than in a rural area with only a few scattered homes. People who lived in rural areas and small towns came to the cities to buy and sell in the occasional markets. As trade continued to increase, these occasional markets gave way to established stores. As modern industries developed, they relied on a large workforce, so workers in those factories tended to live in ever-larger cities where the factories were located.

Cities became the hub of culture and communication for their surrounding areas. New ideas and new ways of doing things spread from the cities into the small towns and settlements nearby. This pattern is why Paul tended to go to cities on his missionary journeys. He went where the people were, and he equipped converts to take the gospel to rural areas. Despite this trend, however, rural populations were more likely to develop and maintain elements of folk culture such as songs, dances, and stories.

During the twentieth century, the world population changed from predominantly rural to predominantly urban. With the spread of mass media, urban ways came to have a large influence, even on rural populations. With modern digital technology, the rate of change has increased significantly.

Today, over half of the world's population lives in urban areas. In some countries, the percentage is even greater. About 80% of the population of the United States lives in urban areas, although the Census Bureau defines any community over 2,500 as urban. In Japan, the urban population is 93% of the total. In Belgium, the figure is 98%.

Cities offer many people much of what they are looking for. According to the World Bank, cities create 80% of gross domestic production around the world. In other words, the vast majority of economic activity takes place in cities. This means that more possibilities for employment and advancement exist in cities. More retail competition means more choices for consumers, often at lower prices. Countries that still have a large majority rural population, such as Malawi and Cambodia, do not have the economic activity that can help large numbers of their people move beyond subsistence living.

Cities can be creative places, where businesses and artisans feed off of each other's ideas and creativity. Cities provide greater access to theaters, museums, and libraries. People who live in cities usually have a diversity of ethnic backgrounds, which others can learn from and relate to.

The growth of urban population has significant consequences. Growing cities require more land, and this has consequences for physical geography. Some areas near cities, such as wetlands or mountains, cannot sustain population growth. Cities therefore tend to grow onto land that was once used for farming. Cities do not only grow out in what is called urban sprawl. Cities also grow up with the construction of high-rise apartments. The increased concentration of people in cities puts greater demands on resources such as water and on services such as police and fire protection.

For instance, the city of Dallas, Texas, built a 12,500-foot-long earthfill dam on the Trinity River east of the city to create a reservoir to supply water for the growing metropolitan area. The city began the project in 1964, and work continued for three years. The water level continued to rise to its desired level for two years after that. The lake drains an area of just over 1,000 square miles. Lake Ray Hubbard transformed the geography and the economy of the area. It provides water to residents as well as boating and other recreational opportunities. However, some cotton farmers lost their land to the project.

Cities also have downsides. More people means more cars, which means more traffic and more pollution. Urban areas generally have a higher crime rate than rural areas. Cities have higher property values, but that also means higher taxes. One irony of city life is that, despite the greater number of people, individuals in a city often feel more isolated and lonely than those living in small towns.

Rural Geography

Rural areas have their own opportunities and challenges. They often provide a beautiful setting in which to live. The cost of living is often lower. It is possible to connect more easily with people who live at a slower pace, although newcomers sometimes find small town society difficult to break into. Small towns usually offer some retail choices, and the prices can be surprising. Our small-town hardware store, for instance, often has better prices than the big box alternative thirty or forty minutes away; and it definitely has better customer service. You might have to travel to take advantage of cultural opportunities, but many small-town residents only seek out such opportunities occasionally anyway.

Rural geography is usually associated with farming. What farmers grow or what livestock they raise depends on the geography of a given area: the soil, terrain, water, weather for the growing season, and the availability of needed transportation. Certain farm products can become associated with particular geographic locations: olives in Italy, cheese in France, dates and figs in the Middle East, fruits and vegetables in southern California.

A phrase that Americans often use is "small-town values." These usually include qualities such as hard work, loyalty to family and friends, honesty and reliability, religious faith, people who know each other and give each other a helping hand when needed, and strong traditions. All of this contributes to what we call quality of life.

The Harvesters *by Pieter Bruegel the Elder (Dutch, 1565)*

Rural settings have their negatives also. A major one is the inconvenience of having to travel relatively far to do much shopping, have access to health care, and enjoy other amenities that cities offer in greater abundance. Rural areas often have significant farming activity, which can involve the use of potentially harmful pesticides and other chemicals in the air, land, and water. With the trend toward urbanization, rural areas have relatively fewer job opportunities. However, an entrepreneur can operate a successful business in a rural setting. For instance, a business that is built on Internet sales just needs a reliable source for an Internet connection and the means for shipping goods to customers.

People who live in extremely isolated rural settings often do not have access to technological development and cultural interchange. Think about the different opportunities available for people who lived in Philadelphia and people who lived in Appalachia in the 1700s, people who lived in London and those who lived in the Scottish highlands during the 1800s, and people who lived in Constantinople and those who lived in the rugged mountain ranges of the Balkan Peninsula in the early 1900s.

Small-town settings such as Spring Hill, Tennessee; Georgetown, Kentucky; and Tupelo, Mississippi, have been transformed when carmakers built automotive plants there. This kind of major change brings the benefits and difficulties of city life to a rural location.

Dependent on Each Other

Rural and urban areas depend on each other. The millions of people who live in apartments and urban neighborhoods depend on rural farmers for their food. At the same time, rural residents receive benefits from innovations in agricultural equipment and communication technology, most of which happen in urban settings.

However, rural and urban areas can also have conflicts. In any given county, state, or national setting, urban residents and rural residents can differ over their ideas of what government priorities should be and how tax revenues should be spent. City dwellers might see a need for better streets and the incorporation of nearby areas into the city limits. Since cities are where more people live, urbanites

Mobile Grocery Store, Bangkok, Thailand

might demand more and better schools and other government services. Rural residents might fear being overlooked as their schools and roads decay and as representatives in government might not notice the need for modernized services even in rural areas.

Needs in both situations are real, and the different constituencies can make their voices heard. Imagine the conflicting priorities of Parisians and those who live in rural France, or residents of the metropolitan area of Chicago and the people who live in the other 80% of Illinois, or those who live in a growing city and farm residents who feel threatened by the growth.

Transportation as Connection

Developments in transportation make both rural and urban living possible for people in modern times. Highways and mass transit systems enable workers to have jobs in cities and live in the suburbs or rural areas. Businesses that were once tied to particular cities can relocate to less expensive places and employ the same people, who simply change their commuting patterns. The downside is that modern transportation has helped to create urban sprawl, which means housing, shops, and manufacturers cover more land area as opposed to being concentrated in the cities.

Modern transportation has made successful city living possible since trucks and trains bring a variety of foods from farming areas to urban markets. Americans just about anywhere can enjoy Florida citrus, Arkansas rice, Vermont maple syrup, and Washington apples thanks to modern transportation. Sea container transport can take prepared foods from many countries to many other places around the world. These goods eventually make their way even into small-town markets. Developments in transportation have not done away with the importance of geography, but they have certainly changed the equation.

Planning

One discipline that considers rural and urban issues together in the context of ever-occurring change is urban planning. Planners consider the land, its characteristics, and its possible uses along with developing trends in population and economics to attempt to forecast the best future activity that will benefit the most people. For instance:

- Will the land proposed for a new residential subdivision perk, or accept septic tanks?
- Does a given tract have wetlands?
- Will a hilly area make development difficult or expensive?
- How easily will a new development connect with existing development?
- Will the land support a large new factory structure?
- Might oil or natural gas be discovered on a tract?

Those in government must also balance the different and often conflicting priorities of concerned groups. A state or local business development organization will likely not see a land-use issue the same way that a conservationist organization does. Is it possible to satisfy both, or will saying yes to one automatically make an enemy out of the other?

Which Do You Choose?

Would you rather live in the country and make occasional forays into the city? Or would you rather live in the city and occasionally go out into the country? Your priorities for living will help determine your answer.

If you want to be in farm implement sales and service, you are likely going to live and work in a rural or small-town setting. If your heart is set on being a civil engineer who helps build major freeways, that work will likely be in urban settings. If an elderly parent needs a higher level of care, someone—the parent or the child—will likely have to move.

The realities of human geography in rural and urban settings help to define our lives, show us what we need to overcome, and open up possibilities that we can accomplish. Wherever we live, we must let God's will guide our lives.

God directed Paul to fulfill an important ministry in Corinth, which was a large city in the Roman world:

And the Lord said to Paul in the night by a vision,
"Do not be afraid any longer, but go on speaking and do not be silent;
for I am with you, and no man will attack you in order to harm you,
for I have many people in this city."
Acts 18:9-10

Assignments for Lesson 143

Worldview Copy this question in your notebook and write your answer: What great good and what great evil are people capable of?

Project Continue working on your project.

Student Review Answer the questions for Lesson 143.

Sign in Colorado

144 Who Owns the Land?

Who owns Mount Everest?

Who owns the Grand Canyon?

Does anyone own the Atlantic Ocean or any part of it?

Who owns the land on which you live?

Perhaps just as important, who gave the owners of any geographic place the right to own it?

Ownership Is Power

The ownership of land is power: power to control what happens on that land and who lives on it; power to create economic goods such as crops, lumber, minerals, and manufactured goods; power to create jobs; power to exert political influence.

Geography is not just the description of various physical features of the earth. People such as homesteaders on the Great Plains have gone to great trouble to own a little piece of geography. People have killed and been killed in wars over the right to claim portions of geography. For instance, a claim that Nazi Germany used to justify aggression against its neighbors was that it needed *lebensraum* (living space) for its people.

Who owns the land? Who should own the land? What is the authority for owning land? What is the best system of land ownership?

Provided by God

"The earth is the Lord's, and all it contains" (Psalm 24:1). Christians understand that the Lord owns the whole earth. Further, in Acts 17:26, Paul says that the Lord "made from one man every nation of mankind to live on all the face of the earth, having determined their appointed times and the boundaries of their habitation"—in other words, God decides when and where we all should live. If God owns it all, how should we view private ownership of land or government claims of authority over the land?

God gave Adam and Eve a garden in which to dwell. They obtained food from the garden, but the Bible nowhere indicates that they thought of themselves as owning the land. After Adam and Eve sinned, God drove them out of the garden; and they took up residence on other land. In turn, Adam and Eve had descendants who took possession of land. Cain, for instance, built a city (Genesis 4:17). What might have been his understanding of who owned the land on which the city stood?

After God scattered people from Babel, they spread out over the earth and took possession of the lands on which they lived. Early on, various locations came to have names, such as the land of Nod and the land of Shinar (Genesis 4:16, 10:10). The sons of Ham and Shem were identified "according to their languages, by their lands, by their nations" (Genesis 10:20, 31). Apparently, tracts of real estate came to be identified with certain people groups. By the time of Genesis 14, identifiable nations lived on specific lands and kings ruled over these lands.

The archaeological record indicates that from earliest times, people believed that God or a god provided the land, that the king as God's representative exercised ultimate human authority over all the land, and that the king parceled it out as he saw fit. People recognized common hunting and grazing grounds. However, records from Egypt and other Middle Eastern nations indicate that people there also recognized private, deeded land ownership. For instance, Abraham had to go through an elaborate procedure to purchase the cave of Machpelah from its owner in order to have a place in which he could bury Sarah. The terms of that sale were very detailed (Genesis 23).

Royal Boundary Marker in Rambouillet, France

God's Promise of Canaan

If the earth is the Lord's, and if He decides when and where people live, then it follows that He can use the land as He wishes. In Genesis 12, the Lord commanded Abram to leave his country for another land that He promised to give to Abram and his descendants. God identified this piece of geography as the land of Canaan (Genesis 13:14-15), land on which other people already lived.

A key part of the covenant that God established with the nation of Israel at Mt. Sinai was His fulfillment of the promise to give Israel a geographic place to live. God did this by enabling Israel to conquer Canaan. The Israelites divided this land among their tribes, and the book of Joshua gives detailed descriptions of the tribal lands and their boundaries. The leaders in each tribe then parceled out land to each family.

The Israelites could buy, sell, or give land; but in the Jubilee year (every fifty years) the land was to revert to the families who owned it originally (Leviticus 25:10-13). The amount of time remaining until the Jubilee was to influence how much a seller could charge for a parcel of land because of the number of crops he could produce on the land (Leviticus 25:15-16). The message that the Lord apparently wanted to send to the Israelites was that they could use the land to provide for their families but that the land was not ultimately theirs. Israelites could build wealth, but not on the basis of the land they possessed. The land belonged to God and only through Him did it belong to the tribes of Israel (Leviticus 25:23).

Kings Owned the Land

Since many societies believed that the king owned the land, the king could parcel out his lands to the nobility, in exchange for which the nobles paid taxes or rendered service to the king. The nobles then collected taxes or labor from those who lived and worked on their lands in exchange for providing them a place to live. The noble and the workers each received a portion of the crop the workers raised. Notice that those on the bottom rung of this pyramid had no one from whom they could collect anything. This system of land ownership characterized medieval Europe.

Kings were often not satisfied with what they already ruled. Kings understood the principle that the ownership of land was power. A king often sent out a conquering army to take the lands of another king. Possessing more land meant that the king was more wealthy and powerful. He had more geography to defend, but he ruled more people whom he could enlist to defend that land. In the Old Testament, the kings of Israel and Judah were often involved in aggression toward other countries and in defending their own land against aggressors. This has been the story of history in every part of the world.

Dutch Colonists in Cimahi, Java (now part of Indonesia), 1902

The Colonial Experience

We have noted that for centuries people—usually nobility or another select few—have owned land. This was true for example in Spain and England when those countries sponsored explorers to come to the New World. The explorers claimed new lands in the name of their rulers, even though people were already living there. Remember: land ownership is power. To possess colonies meant the possibility of more sources of wealth, more areas under the king's control, and more places for his people to live. The kings of colonizing countries held authority over the new lands, but how Spain and England handled the ownership of colonial land in the New World differed. In Spanish colonies, the king granted land to a select few, who ruled large estates. Colonists and indigenous people lived and worked on these estates, but did not own property themselves.

In England's American colonies by contrast, individuals could own small tracts of land. This was one of the attractions of coming to America. People who were not nobles and who knew they would never have a chance to own property in England could own their own land in America. Some owned land when they came, while others worked for a few years as indentured servants in order to receive land when their period of service was finished. Then a person could say, "This piece of land is mine."

Countries continued to establish and govern foreign colonies into the early twentieth century. Again, when those countries did so, people were already living in the places they colonized. Colonizing governments took control, sometimes after doing battle with the indigenous people.

Colonizing arrangements sometimes changed. Germany had to give up its colonies in Africa after losing World War I. The Ottoman Empire had to relinquish the lands that it held at the same time and for the same reason. The League of Nations established mandates by which certain countries oversaw these former colonies (for example, Great Britain oversaw Palestine and France oversaw Syria). The British Empire at one time controlled about one-fourth of the land area of the globe, but the pattern of the twentieth century was for colonizing nations to grant independence to places that had once been colonies. Today only a few countries still hold tiny areas as territories, such as the United States with Puerto Rico, Great Britain with the Falkland Islands, and France with French Guyana.

The American Experience

When the United States became a sovereign nation, the government claimed the land between Canada and the Gulf of Mexico and between the Atlantic Ocean and the Mississippi River, except what private parties owned. The United States subsequently purchased the Louisiana Territory from France, the area that became the southwestern United States from Mexico, Oregon from Great Britain, Alaska from Russia, and so forth. What gave France and the other countries the right to possess land and the right to sell it? The right existed in the stated claims of ownership that were not successfully challenged; in other words, the recognized law of nations. The Northwest Ordinance and the Homestead Act gave individuals the right to purchase property from the government, which claimed the right to sell it. Between 1863 and 1920, over 2.7 million private parties filed claims with the U.S. government for almost 438 million acres of land, an area greater than the size of Alaska.

The western plains saw conflicts over land for a period of time. Ranchers understood that the grasslands were common property on which they could feed their herds and drive them to market. When farmers moved in, they fenced off their individual farms, which changed the dynamic for cattle owners and made their lives more difficult. Armed conflict erupted in a few places. Eventually the right of private ownership won out, and the great cattle drives over open land became a thing of the past.

Native nations in America generally took a different view of land ownership from that held by Europeans. In the view of indigenous people, the land belonged to all people for hunting and living. From their perspective, the Great Spirit provided the land. How could you own something that you didn't make? For Europeans to come in and clear off trees, build permanent structures, and claim land as exclusively their own was an affront, a wrong. The native nations claimed no greater right to the land than the Europeans, but they admitted to no less.

County government offices across the United States have deed books that document property transfers as a public record of ownership.

On the other hand, a few native nations in America did recognize individual ownership of land.

Some colonists, such as Roger Williams and William Penn, arranged to buy land from the native nations living there, whom they saw as having prior rights to the land. Treaties that the United States and individual states negotiated with native nations extinguished or ended indigenous claims to land the Americans wanted in exchange for giving the native people land elsewhere, land that we now call reservations. Unfortunately, the United States often violated these treaties when the federal government later took lands that it had said would belong to the native nations perpetually.

Public Ownership vs. Private Property

The institution of private property (including but not limited to land) is one of the key aspects of free market capitalism. People understand that land ownership means power. If I own a piece of land (or a piece of machinery), I can do with it as I wish and reap the benefits I can derive from it. I have an incentive to take care of it and use it wisely. If I don't, it won't take care of me.

The right of individuals to own property created the possibility of individuals obtaining great wealth by hard work, inheritance, negotiation, or some other legitimate means. Royalty and nobility, of course, already had the power to become wealthy; but private ownership by everyday people opened up significant new possibilities for a larger part of the population. The institution of private property enabled property owners to employ workers, who could become property owners themselves. In the feudal system, people didn't change economic class very often. With the right to own private property, such change was possible—both up the economic scale and down the economic scale.

Socialists and Communists also understand that property ownership means power. This is why they want the state to own all property and why Communist revolutions always involved the seizure of private property. The complaint of socialists is that private property allows property owners to benefit from the labor of others while denying benefits to those people who do the labor. This does happen; think of slavery. Think of tenant farmers and sharecroppers. Think also of workers who earned very low wages for years in factories that wealthy industrialists owned. These outcomes happened because people are sinners, and some people have misused the system of private property. Used well, however, the system of private property distributes wealth more widely than the socialist system does.

Karl Marx advocated the abolition of private property, with all property owned by the state. In his theory, this would create a classless society, without landowners and laborers. The state would then wither away.

The claim of socialism is that, since all the people own the land, the land will therefore be distributed fairly. However, as much as socialists see

Government officials in the Soviet Union forced individual farmers to join collective farms in the early 1930s. Mismanagement led to famine and contributed to the deaths of millions of people.

the failures of private ownership, they do not see the potential injustices of public ownership. Under state control, (1) invariably some individuals come to own more than others—often much more—and (2) distribution takes place by the decisions of a few, who can use deceit, the demand for bribes, or other means to obtain favors. Moreover, in no socialist or Communist system has the state ever withered away. That is a false promise and a pipe dream.

Socialism results in inequities because people are sinners. What is inherent in socialism, however, is the inability of the individual to chart his own course. Instead of being the pawn of the king, under socialism the individual is the pawn of the elite group in charge. What the individual deserves as part of his inalienable rights is to be able to decide for himself what he wants to do with property. Because land ownership is power, it is better for individuals to be able to own property as opposed to only a few holding it and distributing it as they see fit.

However, public ownership of some land does have benefits. It exerts control over the land and prevents private parties from using the land for illegal purposes or from extorting ridiculously large amounts from people who want to purchase it or use it. Government ownership also enables public goods, such as parks and roads, to be available to all citizens and not just to those who would control access or who can own large tracts of land. Public ownership also protects land of exceptional value from exploitation, such as might happen if the Grand Canyon were privately owned and the owner sold the right to build a casino or a Motel 6 on the edge of it.

Now to answer the questions at the beginning of this lesson . . .

Pinnacles Desert, Nambung National Park, Australia

Mt. Everest sits on the border between Tibet and Nepal, so each claims part of it. Most climbers approach the summit from Nepal.

The United States claims sovereignty over the Grand Canyon and administers it through the U.S. Park Service, an agency of the Department of the Interior.

Ray and Charlene Notgrass claim ownership of the property on which they live because they bought it from the previous owner, who had bought it from the owner previous to them, who had bought it from the previous owner, who had received it as property passed down through several generations. In the early 1800s the U.S. government awarded it as a land grant to an American soldier as payment for his service in the Continental Army during the American Revolution. All of these transactions were in keeping with the laws of the United States and the state of Tennessee. Before that, native nations lived and hunted on the land.

Who owns the land? Under what terms can anyone own part of the geography that God created? The worldviews that people have affect how they make laws, how they apply those laws, and how they understand land ownership.

Now these are the territories which the sons of Israel inherited in the land of Canaan, which Eleazar the priest, and Joshua the son of Nun, and the heads of the households of the tribes of the sons of Israel apportioned to them for an inheritance. . . .
Joshua 14:1

Assignments for Lesson 144

Project Continue working on your project.

Student Review Answer the questions for Lesson 144.

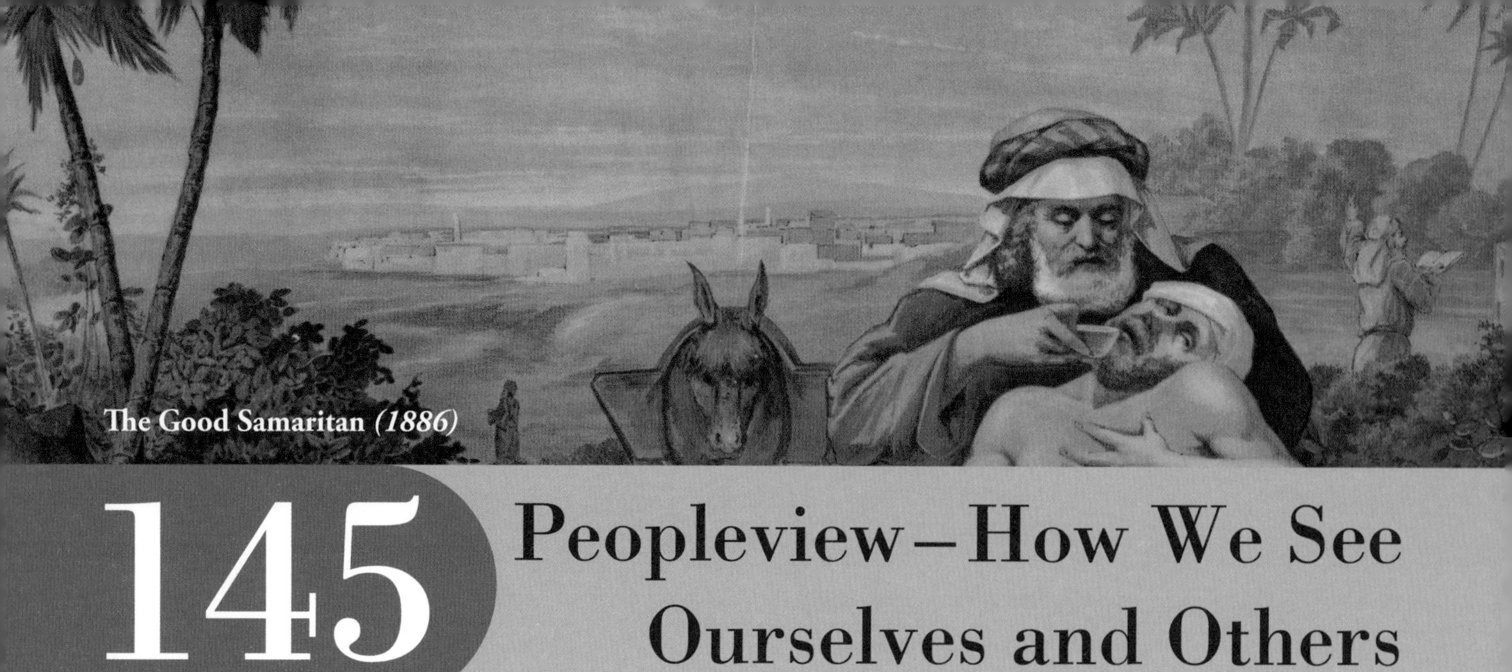
The Good Samaritan *(1886)*

145 Peopleview – How We See Ourselves and Others

You know the story. A man is beaten, robbed, and left for dead along the road—something that happens daily in any number of places in our world. Two religious men see the injured man and hurry by on the other side of the road, not wanting to get involved. Then a man from another ethnic group and a different nationality stops to help the victim. This man, who was far away from his own home, spares no effort to help the assault victim recover from what had happened to him.

Who had the healthier worldview of people: the religious men who made excuses for not helping or the stranger who chose to help? Obviously, the latter. This story is how Jesus explained what it means to love your neighbor as yourself. How you define who your neighbor is depends on how you see yourself. If you see yourself as a neighbor to others, your neighbor will be anyone who needs your help (Luke 10:25-37).

This lesson is about how you see yourself and, as a result, how you see others.

The Way God Made Us

When God created the world, He created material substances. Among other things, He created rocks, trees, planets, and living beings. Human beings have a physical nature, but they also have something more. God made people in His image (Genesis 1:26-27). This means they have an eternal spirit. First Thessalonians 5:23 mentions each person's "spirit and soul and body." This describes who we are: physical body, physical life (98.6° and breathing, blood flowing; this is the meaning of "soul" in this passage), and spiritual identity ("spirit" in this verse; people sometimes call this the soul when just contrasting it with the physical body). Humans are the only part of creation that worship, think about thinking, and in many other ways demonstrate their differences from the merely physical.

If you deny that people have a spiritual, eternal nature, then not only do other people become less than truly human to you, but you also become less than truly human yourself. To say that people have no spiritual identity and are only physical objects—that they are no different from any other created thing—is to say that people are no more valuable than a goat or a radio.

Secular society in a given time and place might decide (however society decides, probably through the people who make its laws) that certain people have special value; but society might also decide that certain people don't have special value, such as the elderly, the infirm, or the unborn.

The Height and Depth of Human Existence

"I am fearfully and wonderfully made," says the psalmist (Psalm 139:14). In Psalm 22:6, the psalmist feels like a worm, but then he realizes his true value when he remembers that he is not a worm but that he is so precious that God brought him from his mother's womb (Psalm 22:9). These and other verses speak of the dignity and worth of human beings, a dignity and worth that are above and distinct from the rest of Creation. People are the only part of Creation made in the image of God, the only part Jesus became one of, and the only part for whom Jesus died.

Because God created people in His image, and because people are a little lower than He is, people are capable of extraordinary accomplishments. People have amazing abilities. They perform life-saving surgery. They build impressive bridges and skyscrapers. They write symphonies and novels and paint beautiful masterpieces. They perform deeds that are genuinely good. They enrich others' lives with their love, wisdom, and humor. People can engage in acts of great sacrifice, even to the point of laying down their lives for others.

People include your grandparents, parents, siblings, and cousins. They include your fellow church members, your government leaders, and your heroes. We value nothing else in Creation the way we value people.

The Failure and the Way Back

However, the crushing disappointment about humankind is that we don't live up to this treasured identity and high calling. We rebel against the One who made us. Like Adam and Eve, we think we have a better way (Romans 3:23, Isaiah 59:2, Ephesians 2:1-3). Because God values us so highly and equips us to do great things, our failure to live as He wants us to is a great disappointment to God and a great loss for the rest of humanity.

This is the human struggle and quandary. A person of inestimable worth is capable of great good, horrendous evil, and deadening apathy. He or she is capable of monumental achievement and wasteful destruction. A person can bless people but also kill them.

God offers redemption, renewal, and reconciliation to every person through Christ (Ephesians 2:4-9). By His grace, believers take on the identity of God's family, and as a result, Christ is not ashamed to call us brethren (Hebrews 2:11). Christians become "a chosen race, a royal priesthood, a holy nation, a people for God's own possession." We go from being nobodies to being the people of God (1 Peter 2:9-10). We go from being "children of [God's] wrath" to being God's workmanship, "created in Christ Jesus for good works," which God had planned for us to do all along (Ephesians 2:3, 10).

With this new identity, Christians can and should view themselves differently. We should think of ourselves as "dead to sin but alive to God in Christ Jesus" (Romans 6:11). A person does not stop sinning after becoming a Christian; but God's assurance is, "There is now no condemnation for those who are in Christ Jesus." The Word assures us that, "if by the Spirit you are putting to death the deeds of the body, you will live" (Romans 8:1, 13).

A Christian should see himself or herself as neither a worm nor an angel. Neither beating yourself up as worthless nor exalting yourself above others is appropriate. A Christian is a precious child of God who has blown it but whom Jesus has received and made new. The proper response is one of humility, because of your failing and because God has shown you redemptive grace. The proper response to God's grace also includes a commitment to living for the One who saved you.

As Andrew Peterson once said, you are a sinner deep down; but deeper down you are an Image-bearer. This means that, despite our failings, we still have a calling and a purpose.

How You See Others

The truth of God's redemption is good news for people. It applies to everyone. However, sometimes we find it hard to extend our personal grace to people who are different from us. We humans are not all the same. People who live in different countries—and even in different parts of the same country—develop different ways of speaking, different patterns of behavior, and different ways of thinking. We develop political differences and religious differences.

Our human differences can cloud our worldview glasses. The human (sinful) tendency is to be critical of those who look or act differently from ourselves and to think of those other people as less worthy or as not measuring up to our worth and the standard we set. This is prejudice, an attitude of condemnation based on prejudging another person or group without any knowledge of the person or group.

The way you answer the following questions comes from your view of other people.

Do you think that most people are pretty much like you, with their own set of experiences, strengths and weaknesses, habits and inconsistencies, successes and failings, or do you think they are somehow less worthy beings?

Can strangers be trusted?

Do you believe that others are out to get you? If you do, you will hear what others say to you with suspicion and fear.

Do you believe any of the following statements to be true?

A man named Michael, holding this sign in Boston, expresses a desire that all humans have.

- Northerners are generally curt and uncaring.
- Southerners are generally unintelligent and lazy.
- People from California are generally laid back and believe "anything goes, dude."
- People from Italy generally drink wine, eat a lot of pasta, and argue vehemently.
- People who were not born where you live—sometimes, people whose great-grandparents, grandparents, and parents were not born where you live—are odd and not to be trusted.
- Women are, for the most part, [fill in the blank].
- Men are, for the most part, [fill in the blank].

(How you answer these last two questions is significant because in each case you are talking about how you view approximately half of the human race.)

- Teenagers are, for the most part, [fill in the blank]. This is important because you are a teenager, and your answer says something about how you see yourself. Someday, you might be the parent of a teenager.
- Politicians are, for the most part, [fill in the blank]. This is important because the people who run our government and who decide about war, taxes, and laws, for the most part, fall into the category of politicians.

The attitudes that you revealed in your answers to these questions are important because when you interact with someone from any of these groups, your view of that group influences your attitude, words, and actions. You might decide to avoid any interaction with a person of a particular group because of your view of people from that group.

The person in Jesus' parable that we call the Good Samaritan did not act on the basis of prejudice. He was of a different ethnic group from the Jews, he followed different religious practices from what the Jews did, and he was far from his home; but he did the right thing by lavishing help on the (probably Jewish) robbery victim. In the parable, the Samaritan obeyed the Lord's command to treat others the way he wanted to be treated (Matthew 7:12). In 2 Corinthians 5:16, Paul set the standard for the worldview of redeemed people when he said that, instead of looking at people with his pre-conversion worldview, "from now on we recognize no one according to the flesh."

Are those "different" people worthy of your love? Before you answer that, ask whether you were worthy of Christ's love. In terms of your innate worth as someone created in God's image, yes, you were worthy of His love; but in terms of your track record, no, you weren't. Yet Jesus died for you anyway. Since He did that, we must not make worthiness as we define it a standard for showing love to others.

Who You Are (and Who Other People Are)

You are a person made in the image of God. So is every other person. We all have physical bodies and physical life, but that is not all we are. We are not mere physical objects. We also have spirits that will live eternally (see 2 Corinthians 5:1). The worthiness of Christians to stand before God comes not from our own deeds but from the innate worth that God gives us and the redemption that we receive through Christ by faith. How well we live and how well we relate to others depend on our grasp of the realities of who God is, who we are, who other people are, and how we should relate to Him and to other people.

C. S. Lewis wrote:

> It is a serious thing . . . to remember that the dullest and most uninteresting person you can talk to may one day be a creature which, if you saw it now, you would be strongly tempted to worship, or else a horror and a corruption such as you now meet, if at all, only in a nightmare. All day long we are, in some degree, helping each other to one or the other of these destinations. It is in the light of these overwhelming possibilities, it is with the awe and the circumspection proper to them, that we should conduct all our dealings with one another, all friendships, all loves, all play, all politics. There are no *ordinary* people [emphasis his]. You have never talked to a mere mortal.

Someone has said, "I do not see the world as it is; I see it as I am." The lens of how you see yourself has a profound impact on how you see others—and God.

But you, why do you judge your brother? Or you again, why do you regard your brother with contempt? For we will all stand before the judgment seat of God.
Romans 14:10

Assignments for Lesson 145

Worldview	Recite or write the memory verse for this unit.
Project	Finish your project for this unit.
Student Review	Answer the questions for Lesson 145. Take the quiz for Unit 29.

Hiking in the Carpathian Mountains of Europe

30 World Geography and You

In this final unit of our study, we discuss some global issues. The first two lessons talk about language and transportation. Then we examine some international organizations. We compare globalism and nationalism. Finally, we see how only Jesus can make the world truly one.

Memory Verse

Memorize Isaiah 2:4 by the end of the unit.

Books Used

The Bible
Exploring World Geography Gazetteer

Project (Choose One)

1) Write a 250-300 word essay on one of the following topics:
   - Halford Mackinder believed that the geographical pivot of history was Eastern Europe and Central Asia, what he called the "Heartland" of the Asian-European-African landmass. What do you think is the geographical pivot of the world today; in other words, what is the geographic location that most influences world affairs? Is it the United States? Is it China? Is it the Middle East or some other part of the world? What are your reasons for thinking this way? (See *Gazetteer* reading after Lesson 146.)
   - What have you come to understand better about the world through your study of *Exploring World Geography*?
2) Look at the essay you wrote in Unit 3 and see in what way your thinking has changed and in what way it has been confirmed. Write an essay telling what you have come to understand better or differently about your worldview through your study of *Exploring World Geography*.
3) Create a work of art that expresses what you think about the world. It can be a picture, a poem, a song, a statue, or some other work.

146 How Do You Say It?

The island of La Gomera is a beautiful jewel in the western part of the Canary Islands, which lie in the Atlantic Ocean off the western coast of Africa near Morocco. La Gomera is covered with mountains and steep, wide ravines.

When a goatherd on the island wants to communicate with someone across a mountainside or across a ravine, he does not shout his message or use a cellphone. Instead, he whistles—a long, complicated whistle with several different sounds that communicate precisely the message he wants to get across. This is the silbo (or sylbo) gomero whistling language of La Gomera.

When the first Europeans came to La Gomera in the 1400s, the people who lived there—who had probably come from northern Africa—already used a whistling language. We have evidence of whistling languages that people used in Africa. When the Spanish came to dominate the Canary Islands, the people of La Gomera adapted ("translated") their whistling into the Spanish language.

A typical technique for using silbo involves inserting the middle knuckle of the forefinger into the mouth to move the tongue as desired to create the whistled sounds. The current form of the language employs different sounds for two vowels and four consonants. In this way, they get their messages across. People can hear these whistles from two miles away or sometimes even farther, significantly farther than people can hear shouted words.

Not everyone on La Gomera communicates in silbo. In fact, some residents of the island look down on it as something only peasants do. But public schools teach it to children who pick it up easily, so it is likely to remain alive. One popular use is in demonstrations in restaurants that cater to tourists. People who use silbo speak Spanish in face-to-face conversation; but where there are few roads and no cellphones, silbo enables communication over longer distances.

People in a few other places on the earth (one estimate is about seventy) use whistle languages, including a small Greek island, a town in northern Turkey, and a village in the French Pyrenees Mountains. Each location has its own techniques and "vocabulary," but they have one thing in common.

They whistle.

What Language Is

Language is how people communicate. Through language we share information (about ourselves, others, and the world), concepts (such as in mathematics and physics), and beliefs (about our fears and hopes, about other people, and about our faith system). People around the world depend on language to communicate, but since the Tower of Babel, the world population has not used one universal language. Instead, people speak (or whistle) in an estimated 7,000 languages. A few languages are the means of communication for hundreds of millions of people across continents, while sometimes only a few hundred or a few thousand people use a language in just a small area. Where people speak a particular language and where that language originated are matters of geography.

What Language Does

Language serves several functions in human society.

Language is identity. The language that a person learns growing up is part of who he or she is as a member of that language group. For instance, the various ethnic groups in the Balkans take a large part of their identity from speaking Albanian or Bulgarian, and so forth. In addition, as a child learns a language from his elders, he also learns stories, traditions, and values; in other words (so to speak), he learns his culture, who they are as a people.

Language both contributes to worldview and reflects it. Norwegian has many words for snow: falling snow, snow on the ground, snowflakes, snowdrifts, and so forth. Nigerian? Not so many. Ancient Egyptians had a great interest in the afterlife, so they had quite a developed vocabulary about elements of it, preparation for it, how it compared to this world, and so forth. Language expresses a group's shared ideas and experiences, which contribute to the worldview of that group and of persons in that group. When one language replaces another through religious conversion or military conquest, it changes the worldview of those people since they come to think and speak in new terms. However, the old worldview can remain to some extent in their minds and words. This is how folk beliefs can exist alongside Christian beliefs.

Language enables the organization and transmission of knowledge, literature, and history. People can do this orally; but a written language preserves the knowledge, literature, and history in more certain form for a longer time. A language group develops a body of literature as writers choose to write poems, drama, stories, and longer works in that language. This also preserves and enhances that group's identity.

Fefor, Norway

Language is power. Conquering armies set taxes, divide land, and make laws. They often require the people of a country to conduct business with the government in their language. Beyond the use of brute force or the mere communication of information, however, language enables people to convince, motivate, and inspire others. Martin Luther King Jr.'s effective use of persuasive language to change laws and the worldview of many Americans is an example.

Language is opportunity. What chance do you have of economic success if you cannot speak, read, or write the language of employers, commerce, and trade? How could you succeed with a revolutionary new idea or product if you had no way to communicate that idea or product to potential buyers? What if Steve Jobs spoke only Swahili? What would have been the market for Apple's goods and services among Swahili-speakers? Instead, he spoke English; and this gave his company the opportunity to sell goods and services to more people.

The sign of the Embassy of the Philippines in Japan features Tagalog, Japanese, and English.

Language and Geography

We could say much more about language in general, but in this context we want to emphasize the connection between language and geography.

In simplest terms, language is a function of geography. In other words, the main factor that determines what language people speak is where they live. For the most part, people who live in France speak French. People who live in Germany speak German. People who live in Vietnam speak Vietnamese. People do have the ability to learn other languages wherever they live, but this geographic factor applies in most cases.

The connection between geography and language is actually much more complex. Language does not stand still. First, languages move as their speakers move. This can happen through conquest, when speakers of a language invade and take control of a new land. The conquerors' language becomes the language in which interaction with the government must take place. As people adopt the new language and as more speakers of the language immigrate to the new land, the new language becomes the norm.

However, this course of events did not happen in the Philippines. The Spanish ruled there for over three hundred years, but the Spanish language never spread much beyond the ruling class. The people continued to use their native languages, of which Tagalog was the most widespread. After the United States took over governing the Philippines in 1898, English spread widely. Two timing-related factors that assisted this development were the increased use of media, such as radio and newspapers that used English, as well as the more significant role the United States played in Asian and world affairs. As a result, English and Tagalog are now the two most common languages in the Philippines. Only a tiny fraction of Filipinos speak Spanish.

People taken to new areas as slaves had to learn the language of their masters in order to get along. African slaves brought to America had to leave their languages behind, but those languages continued to influence their speech. The distinctive Gullah speech that developed in the geographic area of the Sea Islands and coastal lands of South Carolina and Georgia was the result of slaves combining the English of their owners with the African languages they had known.

Ninety-five percent of Brazilians speak Portuguese as their first language. However, before 1500 not one person who lived in what became Brazil had heard even one word of Portuguese. A combination of factors, including continuing immigration by Portuguese seeking wealth from the land, a decline in the indigenous population, and the lack of any military or cultural power to stop the growing influence of Portuguese, led to this radical change.

Languages also change within themselves, as people start using new words (often borrowed or adapted from other languages), use old words in new ways (for instance, "computer" originally meant a person who did calculations and "suffer" used to mean to permit, as in the King James Bible), and discontinue using other words (such as "trow," to think, and "wot," to know).

A major innovation in the development of language was the printing press. Many languages have local dialects. in the past, spelling could vary widely as people wrote words as they heard them. As printed books and documents spread throughout a geographic region where a language was spoken in all its varieties, the dialect and spelling forms that the documents used became standard for that language. The later publication of dictionaries and grammars hardened those practices into the norm. For instance, the works of Martin Luther, which accounted for one-third of all printed publications in German from 1517 to 1525, had a profound influence in creating a standard form of German. His German translation of the Bible, published in 1534, had an even greater impact.

Tens of millions of immigrants left their native lands—Germany, Ireland, Italy, Poland, India, and many more—and came to the United States. Those immigrants spoke their native languages and slowly picked up enough English to get by. They retained their native culture by using their native languages in their homes and neighborhoods. Many ethnic groups in large cities published newspapers in their native languages that offered news about their own enclaves, U.S. news, and news from the countries they had left behind. The children of these immigrants, growing up in the U.S. and attending English-speaking public schools, learned and used their native languages at home but had a greater motivation to learn English and use it in school and at work. The next generation had little motivation to retain the languages of their ancestral countries and used English almost exclusively. The immigrants did not have the political or economic power to push English aside. Instead, the developed English-speaking culture of the U.S. tended to push the native languages aside, although a few of the later generations enrolled in classes to retain a working knowledge of their ancestral tongues.

With the large immigration of Spanish-speaking people into the United States in recent years, some have continued to speak Spanish with one another but often learn enough English to interact with English-speakers. Sometimes immigrants learn only a little English—just enough to help them get by in their jobs and in society. Sometimes an immigrant's job requires them to learn a great deal of English so they can communicate with clients and co-workers.

The situation requires policy decisions for companies in the U.S. and for local, state, and national governments. Some stores post signs in English and Spanish as a way to serve more customers. The question of whether to teach curriculum in public schools in Spanish (or Vietnamese or any other language that a number of students speak) as well as in English is a much-debated one. Teaching in native languages might ease matters for immigrants in the short run but delay their entry into the broader American (English-speaking) community. Only offering instruction in English would open more opportunities for students (see above), but it might make the first steps in their transition more difficult.

Dialects and accents are definitely functions of geography. A dialect is any variety of a language that distinguishes one group from another. A dialect often develops when a smaller group within

Language Lesson in Gadjievo, Russia (2012)

a language group (a remote village, for instance) has little contact with outsiders, and a specific way of speaking that language becomes standard. One mark of a dialect is vocabulary. For instance, are they hotcakes or pancakes? Are they green beans, string beans, or snap beans? Is it a parking garage or a parking ramp?

An accent is the way one sounds when he or she speaks. In other words (pardon the expression), everyone has an accent! People within a language group have accents, such as English-speakers who have a southern accent or a British accent. Where that person has lived has helped to create that person's accent. Another kind of accent is the way someone speaks a second language, such as the way a native Japanese speaks English or the way someone native to the United States speaks French.

Pidgin is any simplified language that enables people who don't share a language to communicate. Such languages are the result of geographic movement. Pidgin often develops in a port city where sellers from one language do business with buyers from another language, and then spreads when the buyers take the pidgin back to their villages and use it there. Examples of statements in pidgin English are "How much this one cost?" ("How much is this?") and "I go now now" ("I am leaving very soon").

Our Mother Tongue

English is an illustration of the complexities a language can have even in a small geographic area. Gaelic-speaking people lived across the island of Britain until waves of Germanic Saxon invaders took over the main part of the island and pushed the Gaelic peoples into what became Wales, Scotland, and Ireland. Each of those lands developed their own languages based on Gaelic.

The Norman invasion of 1066 brought Norman French onto the island, and that language became the language of the ruling class. However, England did not become a French-speaking enclave. Instead, the interplay of the Norman French and Germanic languages contributed to the development of what we know as English.

But British English was not even that simple. Different areas of England developed their own distinctive accents. Sometimes northerners have a hard time understanding people from the south of England, and so forth. The different accents and dialects in England and Scotland help to explain the different accents we have in the United States and other former British colonies. People from different parts of Britain moved to different parts of America and brought their accents with them. This is why people from New England sound different from people who live in South Carolina, even though they all speak English. Many of those who settled in Australia came from the Cockney-accent-speaking part of London, and a version of that dialect of English now predominates there.

Moreover, this interplay of a multiplicity of languages is a major reason why English has so many synonyms and near-synonyms for the same word or thought. With contributions from Gaelic, Saxon/German, French, and—through academic and church activity—Greek and Latin, you usually have several choices to say the same thing. Consider the noun injury: you might also choose wound, hurt, damage, harm, trauma, or several others. For money, you might say cash, wealth, riches, prosperity, or slang words such as moola, deep pockets, or cashola.

Another fascinating geographic fact about English is that out of that small island north of Europe has come the language that today is spoken around the world and is a common language for business, air travel, diplomacy, and many other aspects of life worldwide.

Otto von Bismarck was the Prussian political leader who brought about the unification of Germany in 1871 and led Germany until 1890. He died in 1898. Supposedly someone asked him late in his life what was the single most important political fact of his day, and he replied that it was the fact that North America speaks English.

Bismarck believed deeply in German greatness and he wanted Germany to be a significant power in world affairs, but he recognized the rising power of the United States and the significance of North America's cultural and linguistic heritage from Britain, then the most powerful country in the world. Bismarck apparently felt some jealousy toward the U.S. He is also credited with saying, "God has a special providence for fools, drunkards,

This sign in Nepal is written in Nepalese, Tibetan, and English.

and the United States of America." Ironically, the English-speaking people of the United States helped stop German aggression in the two world wars of the twentieth century.

Geography influences the language that people speak and the dialect and accent they use. It can even influence how a language sounds. Researcher Caleb Everett found that the spoken languages of people who live in high altitudes (4,900 feet or more above sea level) almost always have a feature called ejective consonants. An ejective is a click sound that a person makes with an intense burst of air. English does not have ejectives; the nearest similar sound in English is an intense k. Everett's hypothesis is that people can utter ejective consonants with less effort at higher altitudes because air pressure is lower there.

Wherever we live in geographic terms, that geography and what has taken place on that geography impact the way we use language. We speak. We write. We click. We whistle. We communicate.

At Babel God confused the languages of the people, who scattered across the face of the earth. At Pentecost, thanks to the miracle of languages that God performed that day, He brought people of many languages together through the one message of the gospel.

And how is it that we each hear them in our own language to which we were born? . . . we hear them in our tongues speaking of the mighty deeds of God.
Acts 2:8, 11b

Assignments for Lesson 146

Gazetteer	Read the comments on and the excerpt from "The Geographical Pivot of History" (pages 341-344) and answer the questions in the *Student Review Book*.
Worldview	Copy this question in your notebook and write your answer: What are some of the ways in which people in our world are divided against each other?
Project	Choose your project for this unit and start working on it. Plan to finish it by the end of this unit.
Student Review	Answer the questions for Lesson 146.

White Pass Scenic Railway, Alaska

147 How Do You Get There?

Ray and Charlene's children had a sweet surprise. They told their parents that they wanted to give them a cruise for their fortieth anniversary. "You pick the place," they told Mom and Dad, "and we'll send you there."

It didn't take Ray and Charlene long to decide that they wanted to go on a cruise along the coast of Alaska. They flew into Anchorage. While they were on the plane they saw a huge area of nothing but snow-capped mountains. They went in late summer, so they enjoyed many late hours of sunlight.

On their first full day in Alaska, Ray and Charlene toured the Alaska Native Heritage Center. This facility has walk-in recreations of five different kinds of houses that Alaska Natives once built. They saw demonstrations of indigenous sports competitions and crafts.

The next day, Ray and Charlene rode a train north to the town of Denali. They saw mountains, marshes, and other geographic features along the way. They heard about the effect of the massive earthquake that struck the area in 1964. The following day, they enjoyed an all-day bus ride to Denali National Park—and they actually got to see Denali, the tallest mountain in the United States! Most tourists who go there don't get to see it because of the frequent cloudy weather. When they stopped for lunch, they panned for gold in a nearby stream. Along the way, they saw bear, elk, wolves, and foxes. They saw snow and glaciers high up on several mountains, even though it was August.

They returned to Anchorage and got ready for their cruise. Because a peninsula juts out from the coast just south of Anchorage, Ray and Charlene had to take a train across the peninsula to reach the cruise ship port to begin that part of their trip. But before they got on board, they took an excursion on a smaller vessel to see Blackstone Glacier.

Ray and Charlene Notgrass with Denali Behind Them

Hubbard Glacier Through a Cruise Ship Window

Finally they got on board the cruise ship. Their first stop was another glacier—Hubbard Glacier. They saw the ice calving from the front edge of it and experienced the immense width of the glacier. Ray and Charlene got up close to two other glaciers on the trip.

They also sailed up several deep fjords along the coast. Most days they got off the cruise ship for several hours and visited towns along the coast. Here they saw how Alaska Natives and pioneers lived in years gone by. They visited the state capitol, which is actually a converted office building from the days when Alaska was a territory. They got a tour of an Orthodox church building, a legacy of the Russian exploration of the area. They saw a large display of totem poles that reflect the stories of various nations. They visited a town that was founded during the 1898 gold rush and took a train ride through a valley pass where gold prospectors walked into the Yukon Territory of Canada.

The last day of the cruise they traveled the Inside Passage, the waterway between the coast and the barrier islands off the coast, and landed in Vancouver, British Columbia. From there they flew home.

Ray and Charlene had a great time seeing breathtaking scenery and experiencing the cultures of Alaska. They traveled by plane, train, cruise ship, excursion boat, taxi, bus, on foot, and pickup truck (when their B&B host took them to a bus stop). They were grateful for their children's generosity. They might not have thought of the word geography on the entire trip, but the trip was pretty much all about geography.

Getting There Is Everywhere

People travel. Whether for business, trade, study, exploration, conquest, changing their place of residence, gathering with others, or vacation, people travel from where they are to somewhere else. When they go, they encounter geography, both along the way and at their destination.

People travel despite the fact that geography often does not make travel easy. Flat ground is easier to traverse than mountains; but even on level ground people have to deal with distances and with issues such as driving a covered wagon through tall prairie grass. Rivers provide avenues of transportation, but the difficulties of crossing them (such as the width of the river and swift currents) can limit travel. Deserts and mountain ranges are difficult to cross. Oceans have tended to limit travel also. For millennia, few people crossed the Atlantic and Pacific Oceans. People got to islands, but not easily. Humans are land-dwelling creatures, so most of our travel has taken place on land, over routes that people have found manageable.

Because people want to travel and because geography can present challenges to travel, transportation infrastructure and modes of transportation have been important issues that people have had to resolve. Transportation infrastructure is what people use to travel: roads, railroads, bridges, airports, subways, and so forth. Modes of transportation are what people ride in or send goods on: wagons, carriages, automobiles, tractor trailer trucks, container ships, cargo planes, and so forth.

Infrastructure is a major issue for the development of a nation because it is expensive and because it involves the use of large amounts of land. Infrastructure that crosses difficult tracts of geography is even more expensive. In the early years of the United States, the question of how to pursue "internal improvements" such as roads and canals was a hot political topic. Should these projects be private efforts or left to the states, or were they important enough to the growth of the country that the federal government should undertake them? Canals were also part of the internal improvements program, and they played an important part in the economic growth of Europe as well.

People have traveled throughout history, and other people have been interested in the accounts of those who have traveled. Ancient geographers traveled to places and reported on what they saw. Travel narratives were an important part of American literature in the 1800s, such as *The Journals of Lewis and Clark*, *A Week on the Concord and Merrimack Rivers* by Henry David Thoreau, and *The Exploration of the Colorado River and Its Canyons* by John Wesley Powell.

The speed of travel has been an issue since people began devising ways to get there faster, more easily, and more safely. For centuries people could travel no faster than they or horses or camels could run. Camels can run up to about twenty-two miles per hour, horses perhaps twenty-five to thirty miles per hour; but they can only sustain those speeds for limited periods of time. Sailing ships can sometimes

The Canal du Midi, completed in 1681, connected the French city of Toulouse to the Mediterranean Sea. It is 150 miles long. The completion of the Canal de Garonne (120 miles long) in 1856, going from Toulouse toward the west coast of France, allowed small boats to navigate from the Atlantic Ocean to the Mediterranean.

Tourists in Lisbon, Portugal (2006)

attain greater speeds for short periods, but they are dependent on wind power to do so.

Steamboats were not terribly fast, but they allowed people to travel upstream much more easily than before. Steam-powered railroad engines attained the incredible speed of thirty miles per hour. There were some who were sure that humans could not travel any faster or they would simply disintegrate! But they didn't disintegrate, and advances in travel continued: automobiles, airplanes, space travel, high speed rail lines, and more.

Tourism Geography

A subfield of geography that has grown in recent years is tourism geography, or the geography of tourism. Geography provides the setting for tourist activity. A primary factor in people deciding to travel to a particular place is the geography there. The most visited U.S. national park year after year is Smoky Mountains National Park on the Tennessee-North Carolina border, far from any major city. Museums have geographic components also because they highlight the people and culture of different parts of the world.

Tourism is geographic in nature; it involves going somewhere over geography to a particular geographic place, and getting back. This involves not only infrastructure and modes of transportation but also accessibility. Generally speaking, a place that is easy to get to will attract more tourists. On the other hand, some people enjoy going to out-of-the-way places, which are such because of their geography.

Tourism is a relatively new human activity. People haven't always traveled on vacations; people didn't always have vacations. For much of history, wealthy people were primarily the only ones who could afford to travel. The growth of what economists call the middle class—in the United States and in other industrialized countries—has led to many more people being able to engage in leisure travel. The tourism industry is now one of the largest industries in the world. It can have a significant impact on a small town that has a rich heritage but little or no productive industry.

Tourism geography examines the importance of tourist resources in a place, such as lodging, dining, stores, and museums. It studies the impact of human travel on places, and the impact of places on humans. It considers the environmental impact of tourism, which can be enormous. Even responsible tourists take a toll on a place; irresponsible ones have an even greater impact.

The meaning and significance of tourist districts is a matter of much discussion. Are districts such as the French Quarter in New Orleans, Chinatown in San Francisco, Soho in London, and Greenwich Village in New York genuine snapshots of local culture, or are they atypical places designed to generate revenue from unsuspecting visitors? Can they be both? One question that provides significant insight into the issue is whether the locals go there or not.

Tourism is often a social activity. Groups of people visit places together and discuss what they experience. Large numbers of tourists from different parts of the country or the world can share the same geographic place on a given day. On such visits people learn about how other people live and have lived.

Tourism broadens one's view of the world—his or her worldview. The tourist learns that not everyone in the world thinks and lives like the people in his hometown. Some observers lament that tourists acquire a superficial knowledge of other places and people. This is inevitable. A one-day visit to a place that focuses on shops and a few stunning scenes cannot compare to living in a place all of one's life, but at least the tourist's knowledge increases to some degree. Whether that person begins to talk like an expert on all things Chinese after a two-hour visit to Chinatown is his issue to resolve. Such a brief encounter can lead to deeper study, however.

Going places can have a genuine and long-term impact on tourists. A visit to a stunningly beautiful setting or a place with deep historical significance, for example, can inspire a person for life. Effective means of transportation enable such travel and such meaningful impact in people's lives.

Jesus said He is the way to the Father (John 14:6). The book of Acts sometimes refers to the Christian faith as the Way, as in this passage about Paul (then called Saul) going to Damascus:

Now Saul, still breathing threats and murder against the disciples of the Lord, went to the high priest, and asked for letters from him to the synagogues at Damascus, so that if he found any belonging to the Way, both men and women, he might bring them bound to Jerusalem.
Acts 9:1-2

Assignments for Lesson 147

Worldview Copy this question in your notebook and write your answer: How have you seen a shared faith in Jesus bring people together who were once separated or distant from each other?

Project Continue working on your project.

Student Review Answer the questions for Lesson 147.

International Committee of the Red Cross Headquarters, Geneva, Switzerland

148 What's Bigger than a Nation but Smaller than the World?

Political borders between nations are part of geography. Sometimes geographic features such as rivers and mountains delineate those borders, while at other times borders are simply agreed-upon lines. For example, a long stretch of the border between the United States and Canada follows the 49th meridian of latitude.

Some interests, issues, and needs that people have transcend national borders. Because of this, individuals, groups, and nations have formed international organizations. These organizations have various purposes and exercise different levels of authority.

Non-Governmental Organizations (NGOs)

An NGO is a private group that is not part of a government which seeks to accomplish a certain goal in one country or in many countries. People who have a common interest form the organization to address a particular issue. Some governments might help fund the efforts of an NGO, but the group's purpose is to work outside normal political and governmental channels. Often an NGO seeks to address a need that governments are not addressing. In terms of policies, an NGO can advocate and persuade, but it has no authority to implement policy. In fact, NGOs have to obtain permission from a national government even to be able to operate in that country.

The International Red Cross is an NGO. Another NGO is the physicians' organization Doctors Without Borders (the official international name is in French, Medicines Sans Frontiers) that supplies physicians to crisis areas in the world and advocates for governments to implement certain health practices. The International Olympic Committee, which oversees the summer and winter Olympic Games, is an NGO, as is the World Wildlife Fund for Nature (WWF).

Intergovernmental Organizations (IGOs)

Governments of various countries have worked together to form many international organizations. The goal of these organizations is often to increase trade among the members or to maintain good relationships among member states and avoid conflict. IGOs have official representatives of the governments of member nations. Member nations generally commit themselves to following the decisions of these groups, but usually they are not strictly bound to do so.

Sometimes representatives of governments come together to address specific issues at specific times. The Congress of Vienna was a meeting of diplomats held to sort out the political arrangements of Europe after Napoleon's wars of aggression. The congress met from September 1814 until June of 1815, just before Napoleon's attempted comeback and his final defeat at Waterloo. The treaty negotiations at Versailles following World War I were a similar international gathering. Nations have sent delegates to meetings to address specific issues such as the establishment of time zones (definitely a geographic issue!).

The Group of 20 or G20 meetings bring together the leaders of the governments and the central banks of 19 countries and the European Union for discussion. These meetings just generate talk, but what these leaders say can lead to actions by the individual countries.

Many international organizations are based on geographic areas of the world. Such groups include the North Atlantic Treaty Organization (NATO), the Organization of American States (OAS) for countries in the Western Hemisphere, the Association of Southeast Asian Nations (ASEAN), the African Union, and the Arab League. European neighbors Belgium, the Netherlands, and Luxembourg have formed a political and economic union called Benelux.

One important factor that brings nations together is economics and trade. Several international organizations address this topic. The Organization of Petroleum Exporting Countries (OPEC) tries to guide the oil production of member states in order to maintain an optimum price on the world market. Many times one or

Leaders at the 2019 G20 Summit in Osaka, Japan

Meeting of the United Nations Security Council (2016)

more member states will ignore the guidelines that OPEC issues in order to generate more income for their own economies. The Organization for Economic Cooperation and Development (OECD) brings together representatives of industrialized democracies to discuss best policies. It also generates studies and statistics that countries use to formulate policies. The International Monetary Fund, which is a United Nations agency, takes actions to help countries that have currency problems. The World Bank is associated with but not answerable to the UN. It uses contributions from member nations to make loans for projects that encourage economic development in various countries. The World Trade Organization (WTO) is another UN agency. It oversees regulations for international trade.

The Commonwealth of Nations (formerly the British Commonwealth) is a relatively loose organization of nations that includes the United Kingdom and nations of the former British Empire. The Commonwealth facilitates trade among member states and also maintains positive cultural relationships among member nations.

The United Nations

The largest international intergovernmental organization is the United Nations. The UN seeks to promote peace, justice, and positive relationships among nations. The UN General Assembly and the UN Security Council debate, discuss, and sometimes pass resolutions on various issues that threaten world or regional stability. The UN has established numerous agencies to carry out various responsibilities. These include the International Court of Justice, the United Nations High Commissioner for Refugees, and the United Nations Children's Fund.

The key issue in understanding the role of international organizations is determining their authority. Do member nations surrender any sovereignty to the organization? Generally speaking they do not. International organizations are dependent on independent sovereign nations for funding and for carrying out the ideas, proposals, and resolutions that international organizations develop.

Other International Groups

The members of some professions have organized international groups, such as the International Federation of Accountants and the Actuarial Association of Europe. Such groups encourage countries to adopt certain standards for their professions and keep members aware of issues that arise and other developments (such as technological advances) in their fields.

Some businesses establish offices and production facilities in more than one country. Automobile manufacturers are a prominent example of this kind of company. People often call these multinational corporations. Even though their operations transcend national borders, these businesses are not usually included in lists of international organizations.

Some international organizations are think tanks or policy discussion forums. Examples of these are the Club of Rome ("an organization of individuals who share a common concern for the future of humanity and strive to make a difference"), the Trilateral Commission (seeking to foster better relations among North America, Western Europe, and Japan), and the Council on Foreign Relations (focusing on U.S. foreign policy and international affairs).

Some people believe that these and other such organizations are trying to take over the world. They certainly want to influence policy in several countries. Their members have the kind of wealth and political clout that would cause government leaders to listen to their ideas, much more than if you and I and thirty of our friends got together and offered our recommendations to the governments of the world. There is nothing illegal about a group doing what these groups do; many groups across the political spectrum seek to influence government policy. We might not agree with the recommendations that these groups make, but members of these groups have not been convicted of bribery, extortion, influence peddling, or any other illegal activity in attempting to achieve their goals, as far as this author knows.

The International Cat Association maintains a genetic registry of pedigreed cats and administers the rules for cat shows around the world. Here a judge is evaluating a cat at a 2015 show in Albuquerque, New Mexico.

Of course, since we have all of these international organizations, someone saw a need to organize them; so we also have the Union of International Associations—an association of associations.

We will discuss the European Union in the next lesson, which deals with the debate between advocates of traditional nationalism and globalism.

Is the world better off by the work of these various international organizations? Overall, we are. It is better to talk to one another than to shoot at one another. Certainly there have been failures. We still have wars. We see prejudice against the United States and capitalism in many of these organizations. The money spent on the bureaucracies of these organizations might well be put to better use feeding the hungry, digging water wells, and providing medical care. But the world has not had another major conflagration like the two world wars of the twentieth century, despite longstanding conflicts such as the Cold War and the threat of Islamic terrorism. Perhaps international organizations have contributed to the sometimes uneasy peace in which the people of the world live.

Should we just forget national borders and national identities and have one government for the whole world? We will discuss that question in the next lesson.

The real Peacekeeper and the one whom we can always trust is the Lord.

He rules by His might forever;
His eyes keep watch on the nations;
Let not the rebellious exalt themselves.
Psalm 66:7

Assignments for Lesson 148

Worldview Copy this question in your notebook and write your answer: What can you do to bring people together in the name of Jesus?

Project Continue working on your project.

Student Review Answer the questions for Lesson 148.

Military Parade, India (2007)

149 Globalism and Nationalism

"USA! USA! USA!" chants the crowd as the American gold medalist takes a victory lap around the Olympic track. Later, when she receives her medal, the American flag ascends the pole and the playing of "The Star-Spangled Banner" provides an emotional climax to her victory. National pride fills thousands of hearts in the arena and millions of hearts in front of televisions at home.

Military personnel break into the house and haul off another family who belong to the minority ethnic group that the government has declared to be an enemy of the nation. No one ever finds out what happens to the family. The next day, the president goes on national radio and declares that his government's actions "will restore our national honor and pride."

From time immemorial, human beings have divided themselves and their world into ethnic groups, nation-states, and other segments of the entire world population. Nationhood has provided identity, security, and legal rights. Nationhood has also been the excuse for heinous crimes against humanity.

We live in an interconnected world in which trade, travel, goods, and information flow back and forth across national borders and around the world at a dizzying rate. Immigrants and political refugees desperate for safety and opportunity leave their home countries and enter other countries. They find refuge, but in doing so they put demands on the resources and social systems of the receiving countries.

As a result of these modern conditions, a relatively small but growing number of people say that national borders are no longer a good idea. Their proposals vary, from continuing the existence of nations but with no limitations on the movement of people, to having only one government for the entire world.

Are these ideas correct? Is nationalism an outdated idea? Is national identity good or bad?

The Case for a Single Global Government

Leaders and armies of many nations have wrought great destruction on the world in the name of nationalism. Napoleon's vision for a French empire, Hitler's Germany with its aggression and persecution of those deemed unworthy to live, Serbia's record of ethnic cleansing in the Balkans during the 1990s, and many other extreme nationalist views and their consequences have filled many sad pages of history. People of many countries have sacrificed themselves to serve a brutal, egomaniacal dictator in the name of national honor. Would the world not be better off without this destructive dynamic?

Conflicting national ideologies have also been responsible for ongoing uneasy tensions between nations, with some situations lasting for centuries. We can think of such examples as Israelis and Palestinians, Indians and Pakistanis, Chinese and Vietnamese, and on it goes. Nationalism often assumes or implies an attitude of superiority of one nation over other nations and groups. Thus we can see that nationalism has often been harmful.

Some say the answer is to eliminate this source of conflict among people by thinking of ourselves merely as citizens of the world, not citizens of particular, separate nations. These advocates say that we should have a concern for all people everywhere, since we are part of the human race in a shrinking world, and since what happens in one place impacts people all around the world. Advocates of globalism say that policies reflecting a global awareness would make resources more widely available and result in a more fair distribution. This approach could eliminate tariffs and trade wars. A world parliament could weigh what is best for all people in determining policies to govern all mankind.

The Case for Individual Nation-States

Recognizing national pride and ethnic identity only acknowledges what is. Communist rulers in the Balkans tried to suppress these realities for decades, and the result was a bursting forth of ethnic conflict when Communism fell there. A nation-state is a way for one people to declare their independence from other countries and to claim sovereignty in a specific geographic area. The government of a nation-state compels obedience to its laws and resists invaders. It provides order and security. A nation-state usually has a common history, language, and culture.

Nationalism offers free people the shared ability to govern themselves. Nation-states offer the best protection for individual rights. An inclusive nationalism can help bring about freedom, equality, and well-being.

Historically, national movements fought colonial rule in Africa, India, and many other places. These movements often led to democracy, self-determination, and economic growth. National programs, such as national parks and the Interstate system, have succeeded because they have come from a sense of "we," from the idea that this is ours and we can take pride in it.

Kosovo declared its independence from Serbia on February 17, 2008, and this monument was unveiled that day in the capital of Pristina. The letters are repainted from time to time. This version from 2013 shows the flags of countries that had recognized Kosovo as an independent country. As of 2020, about 100 countries had done so.

People are willing to sacrifice a great deal, even their own lives, for their country. Nationalism can promote concern for others in practical ways. First responders and utility repairmen from various parts of the United States, for instance, often go to a region where a hurricane or other natural disaster has occurred.

Nationalism can encourage character and high moral standards, as people say, for example, "We don't do things that way; we are Americans." Studies have shown that countries with a higher level of nationalism are generally wealthier and have a stronger rule of law.

Nationalism is not healthy if it excludes others within a nation's borders, as white nationalism and Hindu nationalism do. Those ideas exclude instead of include. Yes, nationalists can be fascists or Communists, but they are not necessarily so.

The Eiffel Tower in Paris, France, features a circle of yellow stars on a blue background, a symbol of the European Union, in this 2008 photo.

Patriotism—a love for and pride in one's country—is not wrong. It can be if taken to an extreme, but patriotism can help people develop character, generosity, and sacrifice. We can respect the strengths of other nations even as we still prefer our own.

We can compare national identity to family identity. Family identity is important to the members of a family. Family identity gives family members a sense of who they are. We see every day in our society the results of children who do not have a strong and healthy family identity. And yet, loving your family does not require you to assume that other families are inferior or a threat to your safety. The same is true with nations. Nationhood contributes to a person's identity. Not having that identity can contribute to aimlessness.

The Pitfalls of a One-World Supranational Government

Such an arrangement can become terribly complicated. The European Union is the closest thing we have to a global government in our world today. The EU is an organization of countries in Europe that plans, makes, and oversees economic, social, and security policies for the member countries. It came into effect with the Maastricht Treaty in 1993.

The European Parliament is the legislative body of the EU. It has over seven hundred members who meet for one-week sessions about once per month in Strasbourg, France. The size of each country's representation in Parliament is by population. Members are elected by popular vote every five years. Members of the European Parliament (MEPs) organize themselves by transnational political parties, not by parties within their respective countries. Although the plenary sessions take place in Strasbourg, the Parliament's committee work and other activities take place in Brussels, Belgium.

The European Commission (EC) is the executive arm of the EU. It proposes laws to Parliament, enforces laws already in place, and represents the

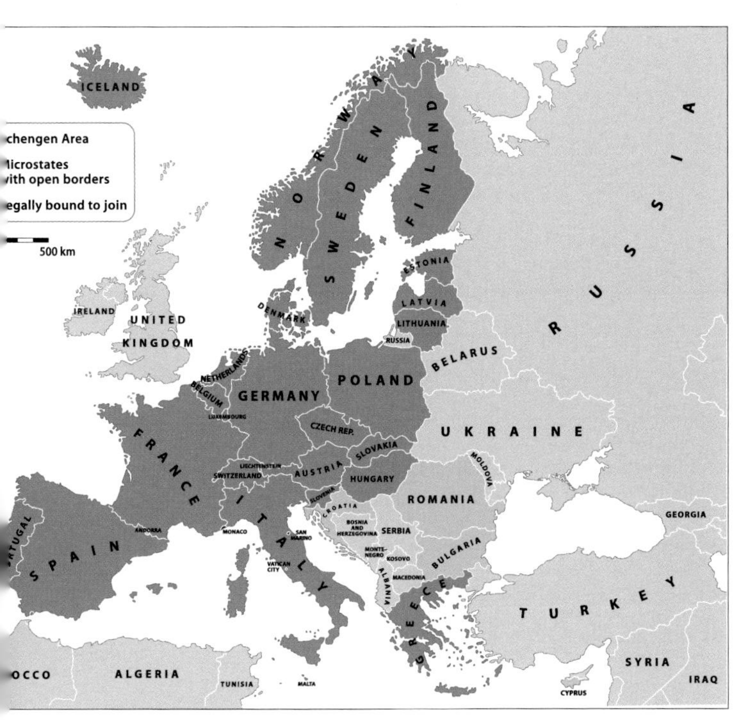

The Schengen Area includes 22 members of the European Union plus three non-members: Iceland, Norway, and Switzerland. Andorra is not a member of the Schengen Area, though Liechtenstein is. Monaco, San Marino, and Vatican City are de facto members since they have open borders with their neighboring states.

EU on the international level. The EC is made up of one member from each country. These people are charged with representing the interests of the EU as a whole and not those of their respective countries. The presidents of the member countries make up the European Council, which selects the president of the Commission. The president then, working with the governments of member nations, selects the representatives from each country for the Commission, subject to the approval of Parliament. The president and Commission serve for the same five year term as Parliament. The president appoints six vice presidents who are responsible for various duties. The Commission is accountable to Parliament, which can dismiss one or all commissioners.

The EU does not oversee all activities in Europe. For example, the Eurozone is made up of countries that use the euro currency (not all EU member countries do). The Schengen Area consists of those countries that have abolished any passport or other border control requirements for travel in and out of these Schengen Area countries (not all EU countries have ended these requirements). The creation of the Schengen Area essentially abolished borders between these countries. It is named for the city of Schengen, Luxembourg, where representatives of the original participating countries signed the agreement. The European Travel and Information Authorization System (ETIAS) permits and keeps track of visitors from countries outside of the Schengen Area who do not need a visa to enter the Schengen Area.

Does all of this sound complicated? IT IS! Added to this is the reality that MEPs and their staff must travel and shift their records between Strasbourg and Brussels repeatedly, plus the fact that staffers must see to the translation of documents and proceedings into the over twenty languages used by EU member states. What happens is that only insiders like representatives and bureaucrats understand the system, so it becomes its own little world that European citizens pay for and have to deal with but can't influence much. In addition, the rulings of the European Court of Justice can affect all of the nations of the EU. This complexity and the threatened surrender of sovereignty were two reasons why a majority of voters in Britain voted to leave the EU in a referendum in 2016.

We can only imagine how much more complicated a world government would be. How many representatives would be adequate? Can you hold an effective legislative session in an athletic stadium?

Other Problems with a One-World Government

Some actions are legal in some countries but illegal in other countries. Christian evangelism is legal in the United States but illegal in Saudi Arabia. How would a single world government sort out these differences? I think we know. Would one-world advocates accept Saudi restrictions on women as the standard for the world? I doubt it.

What would happen if one or more countries don't want to join the plan? Would they have any choice in the matter?

What if the one world government turns sour? Would we just have to accept it, the way Romans had to accept their insane caesars? A one-world government wouldn't guarantee an end to egomaniacs; instead, it might make it easier for them to take over.

How could refugees from one part of the world feel safe from those who are persecuting them if, wherever they fled, they were still subject to the same authority?

Does one-worldism mean that tribes and ethnic groups would lose their identity? Would it mean no national flags, no national anthems? For whom would we cheer at the Olympics? What would be the point of the World Cup?

International government enables corruption. A reliable principle is that local government works best. The closer you can keep governmental responsibility and authority to the voters, the more accountable government will be to the people. Imagine having to contact Brussels or Beijing if you have a Social Security or Medicare problem. In the maze of bureaucracy that a one-world government would require, corruption would be much easier to hide. Law enforcement would be subject to the influence of bribes and prejudice (such as the authorities enforcing a law in one country but letting enforcement slide in another country that the ruling party favors). All people are created with equal value before God, but no country has practiced absolute equality in all things. Even in supposedly egalitarian systems, some people manage to acquire extra wealth and power. The same would be true under a one-world government, and fighting it would be harder to do.

Many of those who promote a one-world government feel dissatisfaction with if not contempt for their own country, for what is near and familiar. They see failings where they are and think things would be better if they didn't have to deal with those failings—not admitting that the failings of a one-world system would probably be worse.

And Then . . .

As we pointed out earlier, we live in a highly-interconnected world. International travel, trade, and communication have become taken for granted. Americans obtain a huge amount of our goods—and some services—from China. Thousands of Chinese students are enrolled in American universities and in universities in other countries. China's Belt and Road Initiative has connected China with many other countries.

Then came the COVID-19 coronavirus. The world practically came to a standstill as the virus, which originated in China and spread around the world by the same means of travel and distribution people used for other things, sickened millions, took the lives of hundreds of thousands, and shut down society and economic activity. All of a sudden, globalism didn't seem like such a good idea.

A Utopian Dream

Simply put, one-worldism is a utopian dream. How many utopian experiments do we need before we admit that utopian visions do not work? Those who would form and run a world government would be those who think they are smart enough to run the whole show and everyone's lives in it, and they would certainly want everyone to conform to one way of life: theirs. How likely would it be that a conservative or a nationalist would be welcome in those halls of power? The one-world ideal might seem like a good idea if your people are in charge; but if people from the other party win a majority, watch out.

A one-world system simply will not work. People would devolve into groups with competing visions and agendas, which we already have. One group or approach would be in, while another group would

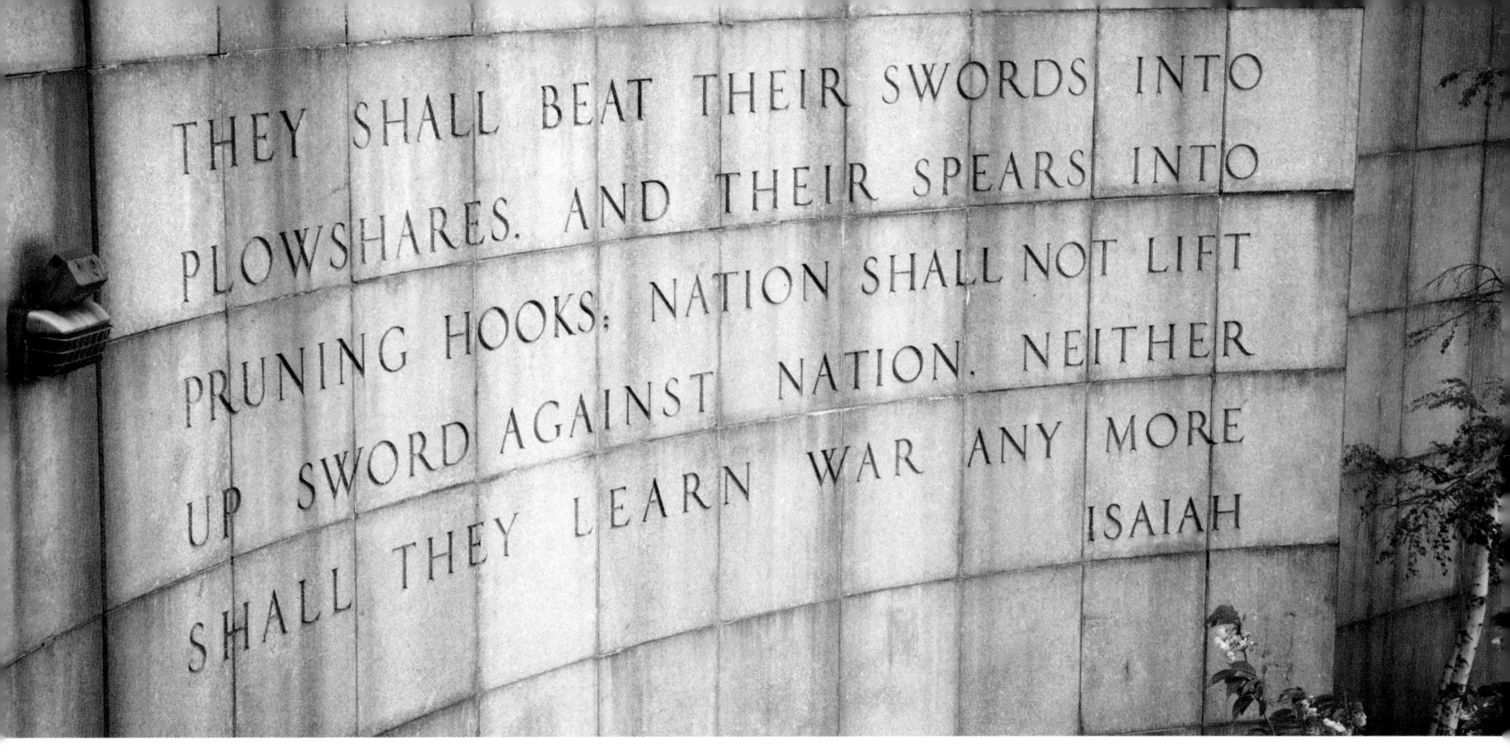

Ralph Bunche Park sits across the street from the United Nations building in New York City. Ralph Bunche (1904-1971) spent his life advocating and working for peace and justice. He received the Nobel Peace Prize in 1950 for his work negotiating a cease-fire between Jews and Arabs in the Middle East, becoming the first person of color so honored. A portion of Isaiah was inscribed on this wall in the park in 1975. It beautifully expresses the dream of all people who seek peace, but it leaves out an important part of Isaiah's message—about God's role in bringing peace to the nations.

be out. The United Nations and the European Parliament haven't solved all of the problems they face; why should we think an even more sweeping approach along the same lines would work?

All political philosophies have weaknesses that lead to problems. What we must do is weigh the various options to see which one has the fewest or the most manageable problems. Bottom line, nationalism, despite its occasional failings, is preferable to one-worldism with its rare successes.

We live in a globally connected world. This creates possibilities for good things, such as more widespread teaching of the gospel and being able to help more people lead better lives. But a globally connected world is not the same thing as a one-world government concept of globalism. As we have seen, this idea is fraught with problems.

A Different Vision

The Lord does not subdivide the church into American, Brazilian, Canadian, Chinese, Danish, Ethiopian, Indian, Kenyan, Korean, and Russian churches. God's ideal for the church is that all Christians around the world be one without division (Galatians 3:28, Colossians 3:11, Revelation 5:9).

On the other hand, the ideal for mankind is not a one-world government but the honoring of national identity that makes us all better people. The difference involves who governs each realm. Christ, the perfect Ruler, guides the church; and Christians need no other ruler over that realm. On the other hand, only flawed human beings rule in this world. What human being would be so completely deserving of our trust as Jesus is for the church?

When Jesus issued the Great Commission to make disciples of all nations (the word for nation in Greek is *ethne*, from which we get the word ethnic; see Matthew 28:19-20), He recognized the reality of the nations. He didn't say we should do away with the nations. The same is true with the passages in Revelation that describe the gathering in heaven of people from every tribe and tongue and people and nation.

Isaiah 2:4, quoted below, expresses the goal of globalists and of all people, that warfare would end and people would not rise up against each other. But in this Messianic prophecy, notice that the verse again does not advocate the disappearance of nations as the ideal but that the nations not fight each other. It is to that end that we should devote our efforts and prayers.

He will judge between the nations
and will settle disputes for many peoples.
They will beat their swords into plowshares
and their spears into pruning hooks.
Nation will not take up sword against nation,
nor will they train for war anymore.
Isaiah 2:4

Assignments for Lesson 149

Project	Continue working on your project.
Student Review	Answer the questions for Lesson 149.

Mount Srd, Croatia

150 Only Jesus Can Make the World One

In this curriculum we have presented the world in its vast diversity of geographic locations and descriptions, ethnic groups, cultures, and belief systems. Unfortunately, people have often used the differences as reasons for conflict. In fact, it didn't take long in the history of mankind for people to become divided.

The sins of Adam and Eve separated them from God.

Cain was so alienated from Abel that he murdered his brother.

Human sin and arrogance at Babel led God to separate the people geographically and by languages.

Then came war.

Genesis 14 records a war among the kings of early city-states in the Middle East.

What we know about the history of the world reveals that city-states and nations fought with each other on every inhabited continent. One group wanted the land or resources that another group had, or one group believed that another group had offended them and wanted to restore their honor.

Dynasties in China rose and fell. Athens and Sparta fought each other several times, as did the Greeks and the Persians. Rome conquered many other peoples all over Europe and into the Middle East and North Africa. Within the late Roman Empire, competing armies and their leaders (who were hoping to become emperor) fought often.

Muslims carried out jihad against Jews and Christians, and Christians fought Muslims in the Crusades.

Protestants and Catholics fought each other in the religious wars in Europe following the Reformation.

European explorers fought native nations in the Americas.

Prejudice

Matters don't have to descend into warfare for people not to get along with each other. People find many reasons to divide one group from another. The poor may resent the rich while the rich look down on the poor. Both conservatives and liberals may think the other side does nothing right and their own side does nothing wrong.

The hatred and distrust between Jews and Gentiles in the ancient world was legendary. Jews saw Gentiles as dirty dog sinners. Gentiles saw Jews as eccentrics who had weird ideas about what they could eat and touch and as people who wouldn't just go along with everybody else.

Jews saw Samaritans as infidel half-breeds. As the Gospel of John put it, "Jews have no dealings with Samaritans" (John 4:9).

This didn't end with the coming of the church. Jewish Christians and Gentile Christians had a hard time accepting each other and getting along with each other as brothers. The attitude of some Jewish brethren toward Gentile believers as portrayed in Acts 15 shows that the Jewish Christians' worldview hadn't changed enough for them fully to accept their new Gentile brothers in Christ. Roman Catholics and Orthodox Christians distanced themselves from each other. The divisions within Christendom are many.

Then we come to ethnic and national prejudices. People with different skin colors have been prejudiced against each other. The English and the French have often not gotten along. The French and the Germans have often not gotten along. The Chinese and the Vietnamese have had conflicts. The Chinese and the Koreans have been at odds.

And have you ever heard of Northerners and Southerners in the United States harboring negative attitudes about each other? How about red-staters and blue-staters?

Jesus Remade Human Geography

Into this maze of conflict, prejudice, suspicion, and hatred, Jesus brought a new way of seeing other people and relating to them. He ministered to a Samaritan woman who needed to turn her life around (John 4:1ff). He rebuked James and John when they wanted to call down fire from heaven on a Samaritan village (Luke 9:54-56). Jesus complimented the faith of a Gentile Roman centurion, saying He had not seen such great faith in Israel (Luke 7:9). He ministered to the daughter of a Gentile Syrophoenician woman (Mark 7:25-30). He spent time in the largely Gentile area east of the Sea of Galilee (Mark 5:1ff).

The apostles and others in the early church recognized this new way that Jesus demonstrated of relating to people who were from other groups. Philip preached the gospel among Samaritans (Acts 8:5). Peter preached the gospel to the Gentile Cornelius (Acts 10). Paul and Barnabas took the gospel to the Gentiles after the Jews at Pisidian Antioch rejected their message (Acts 13:46). This new worldview didn't come easily, but it came.

In his letters, Paul taught Christians how Jesus had brought a revolution in human relations by breaking down the barriers that people had erected. Paul told the Christians at Ephesus that:

> He Himself is our peace, who made both groups into one and broke down the barrier of the dividing wall, by abolishing in His flesh the enmity, which is the Law of commandments contained in ordinances, so that in Himself He might make the two into one new man, thus establishing peace.
>
> (Ephesians 2:14-15)

In Christ there were no longer to be Jews and Gentiles, or even Jewish Christians and Gentile Christians, but one new kind of person, a new humanity: Christians.

In Galatians Paul says that in Christ there is no rank of importance, no second-class citizens of the kingdom:

> For you are all sons of God through faith in Christ Jesus. For all of you who were baptized into Christ have clothed yourselves with Christ. There is neither Jew nor Greek, there is neither slave nor free man, there is neither male nor female; for you are all one in Christ Jesus. And if you belong to Christ, then you are Abraham's descendants, heirs according to promise.
>
> (Galatians 3:26-29)

He makes a similar point in Colossians when discussing the new people they had become:

> . . . a renewal in which there is no distinction between Greek and Jew, circumcised and uncircumcised, barbarian, Scythian, slave and freeman, but Christ is all, and in all.
> (Colossians 3:11)

Other Attempts at Stopping Conflict

People have suggested and tried many approaches to world peace, but they have all failed to one degree or another. In the 1920s, countries signed disarmament treaties hoping to stop war, only to endure the most costly war in history a few years later. The United Nations has attempted to stop fighting in some places in the world, but over seventy years after its founding the world still deals with many trouble spots. When a coalition of nations joined together to oust Iraq from Kuwait in 1990-91, just as the Soviet Union was crumbling and its hold on Eastern Europe had ended, some leaders spoke of a "new world order." Unfortunately, a few years later whatever new world order had come into being looked a great deal like the old world order, just with different antagonists. Where popular movements have overthrown dictators in country after country (as in the Balkans, Libya, and Iraq), old ethnic hatreds and rivalries have emerged and caused renewed violence.

Why Only Jesus?

Since Jesus made us, He should know best how to bring divided people groups back together. Since He died for us, how could we insist on our way above His way when dealing with others? Since He is willing to accept others who come to Him by faith, we should accept them as well. Jesus based His call to follow Him on denying oneself. Self-denial precludes insisting on one's own way.

Jesus taught His disciples to love one another. If we will do that, our divisions will fade. Believers must overcome their divisions in order for the gospel to have more credibility with unbelievers.

The way of Jesus involves a person changing within himself. This enables us to see others in a totally new way, free of the reasons (or excuses) people tend to give for being divided from one another.

Jesus' prayed that His disciples would be one, just as He and the Father are one (John 17:20-21). To honor Him, we should seek to be the answer to His prayer. No one else besides our sovereign Lord can offer such convincing reasons for bringing people together.

This 11th-century illustration by the Spanish monk Facundus depicts the New Jerusalem described in the Book of Revelation. It shows a multi-ethnic group of people surrounding God's throne.

The Redemption of Human Geography

The study of humans' interaction with their geography tells of the division that the Lord brought about at Babel, when He caused them to speak different languages and scattered them across the face of the earth. At Pentecost, the Lord reversed Babel by granting the people who were assembled that day the ability to speak in foreign languages in order to proclaim the one gospel as they ventured to the uttermost parts of the earth (Acts 2).

In the book of Revelation, the Lord gave John the following vision of those who would be assembled around His great white throne. With these appropriate and inspiring words we close this study of people and their geography. May the Lord be honored by our study and our lives.

After these things I looked, and behold, a great multitude which no one could count, from every nation and all tribes and peoples and tongues, standing before the throne and before the Lamb, clothed in white robes, and palm branches were in their hands; and they cry out with a loud voice, saying, "Salvation to our God who sits on the throne, and to the Lamb."
Revelation 7:9-10

Assignments for Lesson 150

Worldview Recite or write the memory verse for this unit.

Project Finish your project for this unit.

Student Review Answer the questions for Lesson 150.
Take the quiz for Unit 30.
Take the sixth Geography, English, and Worldview exams.

Celebrate finishing *Exploring World Geography!*

Mariano Moreno National Library, Buenos Aires, Argentina

Sources

General Sources

Central Intelligence Agency World Fact Book. www.cia.gov/library/publications/the-world-factbook/

Cordesman, Anthony. "Islam and the Patterns in Terrorism and Violent Extremism." Center for Strategic and International Studies, csis.org. October 17, 2017, accessed July 4, 2018.

Encyclopedia Britannica. www.britannica.com

Josephy, Alvin M., ed. *Africa: A History*. New York: Horizon/New Word City (American Heritage), 2016

Kass, Leon. *Leading a Worthy Life: Finding Meaning in Modern Times*. New York: Encounter Books, 2017.

Koran, The. N. J. Dawood, translator. New York: Penguin Books, revised ed. 1993.

Library of Congress Federal Research Division. Country Profiles. https://www.loc.gov/rr/frd/cs/profiles

Lipka, Michael. "Muslims and Islam: Key findings in the U.S. and around the world." Pew Research. pewresearch.org. August 9, 2017, accessed July 4, 2018.

Marshall, Tim. *Prisoners of Geography*. New York: Scribner, 2015.

Melamed, Avi. *Inside the Middle East*. New York, Skyhorse Publishing, 2016

Ofek, Hillel. "Why the Arabic World Turned Away from Science." The New Atlantis, Winter 2011. www.thenewatlantis.com. Accessed June 25, 2018.

Sire, James W. *The Universe Next Door: A Basic Worldview Catalog, Fifth Edition*. Downers Grove, Illinois: InterVarsity Press, 2009

Sowell, Thomas. *Wealth, Poverty and Politics: An International Perspective*. New York: Basic Books, 2015.

Willis, John T. *Genesis. Living Word Commentary Series*. Austin: Sweet, 1979.

World Book Encyclopedia. www.worldbook.com

Lesson 76

Barba, Nicola. *Afghanistan*. Mankato, Minnesota: Arcturus Publishing/Black Rabbit Books, 2008.

Latifi, Ali M. "New School Brings Afghan Students In From the Cold." The United Nations Refugee Agency. www.unhcr.org. October 24, 2019, accessed November 22, 2019.

Bostock, Bill. "Afghanistan Shares a 46-mile Border with China—Here's the Intriguing Story of

How the 2 Countries Became Neighbors." Business Insider, www.insider.com. June 27, 2019, accessed November 26, 2019.

Lesson 77

Hodal, Kate. "'My Dream Is Coming True': The Nepalese Woman Who Rose from Slavery to Politics." www.the guardian.com, October 18, 2017, accessed November 22, 2017.

Kumar, Nikhil. "Why Nepal Is Still in Rubble a Year After a Devastating Earthquake." www.time.com, April 24, 2016, accessed November 22, 2017.

Preiss, Danielle. "Why Nepal Has One of The World's Fastest-Growing Christian Populations." www.npr.org, February 3, 2016, accessed November 22, 2017.

Lesson 78

Diouf, Sylviane. "Africans in India: From Slaves to Generals and Rulers." New York Public Library, www.nypl.org. March 11, 2013, accessed September 12, 2019.

Jones, Timothy. "India's Caste System: Weakened, But Still Influential." www.dw.com. July 17, 2017, accessed September 11, 2019.

Nelson, Dean. "India's Lowest 'Untouchable' Caste Rejoices as Ratcatcher Sworn in as Bihar Chief Minister." *The Telegraph*, www.telegraph.co.uk. May 20, 2014, accessed September 11, 2019.

Vallangi, Neelima. "India's Forgotten African Tribe." www.bbc.com. August 4, 2016, accessed September 12, 2019.

Lesson 79

"Edmund Hillary and Tenzing Norgay Reach Everest Summit, May 29, 1953." www.history.com. November 24, 2009, updated July 28, 2019, accessed October 22, 2019.

"Interview with Sir Edmund Hillary—Mountain Climbing." www.media.smithsonianfolkways.org. 1974, accessed October 23, 2019.

Roberts, David. "Everest 1953: First Footsteps—Sir Edmund Hillary and Tenzing Norgay. www.nationalgeographic.com, April 2003, accessed October 22, 2019.

Salkeld, Audrey. "First to Summit." www.pbs.org, November 2000, accessed October 22, 2019

Lesson 81

Chen, Dene-Hern. "Once Written Off for Dead, the Aral Sea Is Now Full of Life." www.nationalgeographic.com, March 16, 2018, accessed October 17, 2018.

"Dried-up Aral Sea Springs Back to Life." France24 video on YouTube, posted September 18, 2017, accessed September 12, 2018.

Qobil, Rustam. "Waiting for the Sea." www.bbc.com, February 25, 2015, accessed October 20, 2019.

Lesson 82

"Alphabet Soup as Kazakh Leader Orders Switch from Cyrillic to Latin Letters." Reuters via www.theguardian.com. October 26, 2017, accessed October 24, 2019.

Lesson 83

Clot, Lloudmila. "Cotton: Tajikistan's Pride and Shame." www.swissinfo.ch. September 25, 2015, accessed November 20, 2019.

"Country Profile: Tajikistan." Library of Congress Federal Research Division. www.loc.gov. January 2007, accessed November 20, 2019.

Pannier, Bruce. "Tajikistan's Civil War: A Nightmare the Government Won't Let Its People Forget." Radio Free Europe/Radio Liberty. www.rferl.org. June 23, 2017, accessed November 20, 2019.

Smith, Jake. "Good Manners in Central Asia: At the Table." www.mircorp.com. Accessed November 20, 2019.

Sobiri, Bakhtiyor. "The Long Echo of Tajikistan's Civil War." www.opendemocracy.net. June 23, 2017, accessed November 20, 2019.

"Tajikistan Starts Up First Turbine in Dam Set To Be World's Tallest." www.reuters.com. November 16, 2018, accessed November 19, 2019.

"This Is How Bread Takes Shape in Tajikistan." www.smithsonianmag.com. August 12, 2015, accessed November 20, 2019.

Ubaidulloev, Zubaidullo. "The Russian-Soviet Legacies in Reshaping the National Territories in Central Asia: A Catastrophic Case of Tajikistan." Journal of Eurasian Studies, January 2015. www.sciencedirect.com, accessed November 20, 2019.

Lesson 84

Geiling, Natasha. "This Hellish Pit Has Been On Fire for More Than 40 Years." www.smithsonianmag.com. May 20, 2014, accessed November 19, 2019.

Pillalamarri, Akhilesh. "The Epic Story of How the Turks Migrated from Central Asia to Turkey." www.thediplomat.com. June 5, 2016, accessed November 19, 2019.

Pope, Hugh. *Sons of the Conquerors: The Rise of the Turkic World*. New York: Overlook Duckworth, 2005.

Lesson 86

Albert, Eleanor. "China-Taiwan Relations." Council on Foreign Relations, www.cfr.org. Updated October 4, 2019, accessed October 27, 2019.

Chen, Fang-Yu. "The Taiwanese See Themselves as Taiwanese, Not as Chinese." www.washingtonpost.com. January 2, 2017, accessed October 27, 2019.

Jennings, Ralph. "Isolated Diplomatically by China, Taiwan Is Finding Friends in Europe." Voice of America, www.voanews.com. October 18, 2019, accessed October 27, 2019.

Taylor, Alan. "Taiwan's Kinmen Islands, Only a Few Miles from Mainland China." www.theatlantic.com. October 8, 2015, accessed October 27, 2019.

Lesson 87

Fifield, Anna. "Life Under Kim Jong Un." www.washingtonpost.com, November 17, 2017, accessed November 9, 2019.

Kim, Eunsun and Sebastien Falletti. *A Thousand Miles to Freedom: My Escape from North Korea*. New York: St. Martin's Press, 2012, 2015. (Note: Every account I have read of life in North Korea and the experiences of those who escape has a little harsh language and some very troubling scenes. This is because of the nature of the situation that these people have faced.)

Lee, Hyeonseo. *The Girl with Seven Names*. New York: HarperCollins, 2015 (and TED Talk)

Smith, Samuel. "Ex-North Korea Military Officer Once Loved Regime More Than Christ, Now He's Helping Victims Escape." www.christianpost.com. May 13, 2019, accessed November 9, 2019.

Lesson 88

www.ricepedia.com (a resource of the International Rice Research Institute, AfricaRice, and the International Center for Tropical Agriculture), accessed September 28, 2019.

Lesson 89

"How to Host a Japanese Tea Ceremony." www.republicoftea.com, July 18, 2016. Accessed January 23, 2019

Eplett, Layla. "In the Japanese Tea Ceremony, Politics Are Served with Every Cup." www.npr.com, June 23, 2015. Accessed January 23, 2019.

Willmann, Anna. "The Japanese Tea Ceremony." Heilbrunn Timeline of Art History, Metropolitan Museum of Art. www.metmuseum.org, April 2011. Accessed January 23, 2019.

Heiss, Mary Lou and Robert J. Heiss. *The Story of Tea: A Cultural History and Drinking Guide.* New York: Ten Speed Press, 2007.

Surak, Kristin. "Making Tea, Making Japan." Lecture, UC Davis College of Letters and Science, May 12, 2016. Published on YouTube June 7, 2016; accessed January 24, 2019.

Lesson 91

Senthilingham, Meera. "How Quickly Can China Come Back from its One-child Policy?" www.cnn.com, October 13, 2016.

Clarke, Aileen. "See How the One-child Policy Changed China." www.nationalgeographic.com, November 15, 2015.

"China sees millions travel home for Lunar New Year." cbsnews.com, February 14, 2018, accessed September 19, 2018

Lesson 92

Byler, Darren. "China's Hi-Tech War on Its Muslim Minority." www.the guardian.com. April 11, 2019, accessed November 29, 2019.

"China Spiriting Uyghur Detainees Away from Xinjiang to Prisons in Inner Mongolia, Sichuan." Radio Free Asia, www.rfa.org. February 21, 2019, accessed November 29, 2019.

Ng, Eileen and Ken Moritsugu, "Hong Kong Leader Vows to 'Seriously Reflect' on Election. Associated Press story in The Daily Herald, Columbia, Tennessee, www.columbiadailyherald.com. November 25, 2019, accessed November 26, 2019.

Pillalmarri, Akhilesh. "Troubled Today, China's Xinjiang Has a Long History." www.the diplomat.com. July 30, 2015, accessed November 27, 2019.

Schmitz, Rob. "Uighurs Held for 'Extremist Thoughts' They Didn't Know They Had. www.npr.org. May 7, 2019, accessed November 29, 2019.

Samuel, Sigal. "China Is Going to Outrageous Lengths to Surveil Its Own Citizens." www.theatalantic.com. August 16, 2018, accessed November 29, 2019.

"The Uyghur Emergency." www.nationalreview.com. November 27, 2019, accessed November 27, 2019.

"Why Is There Tension Between China and the Uighurs?" www.bbc.com. September 26, 2014, accessed November 27, 2019.

Lesson 94

Edmons, John. "The History of the Shipping Container." www.freightos.com. April 14, 2016, accessed December 3, 2019.

"The History of Containers." www.plslogistics.com. January 21, 2015, accessed December 3, 2019

"JingJinJi, a Chinese Megalopolis in the Making." CKGSB Knowledge. www.knowledge.ckgsb.edu.cn. August 20, 2015, accessed December 3, 2019.

Levinson, Marc. *The Box: How the Shipping Container Made the World Smaller and the World Economy Bigger. 2nd ed.* Princeton, New Jersey: Princeton University Press, 2016.

"The People's Republic of China." Office of the United States Trade Representative. www.ustr.gov. Accessed December 3, 2019.

"Shipping Container History: Boxes to Buildings." www.discovercontainers.com. Accessed December 3, 2019.

"U.S. Trade with China: Selected Resources." Library of Congress Resource Guides. www.loc.gov, accessed December 3, 2019.

Lesson 95

Dennis Prager. "Leftism as a Secular Religion." Heritage Foundation, www.dailysignal.com, August 28, 2018, accessed August 31, 2018.

Bertrand Russell, *Why I Am Not a Christian* (New York: Simon and Schuster, 1957), p. 106.

Lesson 96

"Maritime Zones and Boundaries." Office of General Counsel, National Oceanic and Atmospheric Administration. www.gc.noaa.gov. Updated October 17, 2018. Accessed October 22, 2018.

"Why Is the South China Sea Contentious?" www.bbc.com, July 12, 2016, accessed October 8, 2018.

Jennings, Ralph. "Japan Dials Up Pressure on China Over Southeast Asian Sea." voanews.com, October 12, 2018, accessed October 17, 2018.

"U.S. Navy: Chinese Warships Maneuvered in 'Unprofessional' Manner." www.cbsnews.com, May 29, 2018, accessed May 29, 2018.

Jennings, Ralph. "French, British Ships to Sail Disputed Asian Sea, Rile China." www.voanews.com, June 15, 2018, accessed August 25, 2018.

Lesson 97

Chamberlain, Samuel. "ISIS Claims Responsibility for Philippines Church Bombing That Killed At Least 20." www.foxnews.com, January 27, 2019, accessed January 28, 2019.

Cornelio, Jayeel. "How the Philippines Became Catholic." www.christianitytoday.com. March 9, 2018, accessed January 28, 2019.

Jennings, Ralph. "Energy Deal Deepens Once Unimagined Sino-Philippine Friendship." www.voanews.com, November 30, 2018. Accessed January 2, 2019.

____________. "Sino-Philippine Fishing Deal Would Go Long Way, Experts Say." www.voanews.com, June 18, 2018, accessed June 25, 2018.

Lema, Karen. "Philippine Congress Passes Autonomy Bill for Volatile Muslim Region." www.reuters.com, May 30, 2018, accessed June 24, 2018.

Naholowa'a, Leiana S. A. "Stops Along the Manila Galleon Trade Route." www.guampedia.com, June 26, 2018, accessed January 29, 2019.

Lesson 98

Bisharat, Andrew. "Explore the World's Biggest Cave From Your Couch." www.nationalgeographic.com. May 2, 2018, accessed December 6, 2019.

Rosen, Elisabeth. "How Young Vietnamese View the Vietnam War." www.theatlantic.com. April 30, 2015, accessed December 6, 2019.

Salem, Jarryd. "Explore Hang Son Doong, the World's Largest Cave." www.cnn.com/travel. August 15, 2017, accessed December 6, 2019.

Sullivan, Michael. "Ask the Vietnamese About War, And They Think China, Not the U.S." www.npr.org. May 1, 2015, accessed December 9, 2019.

"The World Bank in Vietnam." www.worldbank.org. October 18, 2019, accessed December 6, 2019.

Lesson 99

Bendix, Aria. "Indonesia Is Spending $33 Billion to Move Its Capital from a Sinking City to an Island Where Forests Have Been Burning." www.businessinsider.com. August 27, 2019, accessed January 2, 2020.

Chappell, Bill. "Jakarta Is Crowded and Sinking, So Indonesia Is Moving Its Capital to Borneo." www.npr.org. August 26, 2019, accessed January 2, 2020.

Rooney, Anne. Tsunami! Mankato, Minnesota: Arcturus Publishing/Black Rabbit Books, 2006.

Sturdy, E. W. "The Volcanic Eruption of Krakatoa." www.theatlantic.com. September 1884, accessed December 11, 2019.

"Volcano World: Describe the 1883 Eruption of Krakatau." www.volcano.oregonstate.edu. Accessed December 11, 2019.

Lesson 101

"The Arrival of Maori." www.newzealand.com. Accessed December 17, 2019.

"History of the Maori Language." www.nzhistory.govt.nz. Accessed December 17, 2019

Lesson 102

Heiss, Anita and Melodie-Jane Gibson. "Aboriginal People and Place." www.sydneybarani.com.au. Accessed January 2, 2020.

White, Dee. "The Biggest Natural Harbours in the World." www.theyachtmarket.com. July 25, 2018, accessed January 2, 2020.

Lesson 103

Hinson, Tamara. "Welcome to Coober Pedy, the World's Strangest Town." www.telegraph.co.uk. March 21, 2019, accessed January 14, 2020.

"History of Uluru-Kata Tjuta National Park." www.parksaustralia.gov. Accessed January 10, 2020.

Lerwill, Ben. "The Strange Story of Australia's Wild Camel." www.bbc.com. April 11, 2018, accessed January 10, 2020.

"The Modern Outback." www.pewtrusts.org. Accessed January 9, 2020.

Nalewicki, Jennifer. "Half of the Inhabitants of This Australian Opal Capital Live Underground." www.smithsonianmag.com. March 2, 2016, accessed January 14, 2020.

Lesson 104

"Christmas Island Environment and Heritage." www.regional.gov.au. Accessed January 14, 2020.

Geoscience Australia (information about several of the External Territories). ga.gov.au. Accessed January 14, 2020.

"Great Barrier Reef—UNESCO World Heritage Centre." www.unesco.org. Accessed January 14, 2020.

"History of the Great Barrier Reef." www.greatbarrierreef.org. Accessed January 14, 2020.

"Norfolk Island National Park." www.parksaustralia.gov.au. Accessed January 14, 2020.

"Red Crab Migration." www.parksaustralia.gov.au. Accessed January 14, 2020.

Lesson 105

"Cactus proved to be a prickly problem." www.examiner.com.au, September 23, 2016. Retrieved January 12, 2018

Evins, Brittany. "Hungry cochineal bug a sharp solution to invasive wheel cactus problem." www.abc.net.au, May 29, 2017. Retrieved January 12, 2018.

"Famous Creation Scientists: Interview with John Mann, M.B.E." www.answersingenesis.org. Retrieved November 22, 2017.

Queensland Historical Atlas, s.v. "Prickly Pear." www.qhatlas.com.au/content/prickly-pear. Retrieved January 12, 2018

Lesson 106

Paine, Lincoln. *The Sea and Civilization: A Maritime History of the World*. New York: Alfred A. Knopf, 2013.

Sobel, Dava. *Longitude: The True Story of a Lone Genius Who Solved the Greatest Scientific Problem of His Time*. New York: 2010.

Lesson 107

Emslie, Karen. "These People Have a Mind-Bending Way to Navigate." www.nationalgeographic.com. April 13, 2016, accessed January 16, 2020.

Lyons, Kate. "Bougainville referendum: region votes overwhelmingly for independence from Papua New Guinea." www.theguardian.com. December 10, 2019, accessed January 17, 2020.

"Papua New Guinea Profile—Timeline." www.bbc.com. February 14, 2018, accessed January 17, 2020.

Westerman, Ashley. "Trying to Form the World's Newest Country, Bougainville Has a Road Ahead." www.npr.com. December 30, 2019, accessed January 16, 2020.

Lesson 108

Herman, Doug. "A Brief, 500-Year History of Guam." www.smithsonianmag.com. August 15, 2017, accessed January 20, 2020.

Lanchin, Mike. "Shoichi Yokoi, the Japanese Soldier Who Held Out in Guam." www.bbc.com. January 24, 2012, accessed January 20, 2020.

Tolentino, Dominica. "WWII: Sgt. Shoichi Yokoi, Last Straggler on Guam." www.guampedia.com. Accessed January 20, 2020.

Lesson 109

Bailey, Anthony. "The Muse of the South Pacific: Literary Encounters with the South Pacific." Island Life Magazine, www.islandlifemag.com. August 4, 2017, accessed January 30, 2030.

Rampell, Ed. "Melville's Paradise." www.chicagotribune.com. March 21, 1993, accessed January 20, 2020.

Lesson 111

Hambling, David. "Does the U.S. Stand a Chance Against Russia's Icebreakers?" www.popularmechanics.com. April 4, 2018, accessed February 6, 2020.

"Northwest Passage." History Channel. www.history.com. Updated September 25, 2019, accessed February 6, 2020.

"Roald Amundsen North-West Passage Expedition 1903-06." Royal Museums Greenwich. www.rmg.co.uk. Accessed February 6, 2020.

Lesson 112

The Canadian Encyclopedia. https://www.thecanadianencyclopedia.ca

Zajac, Ronald. "St. Lawrence Seaway at 60: The project that changed the region." https://montrealgazette.com/news/local-news/seaway-at-60/st-lawrence-seaway-at-60-the-project-that-changed-the-region

Discover Canada - Canada's History. https://www.canada.ca/en/immigration-refugees-citizenship/corporate/publications-manuals/discover-canada/read-online/canadas-history.html

Dictionary of Canadian Biography. http://www.biographi.ca

Lesson 113

Remini, Robert V. *The Battle of New Orleans.* New York: Viking, 1999.

Lesson 114

Aldrich, Marta. "$83 Million Later, Unfinished Dam Being Dismantled." Associated Press story at www.seattletimes.com. October 10, 1999, accessed January 31, 2020.

"The Geopolitics of the United States, Part 1: The Inevitable Empire." www.worldview.stratfor.com. July 4, 2016, accessed January 31, 2020.

"Inland Waterway Navigation: Value to the Nation." U.S. Army Corps of Engineers Institute for Water Resources. www.corpsresults.us. May 2000, accessed January 31, 2020.

"Mississippi River Facts." National Park Service. www.nps.gov. Updated November 24, 2018, accessed January 31, 2020.

"Places We Protect: The Duck River." The Nature Conservancy. www.nature.org. Accessed January 31, 2020.

Waterman, Jonathan. "The American Nile." www.nationalgeographic.com. Accessed January 31, 2020.

Lesson 115

Lenger, John. "Murray: Surgeon with Soul." The Harvard Gazette. www.news.harvard.edu/gazette, October 4, 2001, accessed August 14, 2019.

Murray, Joseph E. "Joseph E. Murray Biographical." https://www.nobelprize.org/prizes/medicine/1990/murray/facts/. Accessed August 16, 2019.

"Nobel Laureate Joseph E. Murray: There Is No Conflict Between Science and Religion." http://2012daily.com/community/blogs/138. January 15, 2012, accessed August 14, 2019.

Tullius, Stefan and Michael Zinner. "Joseph E. Murray" https://fa.hms.harvard.edu/files/memorialminute_murray_joseph_e.pdf, accessed August 14, 2019.

Lesson 116

Kevin J. Avery. "The Hudson River School." Heilbrunn Timeline of Art History. Metropolitan Museum of Art, metmuseum.org, October 2004, accessed May 16, 2018.

Lesson 117

www.babcockranch.com

"Inside Florida's Solar-powered Babcock Ranch." CBS This Morning, January 16, 2018. Accessed on YouTube January 21, 2019.

"Can an Entire Town Run on Solar?" PBS Newshour, November 11, 2018. Accessed on YouTube January 21, 2019.

"GM's Spring Hill Manufacturing Plant to Run on Solar Power by 2023." www.columbiadailyherald.com, May 15, 2020, accessed May 17, 2020.

Pasricha, Anjana. "Southern India Boasts World's First Fully Solar Powered Airport." www.voanews.com, January 27, 2019, accessed January 28, 2019.

Lesson 118

Hawksley, Humphrey. "The Ice Curtain That Divides U.S. Families From Russian Cousins." www.bbc.com. August 31, 2015, accessed February 3, 2020.

Watts, Simon. "Swim That Broke Cold War Ice Curtain." www.bbc.com. August 8m 2012, accessed February 4, 2020.

Wigo, Bruce. "30 Years Ago Today: How Lynne Cox Eased Cold War Tensions by Swimming." www.swimmingworldmagazine.com. December 9, 2017, accessed February 4, 2020.

Lesson 119

Anderson, William. *Laura Ingalls Wilder: A Biography.* New York: Harper Collins, 1992.

McDowell, Marta. *The World of Laura Ingalls Wilder: The Frontier Landscapes that Inspired the Little House Books.* Portland, Oregon: Timber Press, 2017.

Lesson 121

Bendix, Aria. "New 3D-Printed Homes in Mexico Will Cost Their Low-income Residents Just $20 Per Month. Take a Look Inside." www.businessinsider.com. December 13, 2019, accessed February 8, 2020.

Delbert, Caroline. "First 3D-Printed Neighborhood Now Has First 3D-Printed Houses." www.popularmechanics.com. December 12, 2019, accessed February 8, 2020.

"Tabasco." www.history.com/topics/mexico/tabasco. Updated August 21, 2018, accessed February 8, 2020.

"World's First Community of 3D Printed Homes Is Set to House Mexico's Poorest Families." www.goodnewsnetwork.org. December 12, 2019, accessed February 8, 2020.

Lesson 122

Argueta, Al. "The Terrain and Geography of Guatemala." www.hatchettebookgroup.com. Accessed February 13, 2020.

Gallucci, Maria. "IBT Special Report." International Business Times, www.ibt.com. November 23, 2015, accessed February 13, 2020.

Guo, Eileen. "The Fight to Protect the World's Most Trafficked Wild Commodity." www.nationalgeographic.com. August 16, 2019, accessed February 13, 2020.

Interview with Keith Pinson, missionary in Guatemala and Honduras. February 13, 2020.

Lesson 123

"Coffee Culture in Costa Rica." www.costarica.com. Updated February 7, 2012, accessed February 18, 2020.

"Coffee in Costa Rica." www.anywhere.com. Accessed February 18, 2020.

"Costa Rica History." www.entercostarica.com. Accessed February 17, 2020.

"Ecotourism in Costa Rica." www.greenglobaltravel.com. Accessed February 19, 2020.

Fendt, Lindsay. "All That Glitters Is Not Green: Costa Rica's Renewables Conceal Dependence on Oil." www.theguardian.com. January 5, 2017, accessed February 18, 2020.

Fetters, K. Alisha. "Why Ecotourism Is Booming." U.S. News and World Report, www.travel.usnews.com. November 16, 2017, accessed February 18, 2020.

"Geography of Costa Rica." www.costarica.com. Updated June 13, 2015, accessed February 17, 2020.

Hickman, Leo. "Shades of Green." www.theguardian.com. May 26, 2007, accessed February 17, 2020.

"History of Costa Rica." www.costarica.com. Updated June 13, 2015, accessed February 17, 2020.

Lesson 124

"Belize History." www.belize.com. Accessed April 6, 2020.

"Country Profile - Belize." New Agriculturalist, www.new-ag.info. September 2005, accessed April 6, 2020.

"Find Out Everything from Currency to Culture Here in Belize." www.travelbelize.org. Accessed April 6, 2020.

Griffin, Jo. "Why Tiny Belize Is a World Leader in Protecting the Ocean." www.theguardian.com. August 14, 2019, accessed April 6, 2020.

Root, Tik. "How One Country Is Restoring Its Damaged Ocean." www.nationalgeographic.com. April 11, 2018, accessed April 6, 2020.

Lesson 125

Eric Weiner, "How Geography Shapes Our Identity." trend.pewtrusts.org. July 5, 2016, accessed August 20, 2018.

Lesson 126

Corion, Kimron. "How Johanan Dujon Built an Agricultural Biotech Company in St. Lucia." www.huffpost.com. November 28, 2017, accessed March 29, 2020.

Daley, Jason. "New 'Living Museum of the Sea' Established in Domincan Republic Waters." www.smithsonianmag.com. December 3, 2019, accessed March 30, 2020.

Ewing-Chow, Daphne. "Caribbean People and Their Ocean: A Story in Images." www.forbes.com. June 8, 2019, accessed March 30, 2020.

Ewing-Chow, Daphne. "Meet St. Lucia's First Indigenous Biotech Company." www.forbes.com. February 1, 2019, accessed March 30, 2020.

Handwerk, Brian. "Deepest Volcanic Sea Vents Found; 'Like Another World'." www.nationalgeographic.com/news. April 13, 2010, accessed March 31, 2020.

Hares, Sophie. "Caribbean Set to Ride 'Blue Economy' Wave, Say Economists." www.reuters.com. June 1, 2018, accessed March 31, 2020.

"Interesting Facts About the Caribbean Sea." www.justfunfacts.com. Accessed March 30, 2020.

"What Is the Sargasso Sea?" National Ocean Service, www.oceanservice.noaa.gov. Updated October 9, 2019, accessed March 31, 2020.

"Who Were the Real Pirates of the Caribbean?" Royal Museums Greenwich. www.rmg.co.uk. Accessed March 30, 2020.

Yong, Ed. "Why Waves of Seaweed Have Been Smothering Caribbean Beaches." www.theatlantic.com. July 4, 2019, accessed March 31, 2020.

Lesson 127

"A Profile of a Hurricane Hunter Pilot." National Oceanic and Atmospheric Administration (NOAA). www.weather.gov. Accessed March 20, 2020.

Ferro, Shaunacy. "Why It's So Hard to Predict Hurricanes." Popular Science, www.popsci.com. August 29, 2013, accessed march 23, 2020.

"How Does a Hurricane Form?" National Oceanic and Atmospheric Administration (NOAA). www.scijunks.gov/hurricane/. Updated February 18, 2020, accessed March 22, 2020.

"Hurricanes: Science and Society." University of Rhode Island Graduate School of Oceanography. www.hurricanescience.org. Accessed March 22, 2020.

Kaufman, Mark. "Today's Hurricanes Kill Way Fewer Americans, and NOAA's Satellites Are the Reason Why." Popular Science, www.popsci.com. April 6, 2017, accessed March 23, 2020.

Korten, Tristram. "The Bahamas and the Caribbean Have Withstood Hurricanes for Centuries." www.smithsonianmag.com. September 17, 2019, accessed March 23, 2020.

"Tropical Cyclone Naming." World Meteorological Organization. www.public.wmo.int. Accessed March 23, 2020.

Lesson 128

Abbott, Elizabeth. Sugar: A Bittersweet History. New York: Duckworth Overlook, 2010.

"Britain and the Caribbean." www.bbc.co.uk. Accessed March 27, 2020.

Castel, Raul Zecca. "Extorted and Exploited: Haitian Laborers On Dominican Sugar Plantations." www.opendemocracy.net. November 14, 2017, accessed March 27, 2020.

"Did Slavery Make Economic Sense?" www.economist.com. September 27, 2013, accessed March 25, 2020.

Ewing-Chow, Daphne. "Can Caribbean Sugar Make a Sweet Comeback?" www.forbes.com. October 31, 2019, accessed March 27, 2020.

Grono, Nick. "The Economic Case for Ending Slavery." www.theguardian.com. August 15, 2013, accessed March 25, 2020.

Johnston, Mark. "The Sugar Trade in the West Indies and Brazil Between 1492 and 1700." University of Minnesota Bell Library. www.lib.umn.edu. Accessed March 27, 2020.

"Sugar Producing Countries 2020." www.worldpopulationreview.com. Accessed March 27, 2020.

Sumner, Scott. "Ending Slavery Made America Richer." www.econlib.org. September 14, 2014, accessed March 25, 2020.

Lesson 129

"Animals of Cuba." www.cubaunbound.com. Accessed March 13, 2020

"Classic Cars and the Cubans That Keep Them Running." www.anywhere.com/cuba. Accessed March 13, 2020.

"Country Profile: Cuba." Library of Congress Federal Research Division. www.loc.gov. September 2006, accessed March 13, 2020.

"Cuba Profile - Timeline." www.bbc.com. May 1, 2018, accessed March 13, 2020.

"Geography." www.cubaunbound.com. Accessed March 13, 2020.

Zurschmeide, Jeff. "'Doing Anything with Nothing': What It Takes to Keep Cuba's 'Yank Tanks' Ticking." www.digitaltrends.com. October 28, 2015, accessed March 13, 2020.

Lesson 131

Benson, Lee. "About Utah: It Wasn't Easy Being a WWII 'Rubber Soldier'"—A Utah Couple Paying Tribute Knows." www.deseret.com. June 4, 2017, accessed April 9, 2020.

"Brazil to Compensate Rubber Workers from World War Two." www.bbc.com. May 15, 2014, accessed April 10, 2020.

Carlowicz, Mike. "Meeting of the Waters." www.earthobservatory.nasa.gov. June 17, 2012, accessed April 12, 2020.

"Facts About the Amazon Rainforest." www.activewild.com. November 4, 2015, accessed April 9, 2020.

Holland, Jackie. "Nature's Pharmacy: The Remarkable Plants of the Amazon Rainforest—and What They May Cure." www,telegraph.co.uk. May 19, 2019, accessed April 9, 2020.

Langlois, Jill. "Researchers Discover the Tallest Known Tree in the Amazon." www.smithsonianmag.com. September 27, 2019, accessed April 9, 2020.

Nunez, Christina. "Rainforests, Explained." www.nationalgeographic.com. May 15, 2019, accessed April 10, 2020.

"Plants of the Amazon Rainforest." University of California, Berkeley. www.berkeley.edu. Accessed April 9, 2020.

"Rubber Tree." www.rainforest-alliance.org. Updated September 15, 2012, accessed April 9, 2020.

Sherwood, Louise. "Brazilian World War II Workers Fight for Recognition." www.bbc.com/news. August 9, 2010, accessed April 9, 2020.

"Tropical Rainforest Plants." www.celebratebrazil.com. Accessed April 9, 2020.

Lesson 132

Barnes, Amanda and Greg Funnell. "Riding with the Gauchos of Argentina: A Photo Essay." www.theguardian.com. April 19, 2019, accessed April 1, 2020.

Forero, Juan. "Jewish Gaucho Tradition Fades in Argentina." www.washingtonpost.com. June 25, 2011, accessed March 31, 2020.

Foster, David William, et al. Culture and Customs of Argentina. Westport, Connecticut: Greenwood Press, 1998.

"Livestock Production in Latin America and the Caribbean." Food and Agricultural Organization of the United Nations. www.fao.org. Accessed April 1, 2020.

Marshall, Sarah. "Meet Argentina's First Female Gauchos." www.telegraph.co.uk. March 8, 2019, accessed March 31, 2020.

"Pampas Biome." www.blueplanetbiomes.org/pampas.php. Accessed April 1, 2020.

Pham, Diane. "Argentina's Last Jewish Cowboys." www.tabletmag.com. Accessed April 1, 2020.

Smith, James F. "National Symbol: Spirit of the Gaucho Calls to Argentines." Los Angeles Times, www.latimes.com. December 2, 1988, accessed April 1, 2020.

Stewart, Harry. "A Brief History of the Gaucho: The Cowboys of Argentina." www.theculturetrip.com. March 31, 2017, accessed April 1, 2020.

Lesson 133

Bloudoff-Indelicato, Mollie. "What Are North American Trout Doing in Lake Titicaca?" www.smithsonianmag.com. December 9, 2015, accessed April 13, 2020.

Foer, Joshua. "The Island People." www.slate.com. February 25, 2011, accessed April 13, 2020.

"A Guide to Isla Del Sol, Lake Titicaca." www.bolivianlife.com. Accessed April 13, 2020.

Hamre, Bonnie." The Floating Islands of Lake Titicaca." www.tripsavvy.com. Updated June 3, 2019, accessed April 13, 2020.

Jaivin, Linda. "Exploring the Manmade Reed Islands of Lake Titicaca." qantas.com/travelinsider. June 23, 2017, accessed April 13, 2020.

"Lake Titicaca." United Nations World Heritage Center. www.whc.unesco.org. Accessed April 13, 2020.

"Lake Titicaca—Bolivia and Peru." Global Nature Fund. www.globalnature,org. Accessed April 13, 2020.

"Totora Reed—A Unique Resource." www.perunorth.com. February 8, 2018, accessed April 13, 2020.

Lesson 134

Abreu, Fellipe and Luiz Felipe Silva. "In Pictures: The World's Largest Salt Flat in Bolivia." www.bbc.com. January 24, 2016, accessed April 5, 2020.

"Bolivia Sea Dispute: UN Rules in Chile's Favor." www.bbc.com. October 1, 2018, accessed April 3, 2020.

Foster, Ally. "The South American Nation That Has a Navy But No Coastline." www.news.com.au. May 15, 2018, accessed April 3, 2020.

Long, Gideon. "Bolivia-Chile Land Dispute Has Deep Roots." www.bbc.com. April 24, 2013, accessed April 3, 2020.

Mallonee, Laura. "Bolivia Is Landlocked. Don't Tell That to its Navy." www.wired.com. May 11, 2018, accessed April 3, 2020.

Nugent, Clara. "Landlocked Bolivia Wants a Path to the Pacific. Here's Why That Won't Be Happening Any Time Soon." www.time.com. October 3, 2018, accessed April 3, 2020.

Sater, William F. *Andean Tragedy: Fighting the War of the Pacific.* Lincoln: University of Nebraska Press, 2007.

Sierra, Jeronimo Rios. "The Complex Relationship Between Peru, Bolivia, and Chile: A Legacy of the War of the Pacific." www.theglobalamericans.org. October 5, 2018, accessed April 3, 2020.

"War of the Pacific, 1879-83." Library of Congress. www.countrystudies.us. Accessed April 3, 2020.

Woody, Christopher. "Bolivia Is Taking Chile to Court Over a War It Lost 143 Years Ago." www.businessinsider.com. March 29, 2018, accessed April 3, 2020.

Lesson 135

"Global Christianity—A Report on the Size and Distribution of the World's Christian Population." www.pewforum.org, December 19, 2011.

Granberg-Michaelson, Wes. "Think Christianity is Dying? No, Christianity Is Shifting Dramatically." www.washingtonpost.com, May 20, 2015, accessed April 2, 2019.

Jenkins, Philip. "Believing in the Global South." www.firstthings.com, December 2006.

Reese, Derran. "Global Christianity Series," Highland Church of Christ, Abilene, Texas. www.youtube.com.

Lesson 136

Angel, Karen. "The Truth About Jimmie Angel and Angel Falls." www.jimmieangel.org. Revised July 15, 2010, accessed April 14, 2020.

"Canaima National Park." www.venezuelatuya.com. Accessed April 15, 2020.

Carroll, Rory. "Hugo Chavez Renames Angel Falls." theguardian.com. December 21, 2009, accessed April 14, 2020.

"Countries and Their Cultures: Pemon." www.everyculture.com. Accessed April 15, 2020.

Ornes, Stephen. "Inside the Lost Cave World of the Amazon's Tepui Mountains." www.newscientist.com. April 25, 2016, accessed April 15, 2020.

Patowary, Kaushik. "Tabletop Mountains or Tepuis of Venezuela." www.amusingplanet.com. May 11, 2013, accessed April 15, 2020.

Lesson 137

Adalbjornsson, Tryggvi. "The Saga of the Reindeer of South Georgia Island." www.atlasobscura.com, December 24, 2018, accessed October 4, 2019.

Freedman, Sir Lawrence. "The Falklands War Explained." www.historyextra.com, accessed October 4, 2019.

Rohter, Larry. "25 Years After War, Wealth Changes Falklands." www.nytimes.com, April 1, 2007, accessed October 1, 2019.

Lesson 138

"Driving the Pan-American Highway? Here's All You Need to Know." www.funlifecrisis.com. April 9, 2019, accessed April 7, 2020.

Etchevers, Pablo. "Whose Hand Is That?" www.welcomeuruguay.com. Accessed April 7, 2020.

Goncalves, Vania. "The Pan-American Highway: From Alaska to Argentina." www.worldconstructionnetwork.com. August 24, 2016, accessed April 7, 2020.

Harmon, Maureen. "The Story of the Pan-American Highway." Pegasus, the magazine of the University of Central Florida. www.ucf.edu/pegasus. Summer 2019, accessed April 7, 2020.

Haugerud, Ralph. "The North American Cordillera." U.S. Geological Survey, www.pubs.usgs.gov. 1998, last modified December 7, 2016, accessed April 7, 2020.

Sewell, Abby and Jeff Heimsath. "This Suspension Bridge Is Made of Grass." www.nationalgeographic.com. Accessed April 17, 2020.

Sherriff, Lucy. "Monument or Mirage? Giant Hand Rises from the Desert." www.cnn.com. Updated February 6, 2019, accessed April 7, 2020.

"The Pan-American Highway: The Longest Road in the World." www.brilliantmaps.com. April 6, 2016, accessed April 7, 2020.

Lesson 139

Johnson, Ben. "The History of Patagonia." www.historic-uk.com. Accessed April 16, 2020.

Dear, Paula. "Was Welsh Settlers' Patagonia Move a Success or Failure?" www.bbc.com. July 30, 2015, accessed April 16, 2020.

Hansen, Kathryn and Jesse Allen. "Melting Beauty: The Icefields of Patagonia." www.earthobservatory.nasa.gov. March 27, 2018, accessed April 17, 2020.

Moss, Chris. "Guide to Patagonia: What to Do, How to Do It, and Where to Stay." www.theguardian.com. December 12, 2014, accessed April 17, 2020.

The Patagonian Foundation. www.thepatagonianfoundation.org. Accessed April 17, 2020.

Lesson 141

Hackett, Conrad and David McClendon. "Christians Remain World's Largest Religious Group, But They Are Declining in Europe." Pew Research Center, www.pewresearch.org, April 5, 2017, accessed April 10, 2017.

www.worldpopulationreview.com.

U.S. Census Bureau, www.census.gov

Lesson 143

Glaeser, Edward. *Triumph of the City*. New York: Penguin, 2011.

Lesson 144

LeFevre, Robert. "Ownership of Land." www.mises.org, March 16, 2010, accessed June 11, 2019. Chapter 8 of The Philosophy of Ownership, Auburn, Alabama: Ludwig von Mises Institute, 2007.

Libecap, Gary D. "The Consequences of Land Ownership." Hoover Institution, Stanford University. www.hoover.org, August 29, 2018, accessed June 11, 2019.

Lesson 145

Lewis, C. S. *The Weight of Glory and Other Addresses.* Originally published 1949. New York: Collier Books, 1980. Address, "The Weight of Glory," pp. 18-19.

Lesson 146

Birner, Betty. "Why Do Some People Have an Accent?" Linguistic Society of America, www.linguisticsociety.org, n.d., accessed February 11, 2019.

Pitt, Laura. "Silbo Gomero: A Whistling Language Revived." www.bbc.com, January 11, 2013, accessed February 6, 2019.

Ostler, Nicholas. *Empires of the Word: A Language History of the World.* New York: Harper Perennial, 2006.

Than, Ker. "Does Geography Influence How a Language Sounds?" www.news.nationalgeographic.com, June 14, 2013, accessed February 6, 2019.

Lesson 148

"International Organizations on the Web." www.washingtonpost.com. Accessed August 23, 2019.

"What Are International Organizations?" Carleton University, www.carleton.ca. Accessed August 23, 2019.

Lesson 149

Singh, Prerma. "Nationalism can have its good points. Really." www.washingtonpost.com, January 26, 2018. Accessed October 11, 2018.

Winter-Levy, Sam, and Nikita Lalwani. "When is nationalism a good thing? When it unites an ethnically diverse citizenry." www.washingtonpost.com, June 26, 2018. Accessed October 11, 2018.

De las Easas, Gustavo. "Is Nationalism Good for You?" www.foreignpolicy.com, October 8, 2009. Accessed October 11, 2018

Barone, Michael. "Nationalism is not necessarily a bad thing." www.washingtonexaminer.com, August 10, 2016. Accessed October 11, 2018.

Fonte, John. "In Defense of Nations," review of *The Virtue of Nationalism* by Yoram Hazony. www.nationalreview.com, October 1, 2018. Accessed October 8, 2018.

Detail from **Lonely Fishermen on the River in Autumn** *by Tang Yin (Chinese, 1523)*

Image Credits

iii Bhutan: Chandy Benjamin / Shutterstock.com
Uzbekistan: Marina Rich / Shutterstock.com
iv Kosrae: Stu Shaw / Shutterstock.com
iv Kangaroo: John Crux / Shutterstock.com
vi South Georgia: Fredy Thuerig / Shutterstock.com
431 Field: Arlo Magicman / Shutterstock.com
433 Afghanistan: Jose_Matheus / Shutterstock.com
434 Kabul 1842: Library of Congress
Kabul 1978: Cleric77 / Wikimedia Commons / CC BY-SA 3.0
435 Mujahideen: Erwin Lux / Wikimedia Commons / CC BY-SA 3.0
436 Pamir: NOWAK LUKASZ / Shutterstock.com
438 Mountains: soft_light / Shutterstock.com
Baptism: Nepal Church History Project
439 Village: Zzvet / Shutterstock.com
440 Yaks: Lumbini / Shutterstock.com
441 Lumbini: Casper1774 Studio / Shutterstock.com
443 Mosque: Bernard Gagnon / Wikimedia Commons / CC BY-SA 3.0
443 Ikhlas Khan: British Library
444 Mumbai: Wikimedia Commons
Gujarat: Kandukuru Nagarjun / Flickr / CC BY 2.0
445 Ambedkar: Prime Minister's Office (GODL-India) / ID 22565 / CNR 25717
447 Everest: dharmaraj ranabhat / Shutterstock.com
448 Hillary and Norgay: Jamling Tenzing Norgay / Wikimedia Commons / CC BY-SA 3.0
449 Namche: UBC Stock / Shutterstock.com
450 Everest: Carlos Pauner / Wikimedia Commons / CC BY-SA 3.0
451 Garden of Eden: Wikimedia Commons
452 Adam's Peak: thetravellergirl / Shutterstock.com
455 Carpets: Vladimir Goncharenko / Shutterstock.com
457 Boats: Matyas Rehak / Shutterstock.com
458 Cotton: Peretz Partensky / Flickr / CC BY-SA 2.0
Kantubek: ninurta_spb / Shutterstock.com
459 1985: NASA; 2020: NASA Worldview
461 Canyon: Lukas Bischoff Photograph / Shutterstock.com
462 Aktau: yevgeniy11 / Shutterstock.com
463 Festival: Cholpan / Shutterstock.com
465 Market: Luigi Guarino / Flickr / CC BY 2.0
Baking: Michal Knitl / Shutterstock.com
467 Mountains: Valerii_M / Shutterstock.com
468 Family: NoyanYalcin / Shutterstock.com
469 Ashgabat: Michal Knitl / Shutterstock.com
470 Music: Olga Nikanovich / Shutterstock.com
471 Tents: AlexelA / Shutterstock.com
Selfies: Matyas Rehak / Shutterstock.com
472 Monument: Thiago B Trevisan / Shutterstock.com
473 Flowers: Anton Petrus / Shutterstock.com
474 Eagle: MehmetO / Shutterstock.com
477 Mongolia: Wolfgang Zwanzger / Shutterstock.com

479 Islands: Claudine Van Massenhove / Shutterstock.com
479 Mao and Chiang: Wikimedia Commons
480 Chiang: Les Duffin / U.S. Army
481 Carter, Nixon, and Deng: NARA
482 Tsai: O.O / Shutterstock.com
483 Tea: DronaVision / Shutterstock.com
485 Park: Mongkol chai / Shutterstock.com
486 Monument: Chintung Lee / Shutterstock.com
487 DMZ: Narongsak Nagadhana / Shutterstock.com
488 Bridge: WaitForLight / Shutterstock.com
491 Thailand: Take Photo / Shutterstock.com
492 Vietnam: JunPhoto / Shutterstock.com
493 China: Tutti Frutti / Shutterstock.com
495 Ceremony: I am Corona / Shutterstock.com
496 Gathering: Los Angeles County Museum of Art (www.lacma.org)
497 Thailand: SasinTipchai / Shutterstock.com
498 Bowl: Los Angeles County Museum of Art (www.lacma.org)
499 Tea Ceremony: Los Angeles County Museum of Art (www.lacma.org)
501 Yin Yang: RPBaiao / Shutterstock.com
502 Temple: R.M. Nunes / Shutterstock.com
503 Statue: LMspencer / Shutterstock.com
504 Temple: Sean Pavone / Shutterstock.com
507 Wulingyuan: aphotostory / Shutterstock.com
509 Lantern: mehmet dinler / Shutterstock.com
510 Zodiac: Dan Hanscom / Shutterstock.com
Family: TonyV3112 / Shutterstock.com
511 National Day: testing / Shutterstock.com
513 Kashgar: Kylie Nicholson / Shutterstock.com
514 Market: Kylie Nicholson / Shutterstock.com
516 Hong Kong: Luciano Mortula: LGM / Shutterstock.com
517 Protest: Lewis Tse Pui Lung / Shutterstock.com
518 Khorgos: Vladimir Tretyakov / Shutterstock.com
519 Djibouti: Vladimir Melnik / Shutterstock.com
519 Bahamas: Eric Glenn / Shutterstock.com
520 Mosque: CHETTOUH Nabil / Shutterstock.com
521 Map: Sean Killen
522 UN: Drop of Light / Shutterstock.com
524 Ship: Catstyecam / Shutterstock.com
525 Ideal-X: Karsten Kunibert / Wikimedia Commons / CC BY-SA 3.0
526 Cosco: Mariusz Bugno / Shutterstock.com
528 McLean: Maersk Line / Wikimedia Commons / CC BY-SA 2.0
529 Dictionary: Lobroart / Shutterstock.com
530 Bezbozhnik: New York Public Library
533 Beijing: VTT Studio / Shutterstock.com
535 Cambodia: Aleksandar Todorovic / Shutterstock.com
537 Drilling Rig: / Shutterstock.com
538 Protesters: BasPhoto / Shutterstock.com
539 Map: Peter Hermes Furian / Shutterstock.com
540 Diagram: WindVector / Shutterstock.com
541 Soldiers: chauha User:Ha petit / Wikimedia Commons / CC BY 3.0
542 Boat: OlegD / Shutterstock.com
544 Palawan: leoks / Shutterstock.com
545 Volcano: oOhyperblaster / Shutterstock.com
546 San Agustin: Leonid Andronov / Shutterstock.com
547 Panay: Frolova_Elena / Shutterstock.com
548 Market: CaveDweller99 / Shutterstock.com
550 Bridge: Tang Trung Kien / Shutterstock.com
551 POW: SSGT Herman Kokojan / U.S. Army
552 Incense: Alex Varninschi / Shutterstock.com
553 Cave: kid315 / Shutterstock.com
555 Padar: Donnchans / Shutterstock.com
556 Mosque: BagusMartada / Shutterstock.com
Waterfall: Creativa Images / Shutterstock.com
557 Banda Aceh: Photographer's Mate 1st Class John D. Yoder / U.S. Navy
559 Statue: maloff / Shutterstock.com
561 Putao: Udompeter / Shutterstock.com
562 Festival: KaiCamera / Shutterstock.com
563 Fishing: saravutpics / Shutterstock.com
565 Church: gracethang2 / Shutterstock.com
567 Lake: Filip Fuxa / Shutterstock.com
569 Hongi: Tech. Sgt. Shane A. Cuomo / U.S. Air Force
570 Sign: Filip Fuxa / Shutterstock.com
571 Trash cans: plumchutney / Shutterstock.com
572 Waitangi Day: ChameleonsEye / Shutterstock.com
573 Empire: Cornell University – PJ Mode Collection of Persuasive Cartography
574 Elizabeth: Cubankite / Shutterstock.com
576 Sydney: Elias Bitar / Shutterstock.com
577 Paintings: State Library of New South Wales
578 Harbor: anek.soowannaphoom / Shutterstock.com
580 Opera House: Taras Vyshnya / Shutterstock.com
582 Camels: Cezary Wojtkowski / Shutterstock.com
Dancers: Philip Schubert / Shutterstock.com
583 Uluru: Corey Leopold / Flickr / CC BY 2.0
584 Kata Tjuta: John Carnemolla / Shutterstock.com
585 Golf: mark higgins / Shutterstock.com
Church: Dmitry Chulov / Shutterstock.com
587 Reef: Coral Brunner / Shutterstock.com
588 Turtle: Regien Paassen / Shutterstock.com
589 Barnacles: daguimagery / Shutterstock.com
590 Macquarie: Samuel Bloch / Shutterstock.com
Fraser: zstock / Shutterstock.com
591 Norfolk: Rawdon Sthradher / Shutterstock.com
Christmas: skyfall4 / Shutterstock.com
593 Top: Adam J / Shutterstock.com
Bottom: Queensland State Archives
594 Chincilla: State Library of Queensland
595 Larvae: CSIRO / Wikimedia Commons / CC BY 3.0
Cover: State Library of Queensland
596 Monument: Saintrain at English Wikipedia / CC BY-SA 3.0
597 Yasawa: Don Mammoser / Shutterstock.com
599 Map: Princeton University Library
600 Sextant: The Field Museum Library
601 Chronometer: Racklever at English Wikipedia / CC BY-SA 3.0
602 Sinking: Australian National Maritime Museum
603 Hawaii: Wikimedia Commons
605 Papua New Guinea: Michal Knitl / Shutterstock.com

606 Fiji: ChameleonsEye / Shutterstock.com
607 Papua New Guinea: Michal Knitl / Shutterstock.com
608 Plane and Lagoon: Alex East / Shutterstock.com
610 Palau: Norimoto / Shutterstock.com
Hideout: Groverva at English Wikipedia
611 Kiribati: Alex Ruan / Shutterstock.com
612 Guam: San Diego Air and Space Museum Archive
Yap: maloff / Shutterstock.com
613 Latte stones: t.a.m.m.y / Shutterstock.com
613 Replica hut: RaksyBH / Shutterstock.com
615 Bora Bora: Lukas Bischoff Photograph / Shutterstock.com
616 Breadfruit: Tanya Keisha / Shutterstock.com
617 Tahitian Women: Wikimedia Commons
618 Easter Island: T photography / Shutterstock.com
618 Rarotonga: ChameleonsEye / Shutterstock.com
620 Samoa: Martin Valigursky / Shutterstock.com
621 Nauru: Robert Szymanski / Shutterstock.com
622 Tuvalu: Romaine W / Shutterstock.com
623 New Caledonia: Matpix / Shutterstock.com
625 Canada: Angelito de Jesus / Shutterstock.com
627 Beaufort Sea: Pi-Lens / Shutterstock.com
628 Sea of Ice: Wikimedia Commons
Icebreakers: knyazev vasily / Shutterstock.com
629 Nordic OIrion: kees torn / Flickr / CC BY-SA 2.0
630 Map: Peter Hermes Furian / Shutterstock.com
632 Ships: Jon Nicholls Photography / Shutterstock.com
Yacht: U.S. Navy
634 Replica: Markus Daams / Wikimedia Commons / CC BY 2.0
635 Stained glass: jorisvo / Shutterstock.com
636 Statue and Notgrasses: Notgrass Family
637 Monuments: Notgrass Family
638 Satellite: Robert Simmon / Landsat 7 Science Team / NASA
Painting: Wikimedia Commons
639 Battlefield: SF photo / Shutterstock.com
640 Library of Congress
641 Katrina: NASA Worldview
642 Top: Gary Johnson, USCGAUX / U.S. Coast Guard
642 Bottom: Petty Officer 2nd Class Kyle Niemi / U.S. Coast Guard
644 Mississippi: Joseph Sohm / Shutterstock.com
645 Ohio: Smithsonian American Art Museum
646 Map: Horace Mitchell / NASA's Scientific Visualization Studio / Reto Stockli (NASA GSFC) / U.S. Geological Survey EROS Center
647 Colorado: Philip Bird LRPS CPAGB / Shutterstock.com
648 Duck: KennStilger47 / Shutterstock.com
649 Hospital: U.S. Army
650 Woods and Graham: David Woods
Hospital: Jim McIntosh / Flickr
651 Transplant Games: John Marino / Flickr / CC BY 2.0
653 North America: NASA Earth Observatory image by Robert Simmon, using Suomi NPP VIIRS data provided courtesy of Chris Elvidge (NOAA National Geophysical Data Center)
655 Hudson: Daryl Haight / Shutterstock.com
656 A Pic-Nic Party: Brooklyn Museum
656 Harbor Landscape: KaDeWeGirl / Flickr / CC BY 2.0
657 Kindred Spirits: Wikimedia Commons
658 Andes: Metropolitan Museum of Art
658 Colorado: Smithsonian American Art Museum
659 Yosemite: Birmingham Museum of Art
660 Brocho Buster: Metropolitan Museum of Art
661 Panels: Babcock Ranch
662 Lake: Babcock Ranch
663 Square: Babcock Ranch
665 Diomedes: Bering Land Bridge National Preserve / Flickr / CC BY 2.0
666 Satellite: NASA-NOAA
667 Village: Petty Officer 3rd Class Richard Brahm / U.S. Coast Guard
669 Prairie: Jacob Boomsma / Shutterstock.com
669 Ingalls: Public Domain
670 Cabin: Notgrass Family
671 Wildfire: Eugene R Thieszen / Shutterstock.com
672 Creek: Notgrass Family
673 Homestead: Notgrass Family
674 Rocky Ridge: Rose Wilder Lane Collection, Herbert Hoover Presidential Library
675 California: Rose Wilder Lane Collection, Herbert Hoover Presidential Library
676 Sea of Galilee: alefbet / Shutterstock.com
677 Sinai: Kochneva Tetyana / Shutterstock.com
678 Palace: Jukka Palm / Shutterstock.com
679 Jordan Valley: Protasov AN / Shutterstock.com
680 Jerusalem: JekLi / Shutterstock.com
681 Ephesus: Boat Rungchamrussopa / Shutterstock.com
683 Lake: Milosz Maslanka / Shutterstock.com
685 Waterfalls: Anna ART / Shutterstock.com
686 Voladores: Faviel_Raven / Shutterstock.com
687 Cars: veracruz / Shutterstock.com
688 Map: Esra OGUNDAY BAKIR / Shutterstock.com
690 Tikal: SL-Photography / Shutterstock.com
691 Festival: Lucy.Brown / Shutterstock.com
692 Panajache: Kyle M Price / Shutterstock.com
693 Rosewood: MARGRIT HIRSCH / Shutterstock.com
695 Park: iacomino FRiMAGES / Shutterstock.com
696 Tourists: Marco Lissoni / Shutterstock.com
Cart: Jarno Gonzalez Zarraonandia / Shutterstock.com
697 Zipline: Wollertz / Shutterstock.com
698 Quetzal: Ondrej Prosicky / Shutterstock.com
700 Monkey: reisegraf.ch / Shutterstock.com
701 Mennonite: akramer / Shutterstock.com
702 Island: Jashley247 / Shutterstock.com
704 Santa Lucia: Omri Eliyahu / Shutterstock.com
705 Market: Radiokafka / Shutterstock.com
706 Crete: Pietro Basilico / Shutterstock.com
707 Gaza: pokku / Shutterstock.com
709 Curacao: Solarisys / Shutterstock.com
711 Sargassum: avilledorsa / Shutterstock.com
712 Collecting: Johanan Dujon / Algas Organics
713 Cruise Ships: Paulo Miguel Costa / Shutterstock.com
714 Fish: Maria Bobrova / Shutterstock.com

Pirates: Wikimedia Commons
715 Johanan Dujon / Algas Organics
716 Cuba: Inspired By Maps / Shutterstock.com
717 Flight Track and Scott Price: NOAA
718 Scale: Captain Cobi / Shutterstock.com
719 Eyewall: Lieutenant Mike Silah, NOAA Corps, NOAA AOC / Flickr / CC BY 2.0
720 Flooding: Alessandro Pietri / Shutterstock.com
722 Sugarcane: Sevenstock Studio / Shutterstock.com
723 Plantation: Tropenmuseum, part of the National Museum of World Cultures / CC BY-SA 3.0
724 Statue: Kristina Just / Flickr / CC BY 2.0
725 Antigua: British Library
India: Hari Mahidhar / Shutterstock.com
727 El Yunque: Matyas Rehak / Shutterstock.com
728 Map: Peter Hermes Furian / Shutterstock.com
729 Uncle Sam: Wikimedia Commons
730 Fleet: U.S. Navy
731 Cars: Diego Grandi / Shutterstock.com
732 Music: Mikko Palonkorpi / Shutterstock.com
733 Protesters: Library of Congress
734 Library: Leonard Zhukovsky / Shutterstock.com
Houses: Eyemotor / Shutterstock.com
735 Festival: J Erick Brazzan / Shutterstock.com
736 Masks: Eve Orea / Shutterstock.com
737 Collection: Notgrass Family
739 Tractor: Jan Schneckenhaus / Shutterstock.com
741 Rubber: Photoongraphy / Shutterstock.com
742 Roosevelt and Vargas: Library of Congress
743 Globe: NASA image courtesy Reto Stöckli and Robert Simmon
744 Layers: NoPainNoGain / Shutterstock.com
Capybara: GTW / Shutterstock.com
745 Confluence: NASA Earth Observatory image by Jesse Allen and Robert Simmon
746 Chief: Joa Souza / Shutterstock.com
Clearing: guentermanaus / Shutterstock.com
748 Pampas: Foto 4440 / Shutterstock.com
749 Guanaco: Ivan Hoermann / Shutterstock.com
Gauchos: Archivo General de la Nación
750 Beef: AlexCorv / Shutterstock.com
Synagogue: FLLL / Wikimedia Commons / CC BY-SA 3.0
751 Festival: sunsinger / Shutterstock.com
752 Cattle: canadastock / Shutterstock.com
753 Lake Titicaca: NiarKrad / Shutterstock.com
754 Rugs: Cezary Wojtkowski / Shutterstock.com
Isla del Sol: Junne / Shutterstock.com
755 Boats: Eleni Mavrandoni / Shutterstock.com
756 Aerial: qualtaghvisuals / Shutterstock.com
Island: lulu and isabelle / Shutterstock.com
758 Museum: Mark Pitt Images / Shutterstock.com
758 Map: Rawpixel
759 Monument: Mark Pitt Images / Shutterstock.com
760 Base: FrenchAvatar / Wikimedia Commons / CC BY-SA 3.0
Marines and Ship: Israel soliz / Wikimedia Commons / CC BY-SA 3.0
761 Rock: travelview / Shutterstock.com
Salt flat: Shanti Hesse / Shutterstock.com
762 Statue: Diego Grandi / Shutterstock.com
763 Chruch: lembi / Shutterstock.com
764 Baptism: Alf Ribeiro / Shutterstock.com
765 Band: Simplus Menegati / Wikimedia Commons
766 Uniminuto: EEIM / Wikimedia Commons / CC BY-SA 3.0
767 Church: Matyas Rehak / Shutterstock.com
769 Trees: javarman / Shutterstock.com
771 Roraima: sunsinger / Shutterstock.com
772 Angels: © Jimmie Angel Historical Project Archive
773 Angel Falls: Vadim Petrakov / Shutterstock.com
774 Plane: Yosemite / Wikimedia Commons / CC BY-SA 3.0
775 Stanley: amanderson2 / Flickr / CC BY 2.0
776 Soldiers: Wikimedia Commons
Minefield: John5199 / Flickr / CC BY 2.0
777 South Georgia: Brian Gratwicke / Flickr / CC BY 2.0
778 Graffiti: Salim Virji / Flickr / CC BY-SA 2.0
779 Andes: Aaron_M / Flickr / CC BY 2.0
780 Bridge: Jolyn Chua / Shutterstock.com
781 Highway: Library of Congress
782 Hand in Chile: Greg Schechter / Flickr / CC BY 2.0
Hand in Uruguay: Coolcaesar / Wikimedia Commons / CC BY-SA 3.0
783 Map: Sean Killen
Meegan: George Meegan / Wikimedia Commons
784 Costa Rica: Ll1324 / Wikimedia Commons.png
785 Gorsedd: Maria Agustinho / Shutterstock.com
786 Newspaper: Museo Regional de Gaiman
787 Teapot: sonyworld / Shutterstock.com
789 Glacier: Petr Meissner / Flickr / CC BY 2.0
790 Path: KinEnriquez Pixabay
791 Chapel: Richard Avis / Flickr / CC BY-SA 2.0
793 People: Franzi / Shutterstock.com
795 Mumbai: sladkozaponi / Shutterstock.com
798 Generations: Inspiring / Shutterstock.com
799 Kuwait: Vladimir Zhoga / Shutterstock.com
800 Malaysia: The Road Provides / Shutterstock.com
800 Namibia: Watch The World / Shutterstock.com
801 Singapore: Maurizio Callari / Shutterstock.com
803 Migrants: Ververidis Vasilis / Shutterstock.com
806 Train: Romanian National Archives
807 Somalia: hikrcn / Shutterstock.com
808 Malthus: Biblioteca Europea di Informazione e Cultura / Wikimedia Commons
810 Moscow: Linda Kubova / Shutterstock.com
811 Paris Street: Wikimedia Commons
813 Harvesters: Metropolitan Museum of Art
814 Truck: praditkhorn somboonsa / Shutterstock.com
817 Sign: Lisa McIntyre / Unsplash
818 Stone: Henry Salomé / Wikimedia Commons / CC BY-SA 3.0
819 Java: Tropenmuseum, part of the National Museum of World Cultures / CC BY-SA 3.0
820 Deed Books: Library of Congress
821 Farmers: Municipal Archives of Trondheim / Flickr / CC BY 2.0
822 Desert: iacomino FRiMAGES / Shutterstock.com

824 Good Samaritan: Library of Congress
826 Michael: Matt Collamer / Unsplash
829 Carpathian Mountains: ArtemD / Shutterstock.com
831 La Gomera: Astridlike / Shutterstock.com
832 Norway: Kurt:S / Flickr / CC BY 2.0
833 Embassy: Eugene Alvin Villar / Wikimedia Commons
835 Russian: VPales / Shutterstock.com
836 Nepal: Wonderlane / Flickr / CC BY 2.0
838 Alaska: Notgrass Family
839 Glacier: Notgrass Family
840 Canal: sailn1 / Flickr / CC BY 2.0
841 Tourists: Pedro Ribeiro Simões / Flickr / CC BY 2.0
843 Red Cross: Torbjorn Toby Jorgensen / Flickr / CC BY-SA 2.0
844 G20: Alan Santos PR Palácio do Planalto / Flickr / CC BY 2.0
845 UN: Astrid Riecken Comprehensive Nuclear-Test-Ban Treaty Organization / Flickr / CC BY 2.0
846 Cat Judging: Andres Munt / Wikimedia Commons / CC BY-SA 3.0
848 Parade: Angelo Giampiccolo / Shutterstock.com
849 Kosovo: Arild Vågen / Wikimedia Commons / CC BY-SA 3.0
850 Eiffel Tower: Pascal Terjan / Wikimedia Commons / CC BY-SA 3.0
851 Schengen Area: Peteri / Shutterstock.com
853 Isaiah Wall: Glynnis Jones / Shutterstock.com
855 Cross: jelm6 / Flickr / CC BY 2.0
857 Illustration: Biblioteca Nacional / Wikimedia Commons
S-1 Library: Diego Grandi / Shutterstock.com
C-1 Fishermen: Wikimedia Commons